Pliocene Carbonates and Related Facies Flanking the Gulf of California, Baja California, Mexico

Edited by

Markes E. Johnson
Department of Geosciences
Williams College
Williamstown, Massachusetts 01267

and

Jorge Ledesma-Vázquez
Facultad de Ciencias Marinas
Universidad Autónoma de Baja California
Ap. Postal 453
Ensenada, Baja California, Mexico 22800

318

1997

Published by The Geological Society of America, Inc.
3300 Penrose Place, P.O. Box 9140, Boulder, Colorado 80301

Printed in U.S.A.

GSA Books Science Editor Abhijit Basu

Library of Congress Cataloging-in-Publication Data
Pliocene carbonates and related facies flanking the Gulf of
California, Baja California, Mexico / edited by Markes E. Johnson
and Jorge Ledesma-Vázquez.
p. cm. -- (Special paper ; 318)
Includes bibliographical references and index.
ISBN 0-8137-2318-3
1. Rocks, Carbonate--Mexico--Baja California. 2. Geology,
Stratigraphic--Pliocene. 3. Geology--Mexico--Baja California.
I. Johnson, Markes E., 1948- . II. Ledesma-Vázquez, Jorge.
III. Series: Special papers (Geological Society of America) ; 318.
QE471.15.C3P58 1997
552'.58'0972209146--dc21 97-16527
CIP

Cover: Aerial view looking north along a portion of the western shore of Bahía Concepción, Baja California Sur, Mexico. The dark, scalloped subtidal areas in the lower right and the dark area to the right of the light band along the east side of the island are living rhodolith beds. These beds produce and replenish most of the white carbonate sediment offshore, in the lagoon, and on the tombolo between the mainland and the island. Similar rhodolith beds contributed to the development of Pliocene limestones on the Gulf Coast of Baja California. Photograph by Michael S. Foster.

10 9 8 7 6 5 4 3 2 1

Contents

Preface v

1. ***Pliocene volcanogenic sedimentation along an accommodation zone in northeastern Baja California: The Puertecitos Formation*** 1
   Arturo Martín-Barajas, Miguel Téllez-Duarte, and Joann M. Stock

2. ***Development and foundering of the Pliocene Santa Ines Archipelago in the Gulf of California: Baja California Sur, Mexico*** 25
   Maximino E. Simian and Markes E. Johnson

3. ***Holocene sediments and molluscan faunas of Bahía Concepción: A modern analog to Neogene rift basins of the Gulf of California*** 39
   Keith H. Meldahl, Oscar Gonzalez Yajimovich, Christina D. Empedocles, Christofer S. Gustafson, Manuel Motolinia Hidalgo, and Timothy W. Reardon

4. ***Upper Pliocene stratigraphy and depositional systems: The Peninsula Concepción basins in Baja California Sur, Mexico*** 57
   Markes E. Johnson, Jorge Ledesma-Vázquez, Mark A. Mayall, and John Minch

5. ***El Mono chert: A shallow-water chert from the Pliocene Infierno Formation, Baja California Sur, Mexico*** 73
   Jorge Ledesma-Vázquez, Richard W. Berry, Markes E. Johnson, and Sonia Gutiérrez-Sanchez

6. ***Stratigraphy, sedimentology, and tectonic development of the southeastern Pliocene Loreto Basin, Baja California Sur, Mexico*** 83
   Rebecca J. Dorsey, K. A. Stone, and Paul J. Umhoefer

7. ***Bryozoan nodules built around andesite clasts from the upper Pliocene of Baja California: Paleoecological implications and closure of the Panama Isthmus*** 111
   Roger J. Cuffey and Markes E. Johnson

8. ***Origin and significance of rhodolith-rich strata in the Punta El Bajo section, southeastern Pliocene Loreto basin*** 119
Rebecca J. Dorsey

9. ***Living rhodolith beds in the Gulf of California and their implications for paleoenvironmental interpretation*** 127
Michael S. Foster, Rafael Riosmena-Rodriguez, Diana L. Steller, and Wm. J. Woelkerling

10. ***Miocene-Pleistocene sediments within the San José del Cabo Basin, Baja California Sur, Mexico*** 141
Genaro Martínez-Gutiérrez and Parvinder S. Sethi

Index 167

Preface

Naturalist Joseph Wood Krutch (1961) referred to Baja California as the Forgotten Peninsula in the title of his renowned essay on the region's botany. The same epitatet may be used to characterize the status of the region's geology for much of the twentieth century. The northern half of the peninsula has attracted much more attention than the southern half, largely as a result of the pioneering work of Gordon Gastil (San Diego State University) and his associates. Their accomplishments include the first geological map of the State of Baja California (Gastil et al., 1973), the first comprehensive summary of the state's geology (Gastil et al., 1975), and more recently a collection of articles on the prebatholithic stratigraphy of the peninsula's northern sector (Gastil and Miller, 1993).

Completion of the transpeninsular highway (Mexican Federal Highway 1) in 1971 helped enormously to make the entire peninsula, including the State of Baja California Sur, much more accessible for businesspeople, tourists, and all manner of naturalists. More than anything else, however, the present volume owes its conception and guiding spirit to the Peninsular Geological Society—established in 1991 as an international group dedicated to the study of Baja Californian geology. The biennial meetings of this organization bring together a diverse group of individuals who appreciate the interrelatedness and thematic value of their independent research.

Initial plans to organize a body of research following the outline of this volume were made during the second International Meeting on the Geology of the Baja California Peninsula, held April 19–23, 1993, in Ensenada, Baja California. The actual volume grew out of a series of papers delivered at the third International Meeting, held April 17–21, 1995, in La Paz, Baja California Sur. All articles in this collection deal with relationships in the Gulf of California bordering the Baja California peninsula, mostly in the State of Baja California Sur. The chapters are arranged in geographic order (see Fig. 1), with the most northerly study area first (Martín-Barajas et al.) and the most southerly last (Martínez-Gutiérrez and Sethi). Most, but not all, of the contributions deal with aspects of Pliocene stratigraphy. The notable exceptions are chapters by Meldahl et al. and Foster et al., concerning aspects of Recent sedimentation and biology that afford better insight on the Gulf's Pliocene strata.

The topics presented in this volume are organized into three interrelated themes. One concerns the origin and development of carbonate and associated facies. Another deals with rocky shorelines as an appropriate boundary marker for the mapping of facies and the determination of relative sea-level changes. The third theme is that of integration of Pliocene stratigraphy and neotectonics. These themes combine in interesting and mutually supportive ways to advance the study of the Pliocene-Recent record along the Gulf Coast of Baja California. The significance of these three themes is briefly elaborated.

STUDY OF GULF CARBONATES

Entirely by chance, in 1991, a field project supervised by one of us (Johnson) at Isla Requesón in Bahía Concepción overlapped with another project on shallow, submarine beds of living rhodoliths (coralline red algae) supervised by Michael S. Foster (Moss Landing Marine Laboratories). A cursory examination of the carbonate shoals and tombolo at Isla Requesón led to the realization that these sediments were mainly derived from coralline red algae. The association is certainly not lost on the local inhabitants, for whom the Spanish word *requesón* refers to the pustular texture of the curds in cottage cheese and thus by extension to the appearance of the bleached, transported, and abraded remains of rhodoliths in beach deposits. The succession of white carbonate beaches juxtaposed against the red Miocene volcanics of the Comondú Group, as seen along Mexican Highway 1 on the west shore of Bahía Concepción (Fig. 1B) makes a strong visual impression. When one takes into account how rarely carbonates occur on shelves defining the eastern boundaries of major oceans (Ziegler et al., 1984), the development of carbonates within the Gulf of California becomes even more interesting.

Chapter 9 by Foster et al. explores the biological background of living rhodoliths in the Gulf of California and considers their bathymetrical significance for the interpretation of paleoenvironments. A specific example of contemporary sedimentation patterns in Bahía Concepción is offered by Meldahl

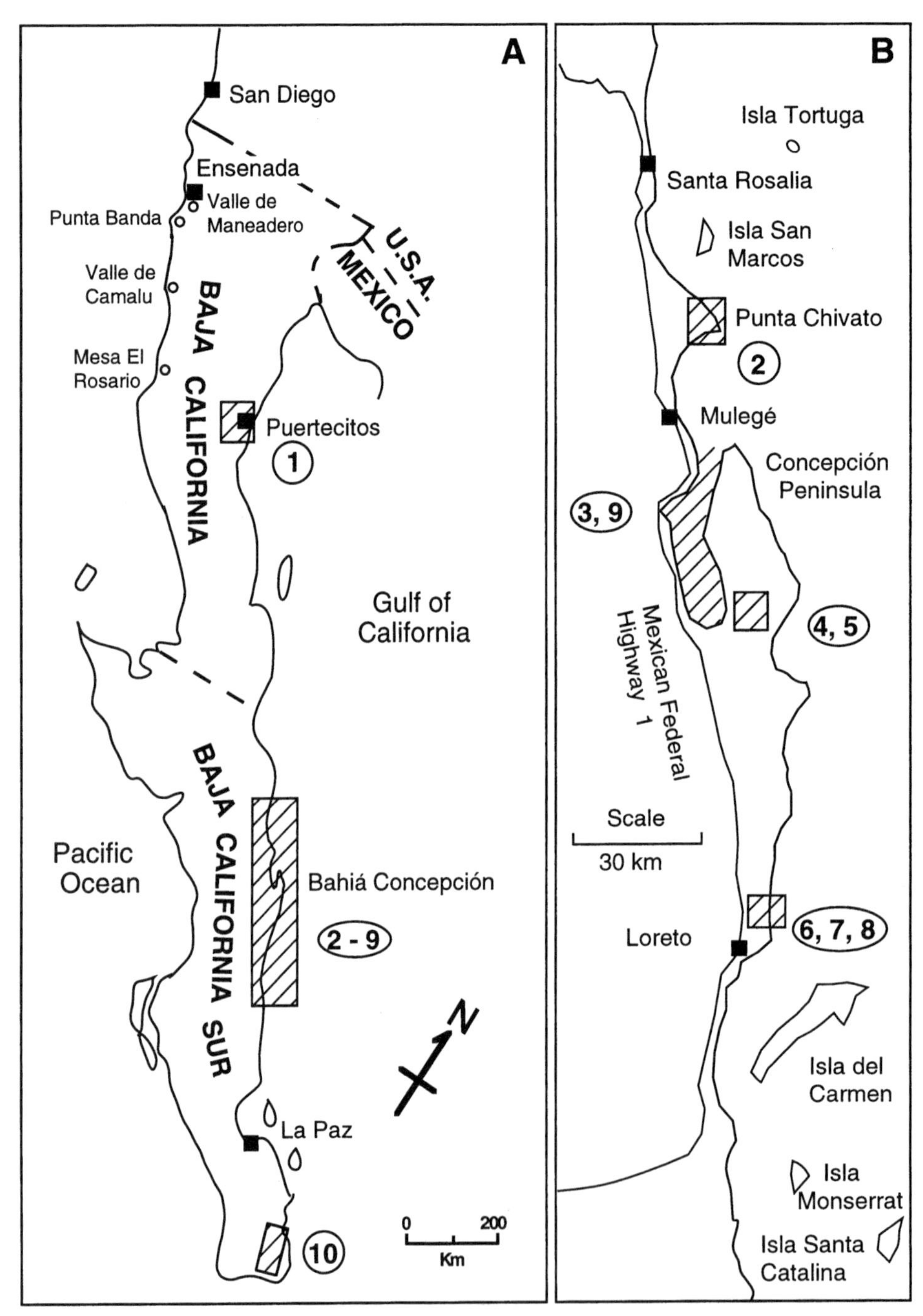

Figure 1. Location of study areas covered in this volume (diagonal lines mark study areas identified by circled chapter numbers; open circles indicate additional areas discussed in the preface). A, map of peninsular Baja California. B, detailed map of the region between Santa Rosalia and Loreto marked by Bahía Concepción in A. The authors of chapters on specific areas indicated by circled numbers on the maps are: Chapter 1: A. Martín-Barajas, M. Téllez-Duarte, and J. M. Stock; Chapter 2: M. E. Simian and M. E. Johnson; Chapter 3: K. Meldahl, O. Gonzalez Yajimovich, C. D. Empedocles, C. S. Gustafson, M. Motolinia Hidalgo, and T. W. Readon; Chapter 4: M. E. Johnson, J. Ledesma-Vázquez, M. A. Mayall, and J. Minch; Chapter 5: J. Ledesma-Vázquez, R. W. Berry, M. E. Johnson, and S. Gutiérrez-Sanchez; Chapter 6: R. J. Dorsey, K. A. Stone, and P. J. Umhoefer; Chapter 7: R. J. Cuffey and M. E. Johnson; Chapter 8: R. J. Dorsey; Chapter 9: M. S. Foster, R. Riosmena-Rodriquez, D. L. Steller, and W. J. Woelkerling; and Chapter 10: G. Martínez-Gutiérrez and P. S. Sethi.

et al. in Chapter 3. Two chapters, by Dorsey and by Simian and Johnson, address Pliocene strata with rich concentrations of fossil rhodoliths.

Other chapters in this collection relate more generally to Pliocene carbonates (Cuffey and Johnson; Johnson et al.; and Martínez-Gutiérrez and Sethi). Another intriguing chapter by Ledesma-Vázquez et al. concerns lagoonal carbonates altered to bedded cherts by silica enrichment through geothermal springs and volcanic ash.

STUDY OF ANCIENT ROCKY SHORELINES

Through their work on contemporary Isla Requesón in Bahía Concepción, Hayes et al. (1993) drew attention to the fossilization potential of Recent rocky-shore biotas occurring in distinct windward and leeward settings. The same principles of faunal analysis are applied by Simian and Johnson in Chapter 2 in their reconstruction of an archipelago of five Pliocene islands preserved in the Punta Chivato area that have distinctly different rocky-shore biotas from windward and leeward environments. In Chapter 8, Dorsey elaborates on a high-energy Pliocene rocky shore in the Loreto area impacted by extensive rhodolith debris. The clast-encrusting bryozoans described by Cuffey and Johnson in Chapter 7 appear to reflect a somewhat more protected environment in the same basin. Midway between Loreto and Punta Chivato is the area where Peninsula Concepción attaches to the mainland (Fig. 1B). Here, surrounding Rancho Rosaliita, is a major Pliocene embayment described by Johnson et al. in Chapter 4. Three peninsulas and four islands subdivide the embayment into four basins interconnected with Pliocene Bahía Concepción. As at Punta Chivato to the north and Loreto to the south, Pliocene coastlines are characteristically shaped by eroded andesites belonging to the Miocene Comondú Group. As a result of the sharp unconformities defining such Pliocene rocky coastlines, they are also significant sources of information regarding the history of relative sea-level changes and tectonic influences.

STUDY OF NEOTECTONICS

Utilization of marine terraces to reconstruct recent crustal movements has been widely attempted on the Baja California Peninsula (Orme, 1973; Ortlieb, 1980, 1991; Rockwell et al., 1989). The results suggest that for the entire post-Pliocene peninsula the mean regional uplift is on the order of 100 mm/ka but that it has diminished during the last few hundred thousand years (Ortlieb, 1991). Punta Banda and the Vizcaino peninsula (on the Pacific Coast) and Santa Rosalia (on the Gulf Coast) are areas where vertical rates of uplift are above the mean for the entire peninsula (Rockwell et al., 1989; Ortlieb, 1991). These higher rates are associated with major tectonic features, such as the Agua Blanca fault in the Punta Banda region or the volcanic structures near Santa Rosalia. Recent crustal shifts are produced by discrete movements of coastal segments through northwest-trending strike-slip faults. These occur not as a single block but as a series of slices delimited by major strike-slip faults, as in the Agua Blanca and San Benito faults. Differential motion between these crustal slices explains the structural and topographical behavior of adjacent areas (Orme, 1973).

Accommodation zones on rift systems produce a distinctive architecture of sedimentary facies. Deposition in the basins clearly emulates the erosion produced in uplifted hinterlands, reflected as thick (Loreto Basin) or thin (Santa Rosaliita, Mulegé, Punta Chivato) sequences, all within the Gulf Extensional Province (Stock and Hodges, 1990). For the Gulf of California, listric-normal faulting with rotations toward the Gulf is the main method by which north-to-northwest normal faults propagate. Associated with these listric faults are antithetic listric-normal and/or normal faults that drop inland to fill grabens created by the listric-normal faults (Dokka and Merriam, 1979).

Ortlieb (1990) shows that the post-Pliocene coasts of Sonora and Baja California have suffered different rates of uplift, with a much quieter east, as opposed to west, coast. Along the Baja California peninsula on the Gulf coast, Ortlieb (1991) suggests that since late Quaternary time the northern and southern areas were less elevated compared to the central region (La Reforma–Santa Rosalia). The central area provides evidence for rates of uplift between 150 and 350 mm/ka.

Another well-documented area that contrasts with the regional mean is Punta Banda, located adjacent to the Agua Blanca fault near the city of Ensenada on the Pacific side of the peninsula (Fig. 1A). Rockwell et al. (1989) estimate the vertical slip rate to be 300 mm/ka, where the highest in a sequence of 14 terraces is situated 298 m above sea level. This example exhibits a horizontal to vertical slip ratio of 25:1 on one of the synthetic faults.

The foremost geomorphological features present along the Gulf Coast of Baja California are marine terraces, combined in many places with rocky-shoreline deposits, at Arroyo Arce (Loreto Basin), Bahía Concepción, Mulegé, Punta Chivato, and Santa Rosalia (Ledesma-Vázquez and Johnson, 1993). Tectonic relationships in the Loreto Basin described in Chapter 6 by Dorsey et al. bear some resemblances to patterns much farther south in the San Jose Del Cabo Basin (Martínez-Gutiérrez and Sethi, Chapter 10). In contrast, post-Pliocene tectonics were much more muted in the Bahía Concepción and Punta Chivato areas (Johnson et al., Chapter 4; Simian and Johnson, Chapter 2).

The Puertecitos area (Fig. 1A), covered by Martín-Barajas et al. in Chapter 1, is defined as an accommodation zone on a rift system recording marine sedimentation and explosive volcanism from late Miocene to early Pliocene time. Divided into two major episodes, the entire sequence is almost 200 m thick. Tectonic subsidence and accumulation rates were much slower in the Puertecitos area, in contrast with other basins farther north where rapid subsidence is associated with major vertical separation on basin-bounding faults.

REGIONAL TECTONIC IMPLICATIONS

During the progression of our studies (Ledesma-Vázquez and Johnson, 1993), we perceived that the Concepción Peninsula is a tectonic block isolated from the west side of Bahía Concepción. The east side of the bay has been uplifted more than the west side of the bay (see also Chapter 3 by Meldahl et al.). The Concepción Bay Fault proposed by McFall (1968) is a feature dividing the Bahía Concepción area into the two blocks. Both are inferred here as topographic lows, but at different elevations.

Previous work, such as that by Umhoefer et al. (1994), shows that the Loreto Basin has been or is being affected by a major tectonic structure, active since the latest Pliocene to the present. This area has a horizontal slip rate that promoted extensive vertical uplift around Arroyo Arce but that diminishes farther north as far as Cerro San Juan. This scenario accommodates the megatectonics of the Gulf of California Fault System in a way similar to that found on the northwestern part of the peninsula on the Pacfic Coast, where Orme (1973, fig. 2) describes Mesa El Rosario and Punta Banda as topographically elevated apexes and Valle de Camalu and Valle de Maneadero as intervening depressions on a 300-km-long series of alternating high and low blocks (Fig. 1A). The rate of uplift for the Arroyo Arce area should be above the mean regional rate for the whole peninsula. Thus, the imprinting of geomorphic patterns in various blocks all along the Gulf Coast, as detailed in this volume, not only accords well with the characteristics of accommodation zones in a rift system but also possesses a tectonic correlation with post-Pliocene structures on the Pacific Coast of Baja California.

Markes E. Johnson, Williams College
Jorge Ledesma-Vázquez, Universidad Autónoma de Baja California

REFERENCES CITED

Dokka, R. K., and Merriam, R. H., 1979, Tectonic evolution of the main gulf escarpment between latitude 31°N and 30°N, northeastern Baja California, Mexico, *in* Abbott, P. L., and Gastil, R. G., eds., Baja California Geology: Geological Society of America Annual Meeting; Field guide and papers: San Diego, California, San Diego State University Geology Department, p. 139–147.

Gastil, R. G., and Miller, R. H., eds., 1993, The prebatholithic stratigraphy of peninsular California: Geological Society of America Special Paper 279, 163 p.

Gastil, R. G., Phillips, R. P., and Allison, E. C., 1973, Reconnaissance geologic map of the State of Baja California: Geological Society of America, 3 maps, scale 1:250,000.

Gastil, R. G., Phillips, R. P., and Allison, E. C., 1975, Reconnaissance geology of the State of Baja California: Geological Society of America Memoir 140, 170 p.

Hayes, M. L., Johnson, M. E., and Fox, W. T., 1993, Rocky-shore biotic associations and their fossilization potential: Isla Requeson (Baja California Sur, Mexico): Journal of Coastal Research, v. 9, p. 944–957.

Krutch, J. W., 1961, The forgotten peninsula: Tucson, University of Arizona Press, 277 p.

Ledesma-Vázquez, J., and Johnson, M. E., 1993, Neotectonica del area Loreto-Mulegé, *in* Delgado-Argote, L. A., and Martín-Barajas, A., eds., Contribuciones a la tectonica del Occidente de Mexico: Union Geofisica Mexicana Monografia 1, p. 115–122.

McFall, C. C., 1968, Reconnaissance geology of the Concepción Bay area, Baja California, Mexico: Stanford University Publications in Geological Sciences, v. 10, no. 5, 25 p.

Orme, R. A., 1973, Quaternary deformation of western Baja California, Mexico, as indicated by marine terraces and associated deposits: Department of Geography, University of California at Riverside, Technical report 0-72-6, p. 627–634.

Ortlieb, L., 1980, Neotectonics from marine terraces along the Gulf of California, *in* Mörner, N. A., ed., Earth rheology, isostasy and eustasy: New York, John Wiley and Sons, p. 497–504.

Ortlieb, L., 1990, Quaternary vertical movements along the coasts of Baja California and Sonora, *in* Dauphin, J. P., and Simoneit, B. R. T., eds., The Gulf and Peninsular province of the Californias: American Association of Petroleum Geologists Memoir 47, p. 447–480.

Ortlieb, L., 1991, Quaternary shorelines along the northeastern Gulf of California: Geochronological data and neotectonic implications, *in* Perez-Segura, E., and Jacques-Ayala, C., eds., Studies of Sonoran geology: Geological Society of America Special Paper 254, p. 95–120.

Rockwell, T. K., Muhs, D. R., Kennedy, G. L., Katch, M. E., Wilson, S. H., and Klinger, R. E., 1989, Uranium-series ages, faunal correlations and tectonic deformation of marine terraces within the Agua Blanca fault zone at Punta Banda, Northern Baja California, Mexico, *in* Abbott, P. L., ed., Geologic studies in Baja California: Society of Economic Paleontologists and Mineralogists, v. 63, p. 1–16.

Stock, J. M., and Hodges, K. V., 1990, Miocene to Recent structural development of an extensional accommodation zone, NE Baja California, Mexico: Journal of Structural Geology, v. 12, p. 315–328.

Umhoefer, P. J., Dorsey, R. J., and Renne, P. R., 1994, Tectonics of the Pliocene Loreto basin, Baja California Sur, Mexico, and evolution of the Gulf of California: Geology, v. 22, p. 649–652.

Ziegler, A. M., Hulver, M. L., Lottes, A. L., and Schmachtenberg, W. F., 1984, Uniformitarianism and paleoclimates: Inferences from the distribution of carbonate rocks, *in* Brenchley, P., ed., Fossils and climate: Chichester, John Wiley & Sons, p. 3–25.

Manuscript Accepted by the Society December 2, 1996

Geological Society of America
Special Paper 318
1997

Pliocene volcanogenic sedimentation along an accommodation zone in northeastern Baja California: The Puertecitos Formation

Arturo Martín-Barajas
Departamento de Geología, CICESE, Ensenada, Baja California, Mexico 22800
Miguel Téllez-Duarte
Facultad de Ciencias Marinas, Universidad Autónoma de Baja California, Ap. Postal 453, Ensenada, Baja California, Mexico 22800
Joann M. Stock
Division of Geological and Planetary Science, California Institute of Technology, Pasadena, California 91125

ABSTRACT

Accommodation zones on rift systems produce distinctive sedimentary facies and facies architecture. The Puertecitos volcanic province in the northern Gulf Extensional Province comprises an accommodation zone and records marine sedimentation and explosive volcanism during upper Miocene and early Pliocene time. The volcano-sedimentary sequence consists of two westward-thinning, wedge-shaped transgressive-regressive marine sequences, each less than 100 m thick, separated by one large pyroclastic flow unit. Northward the lower member is separated from the upper member by an angular unconformity and/or an interval of subaerial erosion. To the southeast, the volcanic units dominate the stratigraphic sections, whereas northward the two marine sequences dominate and contain the distal volcaniclastic facies. We propose the name of Puertecitos Formation to formally name this volcano-sedimentary sequence.

The sedimentary facies crop out in subparallel narrow belts along the present range front to the east of the volcanic province. The fossil assemblages and the sedimentary facies show that this area was a tide-dominated marine embayment and alluvial plain, deepening to the east. Although the environment is favorable for the formation of bioclastic carbonates, the section is essentially devoid of significant carbonate deposits. Instead, the basin was filled with epiclastic and pyroclastic material, and individual volcanic units occur as both subaerial and submarine facies and provide excellent chronostratigraphic markers.

Isostatic sea-level changes and volcaniclastic deposits had a major impact on the facies distribution. Sediment accumulation rates were low. Pre- and postsedimentary deformation was produced by an evenly distributed array of normal faults with small individual offset. This fault pattern produced local facies variations and a distinct facies architecture that contrasts with sedimentary facies associated with basins on rift segments to the north and elsewhere in the Gulf Extensional Province, where rapid subsidence and coarse-grained deposits are associated with large vertical separation on basin-bounding faults.

Martín-Barajas, A., Téllez-Duarte, M., and Stock, J. M., 1997, Pliocene volcanogenic sedimentation along an accommodation zone in northeastern Baja California: The Puertecitos Formation, *in* Johnson, M. E., and Ledesma-Vázquez, J., eds., Pliocene Carbonates and Related Facies Flanking the Gulf of California, Baja California, Mexico: Boulder, Colorado, Geological Society of America Special Paper 318.

INTRODUCTION

Early to late Pliocene marine deposits are extensive and well documented around the Gulf of California depression and farther north in the Salton Trough and the Imperial Valley. These deposits define a late Neogene marine transgression during the early stage of the modern rift system (Fig. 1).

The structure of the Gulf Extensional Province (GEP) has recently been described in terms of rift segments defined by elongated and relatively wide (10 to 20 km) fault-bounded basins (e.g., Laguna Salada, Valle San Felipe–Valle Chico) controlled by alternating polarities of the major bounding faults (Axen, 1995) (Fig 2). These segments are separated by accommodation zones characterized by pre- and synrift volcanism and changes of structural style and topography (e.g., Sierra Las Tinajas–Sierra Pinta, and the Puertecitos volcanic province) (Axen, 1995). In this chapter we examine the sedimentation associated with the eastern part of the Puertecitos accommodation zone. There, marine beds were deposited adjacent to an active Miocene-Pliocene silicic volcanic field (the Puertecitos volcanic province, or PVP) (Fig. 2).

Mio-Pliocene synrift volcanism accompanied marine sedimentation in several places along the western Gulf margin, and volcaniclastic and epiclastic deposits are a major component in some sedimentary sequences in the eastern peninsular margin. However, the relationship between extension, volcanism and sedimentation has been documented in only a few of these localities (see the section "Tectonic and Geologic Setting"). The importance of synrift volcano-sedimentary sequences is that they contain key volcanic horizons that allow detailed reconstruction of the evolution of the basin. Isotopic dating constrains the timing of tectonic subsidence and sedimentation and permits regional correlation with other localities that lack datable volcanic deposits. In particular, intercalated volcanic deposits can be used for absolute calibration of regionally determined biostratigraphic ages. The chronology and volcanic stratigraphy of the eastern and northern part of the PVP are discussed elsewhere (Martín-Barajas et al., 1995; Lewis, 1995; Stock, 1989).

In this chapter we examine the lateral changes of volcanic and sedimentary facies in an accommodation zone characterized by a broadly distributed deformation and concurrent volcanism. We describe the stratigraphy and formally propose the name Puertecitos Formation for the Pliocene volcano-sedimentary sequence that crops out in the coastal zone in this region of northeastern Baja California; we analyze the depositional features of key volcaniclastic deposits that constitute the basis for the stratigraphic correlation; and we propose a model for the lithofacies distribution and paleoenvironments in this accommodation zone during Mio-Pliocene time.

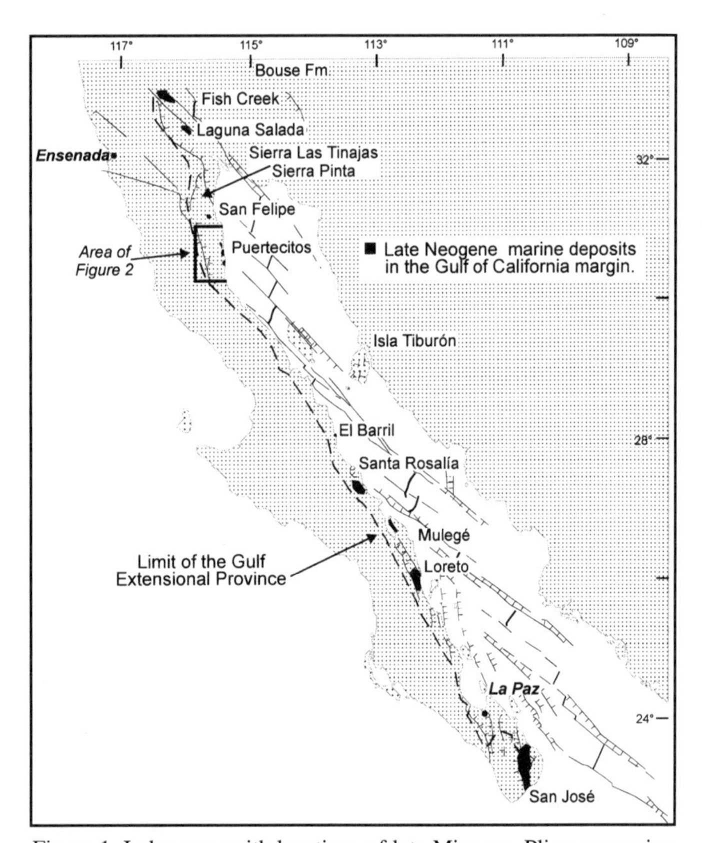

Figure 1. Index map with locations of late Miocene–Pliocene marine deposits along the eastern margin of the Baja California Peninsula.

TECTONIC AND GEOLOGIC SETTING

In northeastern peninsular Baja California (Fig. 2), uplift of the Sierra San Pedro Mártir and the development of the present escarpment at the latitude of Puertecitos began between 11 and 6 Ma, with significant extension prior to 6 Ma (Stock, 1989; Stock and Hodges, 1990). The data we report here suggest that rifting and development of marine basins near Puertecitos, ~30 km east of the escarpment, was underway by the end of this time.

Marine deposition and subsidence at other localities in the northern Gulf of California occurred since middle Miocene time. The oldest known marine deposits are in Isla Tiburón (older than 13 Ma) (Smith et al., 1985). In the Sierra Santa Rosa, southeast of San Felipe, extension and reversals on the drainage patterns are reported to have occurred between 14 and 13 Ma (Bryant, 1986). This period of early-middle Miocene extension and basin subsidence predates the modern rift system and is frequently known as the proto-Gulf stage (Karig and Jensky, 1972). Late Miocene–early Pliocene marine and nonmarine deposits crop out at discrete places along the eastern margin of the Baja California Peninsula (Fig. 1). From south to north, the San José–Los Planes trough includes a late Miocene marine sequence overlain by a >1,200-m-thick sequence of alluvial fan deposits likely accumulated during subsidence of the basin and footwall block uplift (Martínez-Gutiérrez and Sethi, this volume). The Pliocene Loreto Basin contains up to 1,000 m of siliciclastic subaerial to submarine deposits and minor bioclastic limestone beds. Interstratified volcaniclastic units serve as useful chronostratigraphic markers throughout the basin (McLean, 1988; Umhoefer et al., 1994; Dorsey et al., this volume).

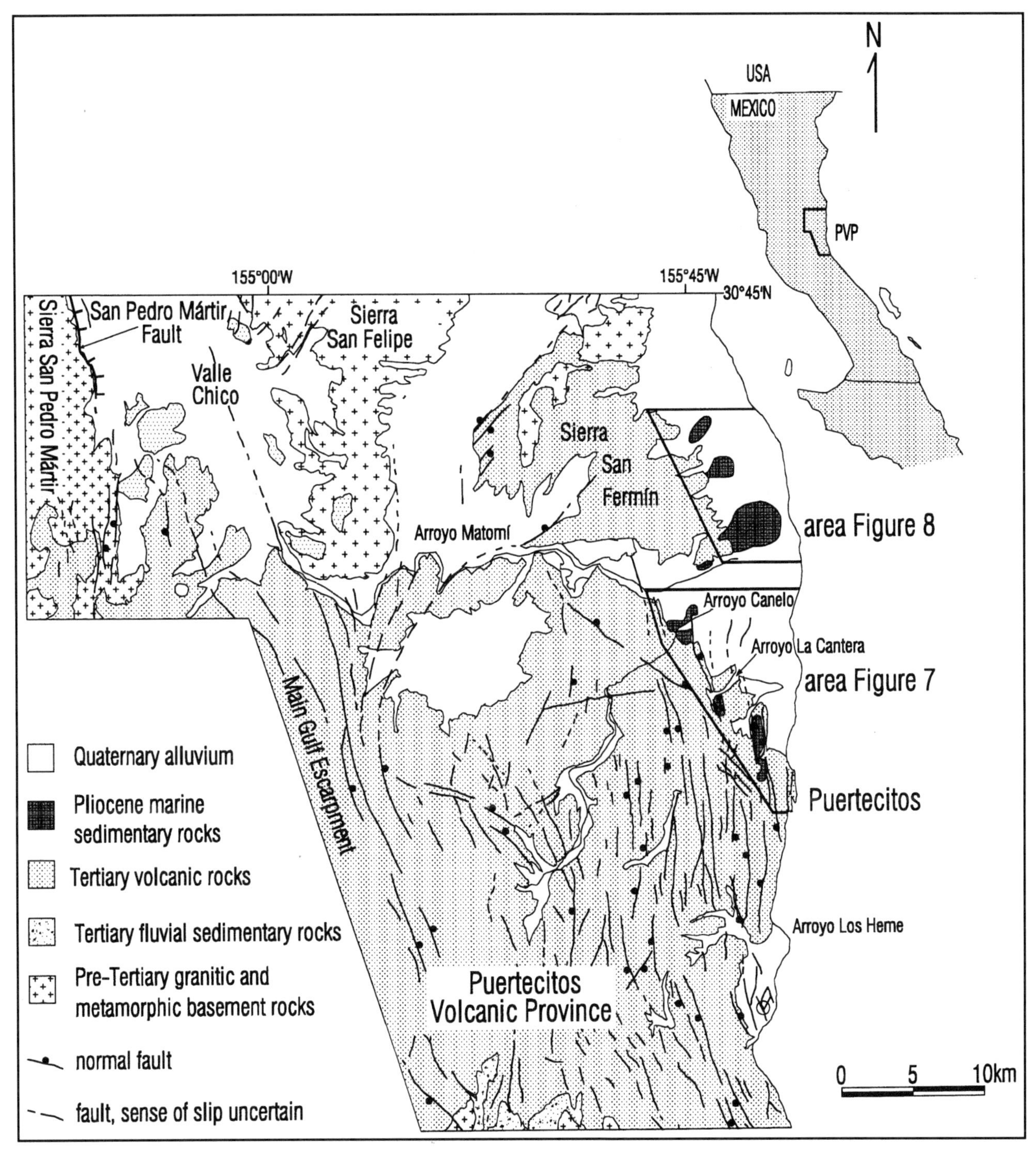

Figure 2. Simplified geologic map of northeastern Baja California and Puertecitos volcanic province (PVP); geologic background modified from Gastil et al. (1975).

South of Bahia Concepción, the upper Pliocene Infierno Formation contains chert beds that likely resulted from alteration of felsic volcaniclastic deposits (Ledesma-Vázquez et al., this volume). In the Santa Rosalía area, synrift volcaniclastic sedimentation is reported in the late Miocene(?)–Pliocene Boleo Formation, which contains primary and reworked volcanic deposits (Wilson, 1948; Carreño, 1982).

In the northern Gulf, 20 km northwest of San Felipe, one of the northernmost exposures of late Neogene marine deposits containing some volcanic influence crops out (see Andersen, 1973; Boehm, 1984). The late Miocene–early Pliocene Llano El Moreno Formation comprises a lower diatomite member and an upper mudstone member (Cañón Las Cuevitas Member) (Boehm, 1984). The clayey sediment in the latter is chiefly smectite derived from alteration of volcaniclastic sediments (Boehm, 1984), and toward the top this unit contains beds of pumice–lapilli-rich sandstone. North of San Felipe, subsequent volcanism in the northwestern Gulf margin is mostly effusive, related to the development of the Cerro Prieto and Salton Trough spreading centers (e.g., Herzig, 1990; Herzig and Elders, 1988); little synsedimentary explosive volcanism has been reported. The Palm Spring Formation in the Fish Creek area includes a thin tuffaceous horizon dated at 2.3 ± 0.4 Ma (Johnson et al., 1983), but the underlying late Miocene–early Pliocene Imperial Formation and the bulk of the Palm Spring Formation in the Laguna Salada and the Fish Creek–Vallecitos areas (Fig. 1) are composed of arkosic sandstone and siltstone derived from local basement rocks and from source rocks from the Colorado River (Vázquez-Hernández et al., 1996; Dibblee, 1984; Kerr and Kidwell, 1991).

The Miocene-Pliocene Bouse Formation exposed in western Arizona and southeastern California may record a marine transgression during the proto-Gulf stage (Buising, 1990). Carbonate deposits at the base have been interpreted as lagoonal and shallow marine, overlain by time-progressive deltaic and prodeltaic deposits. An alternative view holds that these deposits are lacustrine (e.g., Patchett and Spencer, 1995). In the Bouse Formation the basal carbonate sequence unconformably overlies ~10-Ma volcanic rocks, and the basal marine deposits contain a tuffaceous horizon dated at 8.1 and 5.5 Ma (the first age being the more reliable age based on stratigraphic considerations; Damon et al., 1978; Buising, 1990). Up section, the siliciclastic deposits lack a volcanic component. Their age is poorly constrained, but they are likely time correlative with the Imperial Formation, which also contains sediments of the Colorado River delta (Buising, 1990).

In Puertecitos, the marine sequence is interstratified with a series of synrift rhyolite-dacite pyroclastic flows, airfall deposits, and rhyolite domes, ranging in age from 6.5 to 3.0 Ma, that crop out in the northeastern PVP (Sommer and García, 1970; Martín-Barajas et al., 1995; Lewis, 1996). The synrift rhyolite volcanic sequence unconformably overlies arc-related andesite dated at 16 Ma to the south (Martín-Barajas et al., 1995) and 16 to 20 Ma to the west (Stock, 1989). Two major volcanic packages of the PVP in this region are (1) a late Miocene (~6 Ma) Tuff of El Canelo (Tmc) and voluminous late Miocene rhyolite domes (Tmr) and (2) a series of Pliocene (~3 Ma) ignimbrites and reworked volcaniclastic units (refer to Fig. 6). The Tuff of El Canelo and the upper rhyolite flows (Tmru) predate all known exposures of the marine units in this region. However, the Pliocene ignimbrites are synchronous with marine sedimentation to the north and east. Southward, the Pliocene ignimbrite deposits increase in thickness and number of cooling units and their source is inferred to lie east of the present coastline (Martín-Barajas et al., 1995; Martín-Barajas and Stock, 1993).

The modern structural grain of the PVP is dominated by closely spaced north to northwest-striking normal faults, both synthetic and antithetic to the faults of the Main Gulf Escarpment located approximately 30 km to the west (Dokka and Merriam, 1982). The normal faults disrupt the volcanic and sedimentary sequence, and well-preserved northwest-striking fault scarps along the range front between Arroyo Canelo and Arroyo La Cantera suggest very recent fault activity (Fig. 2). The fault pattern in the eastern PVP suggests east-northeast–directed extension and deformation of the Pliocene volcanic and sedimentary rocks (Martín-Barajas and Stock, 1993). Local angular and erosional unconformities, fault-controlled paleotopography, and facies distribution indicate that normal faulting and tilting occurred during deposition of the Pliocene units. Northeast-striking left-lateral strike-slip faults are also present in the Sierra San Fermín and the northernmost part of the Sierra Santa Isabel (south of Arroyo Matomí). These faults cut the Pliocene sedimentary and volcanic rocks, and fault traces are reported to cut Quaternary alluvium across Arroyo Matomí and in the coastal zone east of the Sierra San Fermín (Lewis, 1994; Rebolledo-Vieyra, 1994). The northeast-strike-slip faults may represent boundaries of clockwise-rotated blocks as demonstrated by paleomagnetic studies (Lewis and Stock, 1997; Rebolledo-Vieyra, 1994).

The segment of the Main Gulf Escarpment northwest of Puertecitos is controlled by the San Pedro Mártir fault, a north-northwest–striking, east-dipping, listric normal fault (Fig. 2). From the southern end of the San Pedro Mártir fault to Puertecitos, within the Gulf Extensional Province, the ~6-Ma rocks are severely deformed (Lewis, 1994; unpublished mapping by Stock). The ~6-Ma deformation, the younger rifting and development of marine basins, and the modern structural pattern may all be expressions of a structural zone linking extension at the western edge of the Gulf Extensional Province and slip along the San Pedro Mártir fault to the transform/spreading system of the Gulf of California.

METHODS

Detailed field mapping, along with stratigraphic and lithofacies correlation of the volcanic units and sedimentary rocks, constitutes the basis of this work. We measured several stratigraphic sections in the studied areas and examined the lithological variations of the volcaniclastic and sedimentary units. The sedimentary structures and the fossil assemblages were recorded and used to define the main lithofacies that represent the depositional settings. Part of the stratigraphic data and geologic mapping is published in undergraduate research reports (Rendón-Márquez, 1992; Cuevas-Jiménez, 1994). Regional stratigraphic and lithologic relationships are published elsewhere (Stock et al., 1991; Lewis, 1996; Martín-Barajas and Stock, 1993).

Volcanic and sedimentary rocks were systematically described from thin section. Mineralogical composition of the sand-sized and clayey fractions were determined by point counting analyses and by X-ray diffraction, respectively. The complete results are described elsewhere (Martín-Barajas et al., 1993; Cuevas-Jiménez, 1994). Sediment lithofacies are defined on the basis of sediment color, particle size, mineralogical composition, sedimentary structures, faunal content, and stratigraphic position. Macrofossils were described and compared with the fossil collections of the natural history museum in Los Angeles and San Diego. The proposal of the Puertecitos Formation and its members as formal names follows the North American Stratigraphic Code (North American Commission on Stratigraphic Nomenclature, 1983).

SEDIMENTARY FACIES AND FOSSIL ASSEMBLAGE

The Puertecitos Formation exhibits sedimentological characteristics of a shallow marine, intertidal to subtidal environment, with local progradation of alluvial fan/fan delta deposits derived from the west. Characteristic lithofacies are summarized in Table 1. Muddy subtidal deposits (lithofacies A1) dominate the lower part of the section. Facies A1 likely represents fine-grained deposits on a shallow subtidal setting, below wave base under storm conditions. Its association with fossiliferous muddy sandstone beds (facies B1, Fig. 3) suggests shoaling conditions in the intertidal zone. The presence of coarse-grained massive sandstone and gravelly deposits (facies D1) locally interstratified within facies A1 may indicate sediment gravity flows (slumps) possibly associated with normal faulting. Facies D1 is an uncommon type of deposit and was only observed in the southern part of the Valle Curbina half graben.

Facies C1 is well exposed in the Arroyo La Cantera section and underlies the Tuff of Valle Curbina, a key volcanic unit that limits the top of the lower section (Fig. 4A). Facies C1 typically displays parallel and low-angle planar cross-stratification in sandy to gravelly conglomeratic deposits (Fig. 4A). It probably indicates wave transport in a beach environment. The transitional contact between facies B1 and facies C1 indicates progradation of the coastline from subtidal to intertidal and into subaerial deposits.

In the El Coloradito area, the basal unit is a clast-supported conglomerate (facies E1) with characteristics of both gravity flow deposits and a pebbly to bouldery rocky shoreline. The facies is massive, with <10% matrix. The clasts are poorly sorted and include mainly subangular to sub-rounded pebbles and boulders of welded tuff and rhyolite lava. However, clasts are roughly imbricated and dip to the east, as in modern rocky shorelines in the area. The matrix between clasts is composed of calcareous hash, mostly from echinoderms (sand dollars). These characteristics suggest reworking in a rocky shoreline; however, no typical fossils (e.g., balanus) or boring traces (Tripanites ichnofacies) were observed. Our preferred interpretation is that facies El is an incipient rocky shoreline. Its association with facies B1 suggests a transgressive event and shift into a sandy-muddy substrate in intertidal conditions.

Facies A2 is a reddish to yellowish, poorly laminated, sandy mudstone. It contains contorted laminations and lenses of reworked oyster shells. Coarse sand and gravel are present in the form of irregular lenses or "pods." The matrix is strongly oxidized to limonite-goethite. The association of facies A1 and A2, along with the strong oxidation of the matrix, suggests intervals of subaerial exposure, accompanied by erosion and residual concentration of coarse sediments and diagenetically formed minerals. Facies A2 may mark hiatuses in the sequence (also refer to Fig. 9A and B).

Facies B2 represents an intertidal environment. This facies consists of thin beds of fine-grained sandstone-siltstone, with small-scale ripple cross-stratification and parallel bedding (Fig. 4B). Thickly to thinly laminated, brown to reddish mudstone is commonly associated with the sandstone-siltstone couplets. Also included in this facies are thick beds (10 to 30 cm) of fine-to medium-grained, well-sorted sandstone, with parallel bedding. The well-sorted sandstone beds may represent small-scale sand bars in the intertidal zone. This facies lies in sharp contact with facies C2, which likely represents coarse intertidal channel deposits or channeled deposits in a progradational fan delta body. Facies B2 is similar to facies B1 in that both correspond to intertidal environments. However, two important distinctions are that facies B2 has fewer fossils and less bioturbation and that the lithological composition differs. Facies B2 locally includes bivalve mollusk banks (oysters and clams), whereas facies B1 is typically fossiliferous and highly bioturbated. The mineral composition in facies B2 includes kaolinite and a relatively larger proportion of illite over smectite. The sandy fractions have higher K feldspar/plagioclase ratios and higher amounts of quartz (Martín-Barajas et al., 1993; Cuevas-Jiménez, 1994). These observations suggest a higher contribution of granitic and/or metamorphic rocks to sediments in facies B2.

Facies C2 consists of channeled sandy-to-pebbly conglomerate with planar and trough cross-bedding from the upper flow regime. The laterally discontinuous shape of the beds and the presence of rip-up clasts of muddy sandstone suggest intertidal channel deposits. This interpretation is consistent with association of facies C2 with facies B2, the muddy sandstone facies (see preceding paragraph).

Coarse colluvial deposits (facies D2) interfinger with muddy-sandstone facies in the eastern range front of Sierra San Fermín. The colluvial facies consists of thick beds (10 to 40 cm thick) of clast-supported pebble- to cobble-sized massive breccia and bedded clast-supported conglomerate. The conglomerate shows variable amounts of coarse sand matrix, with parallel and trough cross-stratification. These beds alternate with the massive beds of sedimentary breccia and may result from alternating sheet flood and debris-flow events, in an incipient alluvial fan system. This facies is clearly associated with the faulted range front of the Sierra San Fermín.

The Puertecitos Formation also includes alluvial deposits that consist of sandy to gravelly conglomerates, both matrix supported and clast supported, that normally represent progradation of the shoreline during low sea-level stands.

The most diverse macrofossil assemblage was found in the muddy sandstone facies (facies B1) from the Arroyo La Cantera area. In this area 24 species were identified, of which 11 are bivalve mollusk and 10 are gastropods, one is brachiopod, one

TABLE 1. LITHOFACIES DESCRIPTION AND INTERPRETATION.

Facies	Lithofacies Description	Associated Lithofacies	Interpretation
A1	Yellow-ocher to greenish gray mudstone, massive to thinly laminated. As much as 90% clay and silt, 5 to 10% of fine sand. Mostly glass, lithic volcanic, and plagioclase, traces of pyroxenes, opaques and biotite. Smectite>>illite in the clayey fraction. Gypsum-anhydrite and halite are common, both as a cement in the muddy matrix and as fracture-filling material. Fossils are scarce, worm tubes and bivalve shell molds locally found. Few reworked, and poorly preserved benthic foraminifera.	Commonly in transitional contact with facies B1 and E1.	Shallow, low-energy subtidal deposits below the wave base under storm conditions. Inner to outer shelf.
A2	Reddish, massive to crudely laminated sandy mudstone, with lens (<15 cm thick) of coarse volcanic sandstone and gravel. Abundant iron oxide (limonite and goethite). Lenses and beds of coarse bioclasts (mainly oysters) in a sandy to muddy matrix. The valves are both complete and fragmented.	Commonly in sharp contact with facies A1, facies B2 and C2 in the Sierra San Fermín.	May represent hiatuses or local unconformities within the section. Coarse volcanic and bioclasts probably concentrated as lag or residual material
B1	Yellow to brown-beige, fossiliferous, muddy sandstone. Variable thickness of beds (few cm to 1 m), graded and parallel bedding. Pervasive bioturbation (tubes) in thick strata and lack of distinct bedding. Sand-sized particles are mainly bioclastic fragments, devitrified to fresh volcanic glass and volcanic lithic fragments, and plagioclase. Carbonate and gypsum cement locally important. Both primary and reworked shell banks of molluscan bivalves, gastropods, brachiopods, echinoderms, and arthropods (Table 2).	In transitional contact with facies A1 and E1.	Intertidal to shallow subtidal deposits above wave base. *Panopea* sp. is a good bathymetric indicator living between the intertidal zone and 20 m depth.
B2	Couplets of sandstone and mudstone in thin beds interbedded with thicker, massive beds of muddy sandstone and well-sorted sandstone. The couplets show sharp bases and ripple cross-laminations and flaser structures within laminated mudstone layers. The sandstone beds show poorly planar parallel laminations and thin beds that frequently contain lenses of coarse lithic sand and bioclasts. Transport direction from asymmetric ripple crest in couplets is NE-SW in El Coloradito and Campo Cristina (Fig. 4b).	Associated in sharp contact with facies A2, C2, and D2.	Intertidal plain deposits and sand bank deposits.
C1	Coarse sand and sandy to pebbly conglomerate beds. Parallel stratification, planar-tabular and low-angle cross-bedding (Fig. 4a). Variable bed thicknesses (2 to 20 cm). Large to medium-sized, well-rounded to angular cobbles and pebbles in a poorly cemented, sandy matrix. Rock fragments are mainly rhyolite lava, and minor welded and unwelded tuff and pumice. Sand-sized lithic fragments>>volcanic glass> plagioclase. Shell fragments common.	In gradational contact with facies B1. In sharp contact with coignimbrite ash fall (cf, Fig. 12a).	Sandy to pebbly beach face deposit.

TABLE 1. LITHOFACIES DESCRIPTION AND INTERPRETATION. (continued - page 2)

Facies	Lithofacies Description	Associated Lithofacies	Interpretation
C2	Gray, sandy-to-pebbly conglomerate in laterally discontinuous, lens-shaped beds. Sharp bases, trough and parallel cross-bedding, and normal grading. Contains bioclasts and intraclasts of muddy sandstone.	Laterally interfingers, in sharp contact with facies B2, and more rarely with A2. Associated with Pap (Ptf) in Campo Cristina.	May represent fan-delta channeled deposits, and/or tidal channels.
D1	Yellow to brown color, matrix-supported pebbly conglomerate (diamictite) and poorly sorted coarse sand, massive beds. May show bioturbation. Clayey-silty matrix with sand-sized lithics and minor crystals. Bed thickness from ~0.1 to 1 m. Pebbly to coarse sand–sized volcanic clasts >bioclasts.	In gradational contact with facies B1 in Valle Curbina.	Gravity flow (slump) deposits, probably related to synsedimentary faulting and tilting.
D2	Alternate thick beds (10-40 cm) of clast-supported massive sedimentary breccia, and clast-supported to matrix-supported bedded conglomerate, with planar and low-angle cross-bedding. Rock fragments are angular volcanic rocks and very minor granitic rock fragments. Matrix (if present) is composed of coarse sand.	This facies laterally interfingers with lithofacies B2 and unconformably overlies and is juxtaposed in fault contact with granitic and/or volcanic rocks.	Proximal colluvial deposits associated with faulted range front. Laterally passing into intertidal inlet and plain.
E1	Poorly bedded, clast-supported conglomerate, with imbricated cobbles and boulders. Subangular to rounded clasts, clast imbrication. Matrix, where present, is chiefly calcareous hash of sand dollars. Exclusively volcanic rock fragments.	Basal conglomerate over volcanic ramp-like basement (major unconformity). In gradational contact with facies B1.	Rocky (boulder to pebbly) shoreline deposits, based on clast imbrication and matrix. However, no encrusting or boring organisms present.

echinoderm, and one cirripedium (Table 2). Pervasive bioturbation, encrusting organisms (e.g., balanus), and rock-boring ichnofossils are also common in facies B1. Molds of *Chione* sp. are locally abundant in the yellow mudstone facies, which also contains thin irregular tubes (1- to 2-mm diameter) of bioturbation. The overall fossil assemblage corresponds to a warm shallow marine environment (0 to 20 m) with a muddy to muddy-sandy substrate. In this facies *Turritella* n. sp. (Fig. 5c) is abundant. *Turritella* n. sp. is similar to *Turritella planigyrata*, but the former has a thicker rolling pattern. This fossil may be considered distinctive of the Puertecitos Formation since it was found in both members. Another probable endemic species is *Dosinia* n. sp. (Fig. 5a), which is similar to *Dosinia* cf. *ponderosa* but is smaller in size and has thicker growing lines. *Dosinia* n. sp. was only observed in the upper Delicias Sandstone Member. These two species, *Turritella* n. sp. and *Dosinia* n. sp., may be considered endemic species of the Puertecitos area, since they have not been reported in other circum-Gulf late Neogene sequences (e.g. Smith, 1991; Bell-Countryman, 1984; Hertlein, 1968).

The fossil assemblage of the lower member is compatible with a warm, shallow-water environment, similar to the modern Gulf. *Panopeas* sp. constrains water depth to less than 20 m (Keen and Coan, 1974). The echinodermate *Encope* sp. was found in life position, suggesting that this sandy-muddy lithofacies was below the wave base. However, reworked fossils are abundant in the upper part of lithofacies B1, suggesting shoaling conditions.

Upward the sequence coarsens and contains *Encope* sp. cf. *chaneyi* in life position. This organism is reported to live at a maximum depth of 30 m but is very common in the lower intertidal zone (Morris et al., 1980). Thus, this fossil represents the subtidal-intertidal transition. Its association with encrusting organisms (*Balanus* sp.) and vertical tubes of bioturbation distinctive of the skolithos ichnofacies is also consistent with shoaling conditions and progradation of the coastline.

Calcareous microfossils are sparse in the Puertecitos Formation (Table 3). Microfossils are recrystallized and poorly preserved as a result of diagenesis. However, the specimens that were identified indicate shallow marine environments, specifically the ostracod *Perissocytheridea meyerabichi*, which is typical of tidal flats and eulitoral zones (Ana Luisa Carreño, written communication, 1994). *Perissocytheridea meyerabichi* and *Cyprideis* sp. are relatively common in the upper member.

Only two benthic foraminifera were unequivocally identified

Figure 3. Fossiliferous muddy sandstone bed (facies B1). Note pervasive bioturbation.

Figure 4. A, sandy conglomeratic facies C1 beneath the Tuff of Valle Curbina (Tpvc) in the Arroyo La Cantera section. Note the planar parallel and low-angle cross-bedding. B, ripple cross-stratification in fine-grained sandstone facies. The photo corresponds to the Campo Cristina section (Facies B2 in Table 1).

in the mudstone facies A1 from Valle Curbina: *Bulimina* sp. and *Buliminella* sp. are indicative of a subtidal to intertidal environment. However, the most common benthic foraminifera found in the fine-grained facies of the Puertecitos Formation is *Nonion grateloupi* (Table 3). This fossil is widely distributed in late Neogene to Recent time and has been found in the neritic environment (A. L. Carreño, written communication, 1994). Apparently this calcareous microfossil resisted the effects of diagenesis.

STRATIGRAPHY

The Puertecitos Formation is composed of two formal members, which are chiefly marine, and a coeval volcanic sequence (Fig. 6). The marine sequence is divided into a late Miocene(?)–early Pliocene member, called the Matomí Mudstone Member (Pml), and an upper, early to late Pliocene member, here defined as the Delicias Sandstone Member (Pmu). The Tuff of Valle Curbina is the lowest unit of the volcanic sequence, which also includes three thin, densely welded ash-flow tuffs (the Tuff of Mesa El Tábano, or Tpt) that overlie the Tuff of Valle Curbina in the northeastern volcanic province. The volcanic sequence is entirely coeval with the Delicias Sandstone Member, and some of the pyroclastic flows (e.g., the Tuff of Valle Curbina and one unit of the Tuff of Mesa El Tábano) interfinger with the Delicias Sandstone Member toward the north. The Delicias Sandstone Member also contains three reworked ashy-lapilli layers and several bentonitic horizons not identified within the volcanic sequence. These reworked volcanic units are considered part of the Delicias Member. Because of their importance as stratigraphic markers, their lithologic characteristics are described in the section "Volcaniclastic and epiclastic horizons."

Matomí Mudstone Member

The Matomí Mudstone Member is best exposed along the range front bordering the Arroyo Matomí alluvial plain. The type localities are in Valle Curbina and Arroyo La Cantera (sites 1 and 2, Fig. 7); the latter is unnamed on topographic maps of the region and was previously referred to as Arroyo Los Heme Norte by Stock et al. (1991). Valle Curbina and Arroyo La Cantera are located 5 km northwest of Puertecitos. They form two adjacent alluvial valleys, controlled by north-south–striking, high-angle normal faults that produce narrow horst and graben structures. The eastern graben, Valle Curbina, forms a small closed basin, 0.5 km wide in the east-west direction and 2 km long, open to the southeast (Fig. 7). On the west side of Valle Curbina, the Matomí Mudstone Member overlies and pinches out against a 5.8 ± 0.5-Ma rhyolite dome (Martín-Barajas et al., 1995). We thus infer a maximum age of ~6 Ma for the Matomí Mudstone.

The base of the Matomí Mudstone crops out in the El Coloradito and Santa Catarina areas east of the Sierra San Fermín (Fig. 8). The base includes a 5-m-thick basal conglomerate (facies E1) that unconformably overlies ~6.5-Ma rhyolite tuffs (Lewis, 1995). The conglomerate grades into ~10 m of coarse- to medium-grained sandstone of facies B1; we thus infer a maximum outcrop thickness of 50 m for the Matomí Mudstone Member (Fig. 9A). Nevertheless, because of the wedge shape of this member, its thickness probably increases eastward in the subsurface.

TABLE 2. MACROFOSSIL DESCRIPTION*

Species	VC	LC	AM	EC	Facies
Pelecypoda					
Argopecten sp.	X	X	X		B1
Chionopsis cf. *jamaniana* Olsson		X	X	X	B1, A1
Dosinia cf. *Ponderosa* (Gray)		X			B1
Flabelli pecten sp.	X	X	X		B1
Leopecten bakeri (Hanna and Hertlein)	X	X	X		B1
Lucina lampara (Dall)	X	X			C1
Ostrea angelica Rochebrune	X	X	X	X	B1, C2, A2
Ostrea californica Marcou		X		X	B1
Ostrea erici Hertlein	X	X	X	X	B1, C2, A2
Ostrea vespertina Conrad	X		X	X	B1, C2, A2
Panopea sp.		X			B1
Solen sp.		X			B1
Gastropoda					
Architectonica nolis Bolten		X			B1
Conus sp.		X			B1
Macron sp.		X			B1
Muricanthus sp.		X			B1
Nerita sp.			X		C2
Gastropoda (continued)					
Oliva incrassata (Solander)	X	X		X	B1
Oliva spicata (Bolten)		X			B1, C1
Polinices sp.	X	X		X	B1
Solenoisteira anomala (Reeve)	X	X		X	B1, C1
Strombus sp.		X			B1
Thais sp.		X			B1
Turritella n. sp. aff. *T. planigyrata* Guppy	X	X	X	X	B1, C1
Brachiopoda					
Eogrgryphus sp?		X			B1
Echinodermata					
Encope sp. cf. *E. chaneyi* Durham	X	X	X	X	B1, E1
Cirripedia					
Balanus sp.	X	X	X	X	B1, C1
Chondrichties					
Carcharodon sp.			X		B1

*Key code for localities: VC = Valle Curbina; LC = La Cantera; AM = Arroyo Matomí; EC = Arroyo El Canelo.
X = specimen is present.

Stratotype of the Matomí Mudstone Member. In Valle Curbina the stratotype is a 35-m-thick, coarsening-upward sequence (Fig. 9A). It consists of a yellow-ocher mudstone in the lower 20 m of the section (facies A1). Upsection the mudstone contains beds of coarse sandstone and gravel that grade into a muddy fossiliferous sandstone-siltstone and into sandy to pebbly conglomerate (facies B1 and C1, respectively). The latter underlies the Tuff of Valle Curbina (Fig. 9A). Very rapid east-west lateral variation of lithofacies occurs, and eastward the uppermost marine deposits are fine grained. Westward the top of this section is characterized by sandy and conglomeratic facies that define a paleoshoreline environment. Toward the top the yellow mudstone facies contains a dark-gray lithic tuff, 2.5 m thick, which is a useful marker for local correlation and for fault offset estimation (Ptb in Fig. 6 and 9A).

The coarse-grained facies C1 and B1 are best exposed in Arroyo La Cantera (site 1, Fig. 7). There, the muddy fossiliferous sandstone strata are poorly bedded, with planar, normally graded beds 2 to 10 cm thick and massive beds up to 1.0 m thick, both showing intense bioturbation. The sandstone facies coarsens upward and grades into pebbly sandstone beds and then into sandy to pebbly conglomerate up to 7 m thick (Fig. 9A). In Valle Curbina and Arroyo La Cantera the Tuff of Valle Curbina overlies the coarse-grained facies, but eastward in Valle Curbina the tuff overlies several different facies of fine-grained deposits.

The boundary between the lower and upper members coincides with a change in lithological composition and a distinctive change in color from yellowish to beige-brown in the fine-grained deposits (Martín-Barajas et al., 1993; Cuevas-Jiménez, 1994). Northward, this color change is distinctive of the transition between the lower and the upper members. It is very useful in locations where the Tuff of Valle Curbina either was not deposited or has been eroded.

Delicias Sandstone Member

The Delicias Sandstone Member is best represented in a composite section-stratotype in the marine beds encircling the range front of the Sierra San Fermín, including the Arroyo Canelo section (site 5, Fig. 7) and, above it, the Campo Cristina section (Fig. 9B). The marine beds in this area are disrupted by a series of north-northeast–striking, east-dipping, high-angle normal faults (site 4, Fig. 8). These faults also affect a series of rhyolite flows (Tmr) cropping out in the southern range front of the Sierra San Fermín where tilted domino blocks dip west 20° to more than 30°. The rhyolites are more tilted than the overlying sedimentary sequence, suggesting that some displacement along the normal faults preceded deposition of the marine beds.

The Delicias Sandstone Member consists of a lower, time-transgressive section (El Canelo section) and an upper, time-regressive section (Campo Cristina section) composed of cyclic, fining-upward sequences of green, white, or brown muddy sandstone alternating with sandstone-siltstone massive beds and reworked volcaniclastic deposits and bentonitic layers (Fig. 9B). The sandstone-siltstone beds interfinger with alluvial and coarse marine deposits toward the eastern range front of Sierra San Fermín. Toward the top, the marine sequence unconformably underlies alluvial conglomerates that cap the marine terraces. But locally, in the El Coloradito area, the marine sequence concor-

Figure 5. Macrofossils characteristic of the Puertecitos Formation. a, *Dosinia* n. sp.; b, *Architectonica nobilis* Bolten; c, *Turritella* n. sp. aff. *T. planigyrata* Guppy; d, *Panopea* sp.; e, *Encope* sp. cf. *E. chaneyi* Durham.

dantly grades upward into alluvial conglomerates, and the sequence shows shallower tilting upsection.

Stratotype of the Delicias Sandstone Member. The lower Arroyo Canelo section is a 10-m-thick coarsening-upward sequence that overlies the Tuff of Valle Curbina in its westernmost outcrop (site 5, Fig. 7). The section described here corresponds to the hill located north of the wash. This section begins with thick (10 to 30 cm) beds of fine-grained sandstone, siltstone, and mudstone with indistinct internal layering (facies B2) (Fig. 9B). Abundant bioclasts, mainly oysters and turritelids, are common in the sandstone beds. A greenish bentonitic layer less than 0.7 m thick and a poorly consolidated white-gray pumice-lapilli tuff (Ptg in Fig. 9B), 1 to 1.5 m thick, lie interstratified in the lower half of the section. Ptg is considered the submarine equivalent of unit b of the Tuff of Mesa El Tábano (Stock et al., 1991), and an isotopic date of 3.08 ± 0.04 Ma was obtained on a

TABLE 3. MICROFOSSIL DESCRIPTION*

Calcareous Microfossils	Lithofacies	Member	Relative Abundance†
Benthic Foraminifers			
Amonia beccarii (Linneaus) var.	B2	Delicias	Rare
Bolivina sp.	A1	Matomí	Rare
Bulimina elegantissima (d'Orbigny)	A1	Matomí	Rare
Bulimina subfusiformis (Cushman)	C2	Delicias	Rare
Buliminella subfuiformis (Cushman)	C2	Delicias	Rare
Cibicides pseudoungeriana (Cushman)	C2	Matomí and Delicias	Rare
Elphidium articulatum (d'Orbigny)	C2	Delicias	Common
Loxostomun instabile (Cushman and McCullock)	C2	Delicias	Rare
Nonion costiferum (Cushman)	A1	Matomí	
Nonion grateloupi (d'Orbigny)	A1, C2	Matomí and Delicias	Abundant
Pullenia bulloides (d'Orbigny)	A1	Matomí	Common
Quinqueloculina laevigata (d'Orbigny)	C2	Delicias	Rare
Planktonic Foraminifers			
Globigerina bulloides (d'Orbigny)	C2	Matomí and Delicias	Rare
Globigerina glutinata	A1	Matomí	Rare
Ostracods			
Caudites sp. (juvenil)	C2	Delicias	Rare
Cyprideis currayi (Swain)	C2	Delicias	Rare
Cyprideis sp.	C2	Delicias	Common
Cytherella sp.	C2	Delicias	Rare
Perissocytheridea meyerabichi (Hartmann)	C2	Delicias	Common

*Code for lithofacies as in Table 1.
†Relative abundance of specimens: rare = <5, common = 5–10; abundant = >10.

sample from an outcrop in the southern range front of the Sierra San Fermín (Martín-Barajas et al., 1995).

Upward the marine section contains coarse-grained fossiliferous sandstone and pebble conglomerate in tabular sets with planar and low-angle cross-bedding. This coarse-grained deposit grades into nonmarine bedded sandy conglomerate. This part of the El Canelo section contains a subaerially reworked pumice lapilli tuff (Ptf in Fig. 9B) interlayered in 2.5 m of conglomerate. Ptf has no equivalent in the Tuff of Mesa El Tábano (Stock et al., 1991), but it may correlate with an ashy-pumiceous sandstone, Pap, interstratified within coarse-grained deposits in the lower part of the Campo Cristina Section (Cuevas-Jiménez, 1994; Rebolledo-Vieyra, 1994). Ptf is thus an important stratigraphic marker for correlation between the Campo Cristina and Arroyo El Canelo sections of the Matomí Sandstone Member.

The type locality of the upper section, the Campo Cristina section, is in the marine terraces east of the Sierra San Fermín, 500 m south of the unpaved road that runs east from km 57 of the highway to Campo Cristina (site 4, Fig. 8). There the marine section is cut by a series of closely spaced, west-dipping normal faults that produce tilts up to 30° to the east and cause repetition of the section. It is covered in angular unconformity with up to 14 m of nonmarine, matrix-supported conglomerate grading up into a clast-supported conglomerate. These deposits likely represent alluvial conditions. From base to top, the marine section includes 2.5 to 3.0 m of clast-supported and matrix-supported conglomerate with planar parallel and trough cross-bedding (facies C2). Beds show sharp contact bases and normal grading (section 4, Fig. 9B). This conglomerate is chiefly composed (>85%) of clasts of volcanic rocks, with less than 10% subrounded plutonic (granodiorite) and metamorphic rocks (Cuevas-Jiménez, 1994). It also contains intraclasts of dark-brown sandy mudstone, derived from the underlying fine-grained deposits. The conglomerate grades into a 2- to 3-m-thick white pumiceous sandstone correlated with unit Ptf from the El Canelo section (Rebolledo-Vieyra, 1994; Cuevas-Jiménez, 1994). This volcaniclastic sandstone (Ptf in Fig. 9B) is described in more detail in the following section. It underlies, in sharp contact, a coarse sandstone interbedded with matrix-supported conglomerate, that grades up into a sequence of cyclic-fining upward, thin beds of sandy mudstone and fine-grained sandstone (facies B2). This part of the sequence includes two greenish bentonitic horizons and a reworked greenish lapilli tuff toward the top (Pte in Fig. 9B). Pte is the uppermost volcaniclastic deposit in the sequence.

Volcaniclastic and epiclastic horizons

The volcanic horizons in the sedimentary sequence are both primary and reworked submarine equivalents of some of the subaerial volcanic deposits included in the volcanic section. Detailed descriptions of the geochronology and the petrologic characteris-

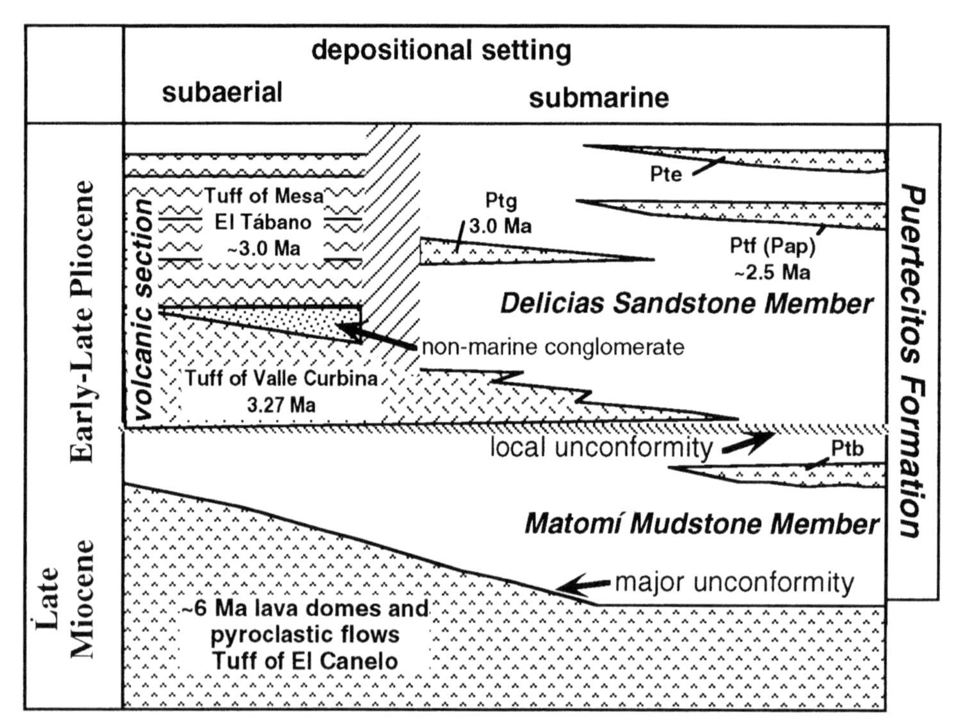

Figure 6. General stratigraphic relationships between the volcanic and marine rocks in the northeastern PVP. Relative thickness not to scale. Abbreviations: Pte = reworked green tuff; Ptg = pumiceous tuff; Ptf = reworked tuff; Ptb = dark-gray lithic tuff.

tics of the subaerial volcanic rocks can be found in Martín-Barajas et al. (1995) and Martín-Barajas and Stock (1993). The volcanic section described here is composed of the Tuff of Valle Curbina and the Tuff of Mesa El Tábano. Southward the Tuff of Valle Curbina underlies a series of pyroclastic flow and fall deposits labeled Tuff of Los Heme, collectively up to 200 m thick, that includes more than 30 cooling units. On the basis of isotopic (^{40}Ar/^{39}Ar geochronology) and paleomagnetic data (Martín-Barajas et al., 1995; T. Melbourne, unpublished data) the Tuff of Mesa El Tábano correlates with the lower 10 units of the Los Heme Tuff, as the number and thickness of the tuffs increase to the southeast toward the inferred source.

A period of ~200 k.y. elapsed between deposition of Tpvc and deposition of the first welded ash flow unit (Tpt-c) within the volcanic member. At the mouth of Arroyo La Cantera (Fig. 7), an alluvial conglomerate more than 10 m thick overlies the Tuff of Valle Curbina (facies D2 on log 1, Fig. 9A). The conglomerate thins to the east and is capped by two welded tuffs of the Pliocene Tuff of Mesa El Tábano. The conglomerate contains only volcanic clasts and shell fragments at the base and records a shoreline progradation following deposition of Tpvc. Farther north, in Arroyo El Canelo and the Sierra San Fermín, marine deposition in a tide-dominated setting continued after the eruption of the Tuff of Valle Curbina. Subsequent marine sedimentation includes several volcaniclastic beds, with a source located east and southeast of Puertecitos.

Tuff of Valle Curbina. Because of its wide distribution, the Tuff of Valle Curbina (Tpvc) is a marker horizon throughout the eastern PVP. Its thickness increases from a few meters in the southern Sierra San Fermín to more than 60 m in Arroyo Los Heme (Fig. 10). Measured sections along the eastern volcanic province suggest a probable source seaward, in the southeastern part of the province, and a minimum volume of 5 km^3 (Martín-Barajas and Stock, 1993). This unit is characterized by a yellow-ocher color and a massive aspect that contrasts with the overlying dark, thin, densely welded ash flows.

Textural and lithological variations along the coastal zone indicate that seawater had little effect on the facies distribution of Tpvc. Southward, the flow was deposited in subaerial conditions and fills paleotopography on ~6-Ma andesite and rhyolite rocks. It includes a slightly welded zone in Arroyo Los Heme, but the massive aspect and poorly consolidated ash and pumice lapilli are distinctive throughout the area (Fig. 11). A deposit of 5- to 10-cm-thick, poorly laminated ash fall is present at the base and may represent a coignimbrite air fall deposit (see Fig. 4a). In Valle Curbina and the Matomí area both the ash fall and the pyroclastic flow deposits appear to have been deposited in shallow water, but no evidence of winnowing is observed. The flow is massive and contains degassing pipes in Valle Curbina (Fig. 11). In Arroyo Canelo, the top of the flow was partially eroded and reworked and is composed of 10- to 20-cm-thick beds of pumiceous lapilli and gravelly lithic sandstone, with planar horizontal laminations and planar low-angle cross-stratification. This facies may represent reworking in subaerial conditions (alluvial ?) or in a beach setting.

In Valle Curbina, Tpvc is a yellow-orange to purplish-gray,

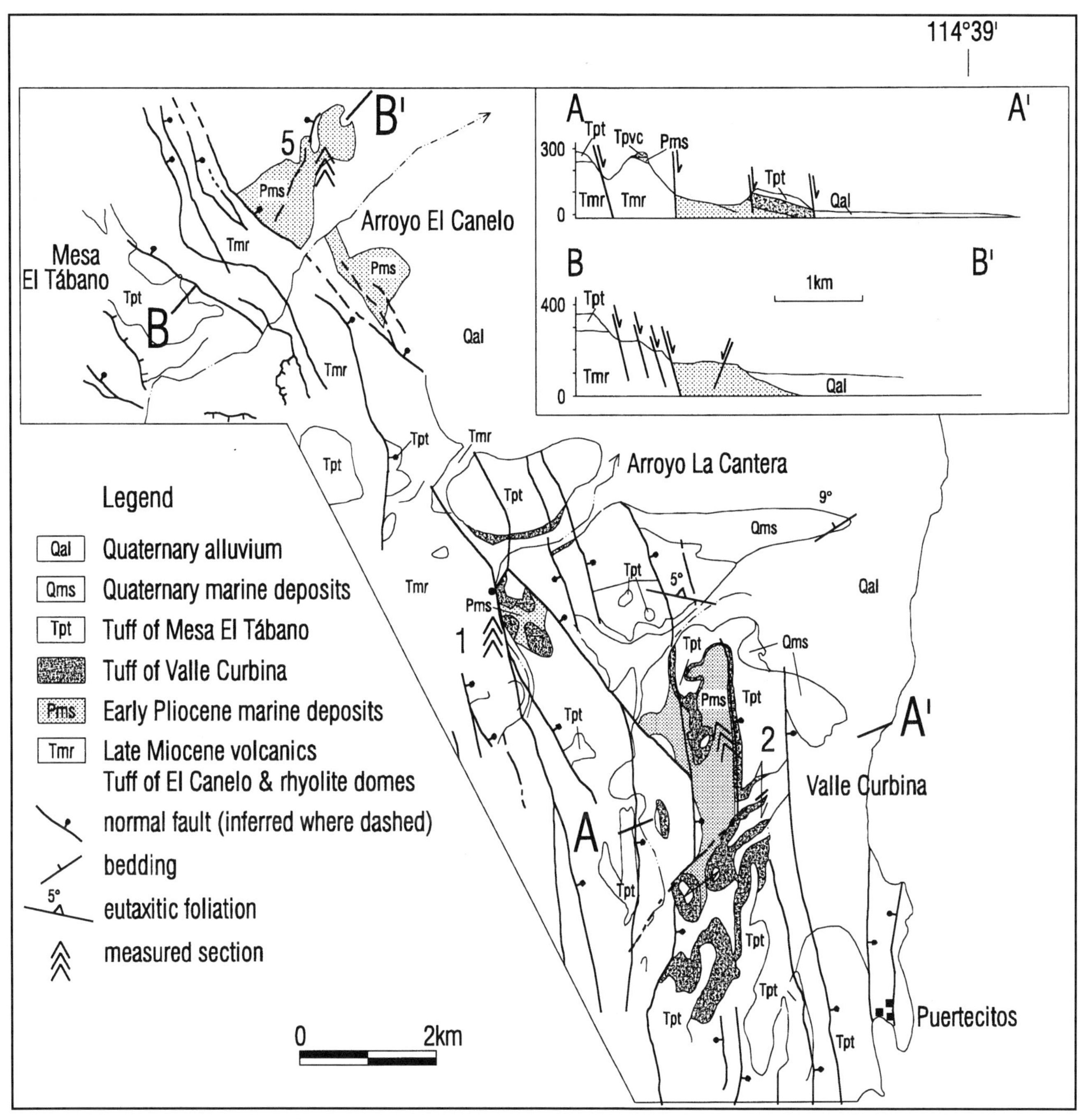

Figure 7. Simplified geologic map of the Valle Curbina–Arroyo Canelo area and cross sections in Valle Curbina (A-A′) and Arroyo Canelo (B-B′). Abbreviations: Tpvc = Tuff of Valle Curbina; Tmr = late Miocene rhyolite domes and Tuff of El Canelo; Tpt = Tuff of Mesa El Tábano, Qms = Quaternary marine sedimentary rocks. Most of Qms is covered by Quaternary colluvial and alluvial deposits not shown here.

matrix-supported, pumice-lapilli pyroclastic flow deposit, with 75 to 80% glass shards and pumice, 15% crystals, and less than 10% lithics. The crystal fraction is mainly plagioclase (andesine), biotite, and opaque minerals, with traces of hornblende (ferro-edenite) and clinopyroxene. Lithic fragments are mostly rhyolite, with some andesitic and quartz-bearing granitic rock fragments. The lithic content clearly decreases between Valle Curbina and the El Canelo area. Another distinction is that Tpvc has both colorless and brown glass shards; microprobe analyses reveal that both types of glass are rhyolitic, but the brown glass has as much as 2% less silica and 0.5 to 1% more iron. The $^{40}Ar/^{39}Ar$ date obtained from plagioclase concentrates from this

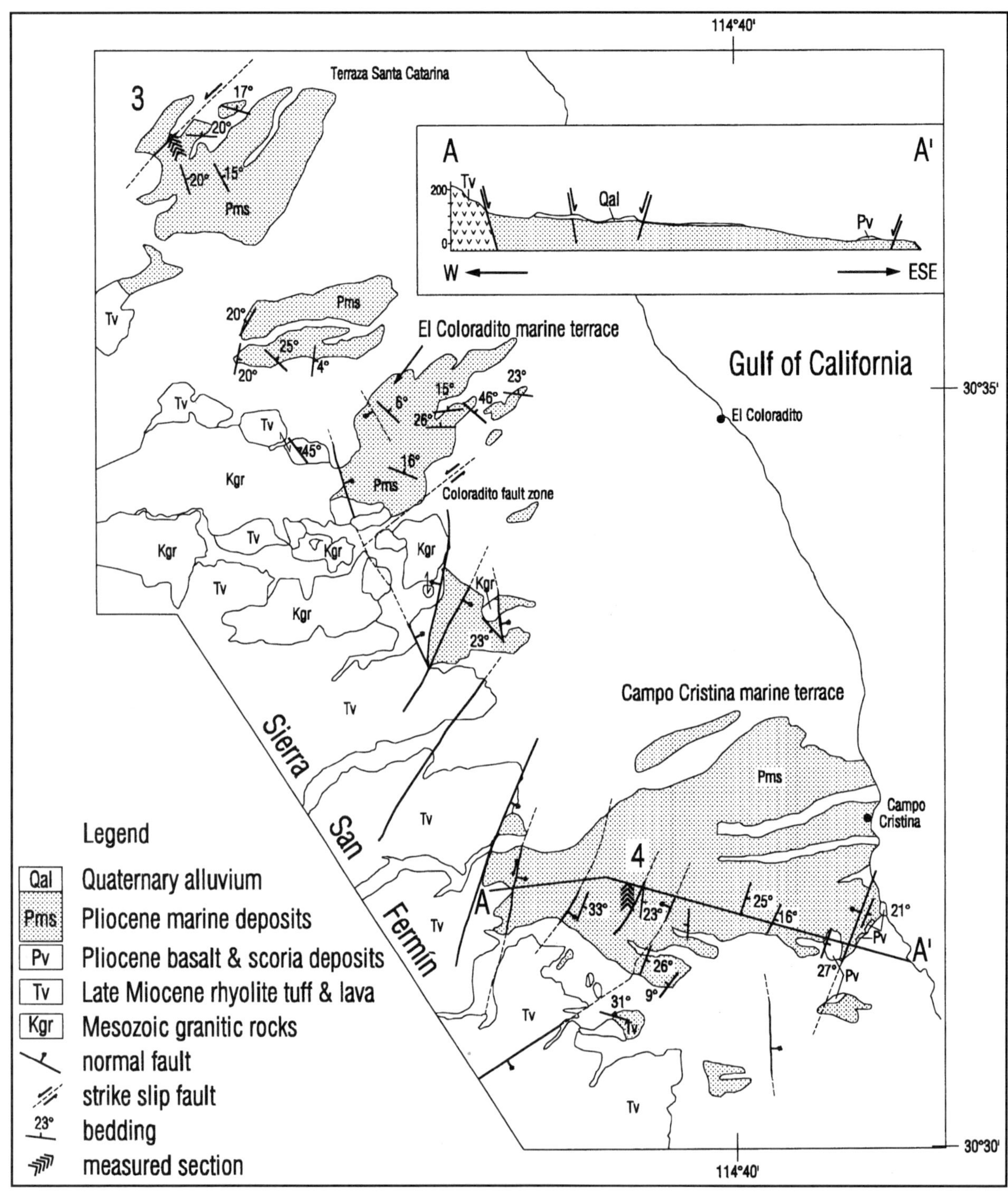

Figure 8. Simplified geologic map of the eastern Sierra San Fermín and cross section from along the Campo Cristina marine terrace.

unit is 3.27 ± 0.04 Ma (Martín-Barajas et al., 1995), and it is reversely magnetized (Tim Melbourne, unpublished data).

In the southern range front of the Sierra San Fermín, the distal facies of the Tuff of Valle Curbina overlies the mudstone facies of the Matomí Mudstone Member. The base of Tpvc contains shell fragments and fine-grained sediments that were incorporated in the flow. The top of the pyroclastic flow deposit is crudely laminated. Farther north, Tpvc is absent between the lower and the upper member of the Puertecitos Formation. The contact, however, is locally an angular unconformity and marked with a distinctive color change in the mudstone facies from yellow to reddish-brown to light brown.

Waterlain pumiceous tuff Ptg. The volcanic unit Ptg crops out in the Sierra San Fermín and in Arroyo Canelo. Here Ptg is stratigraphically located 7.0 m above the Tuff of Valle Curbina within the El Canelo section (see Fig. 9B). Ptg is a 1.8-m-thick, gray to white, crystal-rich pumiceous tuff, with planar bedding and reverse grading at the base and massive bedding upward. The planar bedding and reverse grading suggest reworking of the volcaniclastic ejecta during deposition. Phenocrysts are andesine plagioclase, augite, and hypersthene, with minor opaques and trace hornblende. On the basis of its stratigraphic position, dacitic composition, phenocryst assemblage, and microprobe analyses of phenocrysts, this unit is correlated with unit b, the only dacitic ignimbrite within the Tuff of Mesa El Tábano. The $^{40}Ar/^{39}Ar$ (plagioclase) age of this tuff is 3.08 ± 0.04 Ma (Martín-Barajas et al., 1995), consistent with the age of 3.0 Ma obtained in the upper ash flow unit of Mesa El Tábano (Sommer and García, 1970).

A dramatic change in the thickness and welding may have occurred in contact with water. Preserved thicknesses of unit Tpt-b nearby, where it was deposited in subaerial conditions, range from 10 to 25 m, but less than 2 m of poorly consolidated ashy lapilli are preserved in unit Ptg in the marine section.

Reworked tuff Ptf. The Ptf unit crops out toward the top of the El Canelo section, interbedded with fluvial conglomerates. On the northernmost hill in Arroyo El Canelo, the unit is 2.5 m thick and is tilted 10° to the southeast (Fig. 12). Ptf is a gray-white, poorly consolidated lapillistone, characterized by plagioclase (andesine and oligoclase), biotite, hornblende (ferro-edenite), and minor clinopyroxene. The phenocryst assemblage suggests that this unit has no subaerial equivalent on Mesa El Tábano or in Valle Curbina–El Canelo. Ptf from Arroyo El Canelo correlates with a 2-m-thick volcanic sandstone that crops out along the marine terraces in the east of Sierra San Fermín. This unit shows planar parallel and low-angle cross-bedding and alternating coarse, mostly pumice, layers and fine ashy layers. The correlation is based on mineralogical considerations. In previous work this unit is labeled Pap; it constitutes an important stratigraphic marker for the upper sequence along the coastal zone to the north (Rebolledo-Vieyra, 1994; Cuevas-Jiménez, 1994).

The facies distribution indicates that Ptf was deposited in subaerial and submarine conditions as an airfall deposit. In El Canelo it was partially reworked in a braided stream environment, but a massive basal lapilli unit, 30 cm thick, suggests that the primary deposit was a fallout tephra. Unit Ptf in shallow marine conditions is totally reworked (Fig. 12). It is interbedded with coarse channel deposits in the Campo Cristina area and with fine-grained intertidal deposits eastward and northward. The easternmost outcrop of Ptf is ~10 m thick and has in the lower 2 m planar parallel beds enriched in lithics and crystals. Higher in the outcrop, pumiceous ash and fine-grained pumiceous lapilli with trough and tabular cross-bedding and load bed structures are distinctive. No primary massive volcaniclastic facies, such as those reported for subaqueous pyroclastic flows elsewhere (e.g., Fisher and Schmincke, 1984) are present, a situation that also is consistent with an airfall deposit.

Reworked green tuff Pte. A greenish ash, with lapilli of reddish-black scoria and yellow-beige pumice, occurs near the top of the Matomí Sandstone Member along the coastal plain east of the Sierra San Fermín. It contains variable amounts of epiclastic phenocrysts and lithic fragments. This unit, Pte, includes well-sorted beds of coarse ash and poorly sorted beds of pumice and scoria lapilli, with planar parallel and cross-bedding and normal and reverse grading. Pte is up to 2 m thick in Campo Cristina but less than 0.5 m thick in the El Coloradito area. There, it is interbedded with conglomeratic sandstone that grades into clast-supported conglomerate, which is likely a subaerial deposit. Isopachs suggest that the source of this epiclastic unit may lie to the southeast.

OVERALL GEOMETRY AND SEDIMENTATION

The volcano-sedimentary sequence consists of two westward-thinning, wedge-shaped transgressive-regressive marine sequences each less than 100 m thick, separated by one large pyroclastic flow unit (Tpvc). To the southeast, the volcanic units dominate the stratigraphic sections, whereas northward the two marine sequences dominate and contain the distal volcaniclastic facies.

The sedimentary facies crop out in subparallel narrow belts along the present range front east of the Sierra San Fermín (Fig. 13). The coarse-grained facies extends less than 400 m eastward, is less than 25 m thick, and rapidly interfingers with muddy sandstone facies in the upper member. Fine-grained deposits to the east are thicker than coarse-grained deposits adjacent to the range front. This facies relationship is also observed in the lower member in the Valle Curbina area where rocky shoreline and coarse beach deposits thin and pinch out over a rhyolite dome. To the east the conglomeratic deposits pass rapidly into fine-grained sediments within less than 100 m.

In Valle Curbina, deposition of the volcanic units caused progradation of the coastline toward the east, but northward and westward the marine sedimentation continued after deposition of this pyroclastic flow. The effects of the pyroclastic flow deposits on the redistribution of sedimentary facies may have been more important to the south. Southward, the position of the coastline may have changed dramatically as the thickness and number of pyroclastic flow units increased. In contrast, the effects on the Pliocene shoreline configuration as a result of deposition of the Tuff of Mesa El Tábano in the Matomí–Sierra San Fermín area are negligible.

Facies distribution in the Matomí–San Fermín area suggests that the position of the coastline in early Pliocene time was controlled by paleotopography previously developed on ~6-Ma rhyolite domes and flows. In the Arroyo Canelo area, a ~6.5-Ma pyroclastic flow sequence was severely deformed and tilted as a result of extensional faulting prior to deposition of the lower marine sequence. The contact between the volcanic basement and the marine sequence is, however, a ramplike surface dipping from 5 to 15°; locally it is a high-angle fault contact. No buttressed sedimentary sections were observed in the lower marine member.

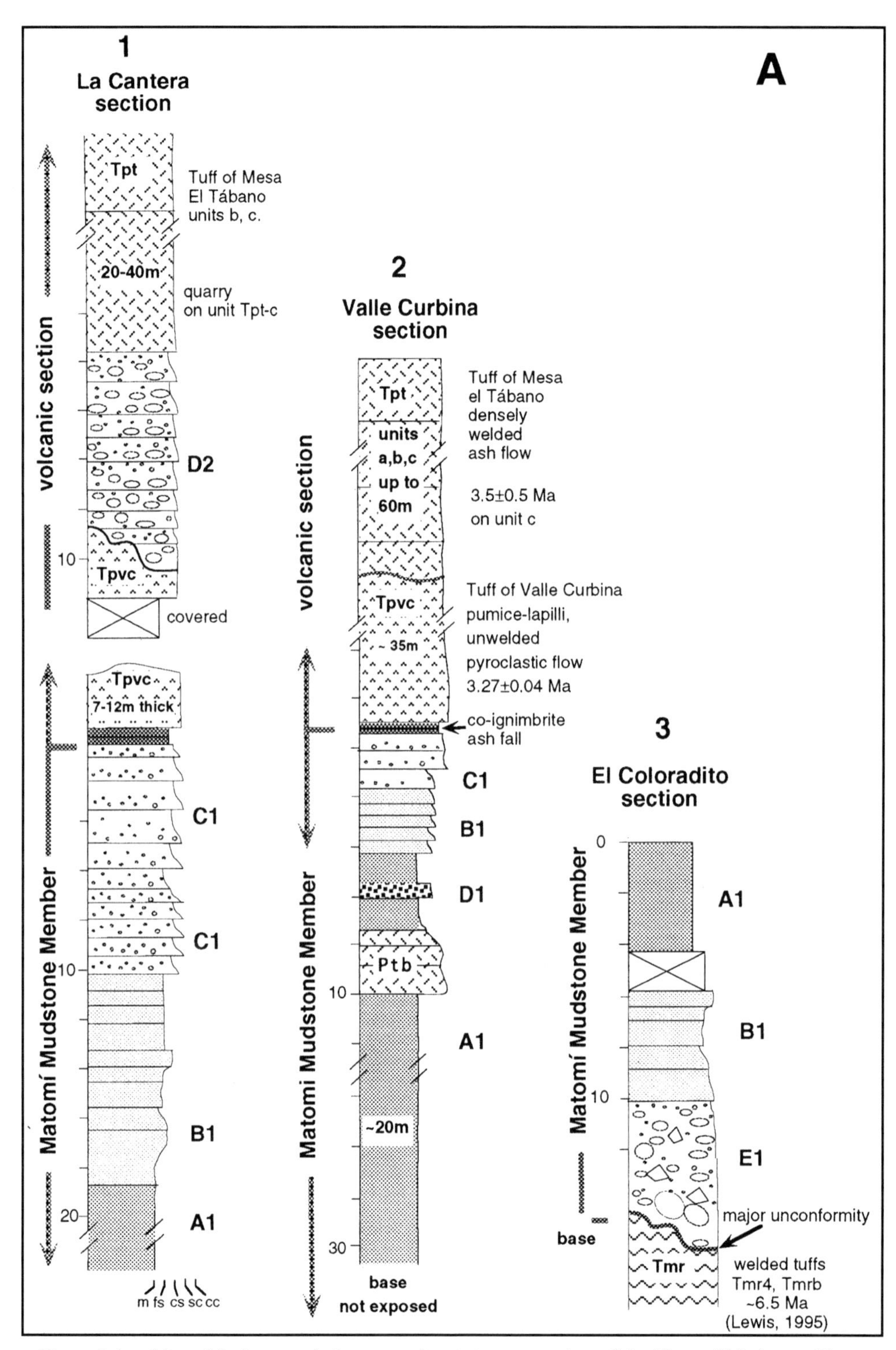

Figure 9 (on this and facing page). A, composite stratotype section of the Matomí Mudstone Member and the volcanic section of the Puertecitos Formation. Abreviations: m = mud, fs = fine sand, cs = coarse sand, sc = sandy conglomerate, cc = conglomerate. Other abbreviations as in Figures 6 and 7. (See location of the stratigraphic logs in Figs. 7 and 8.) B (on facing page), composite stratotype section of the Delicias Sandstone Member of the Puertecitos Formation. Abbreviations as in Figures 6 and 7. (See location of the stratigraphic logs in Figs. 7 and 8.)

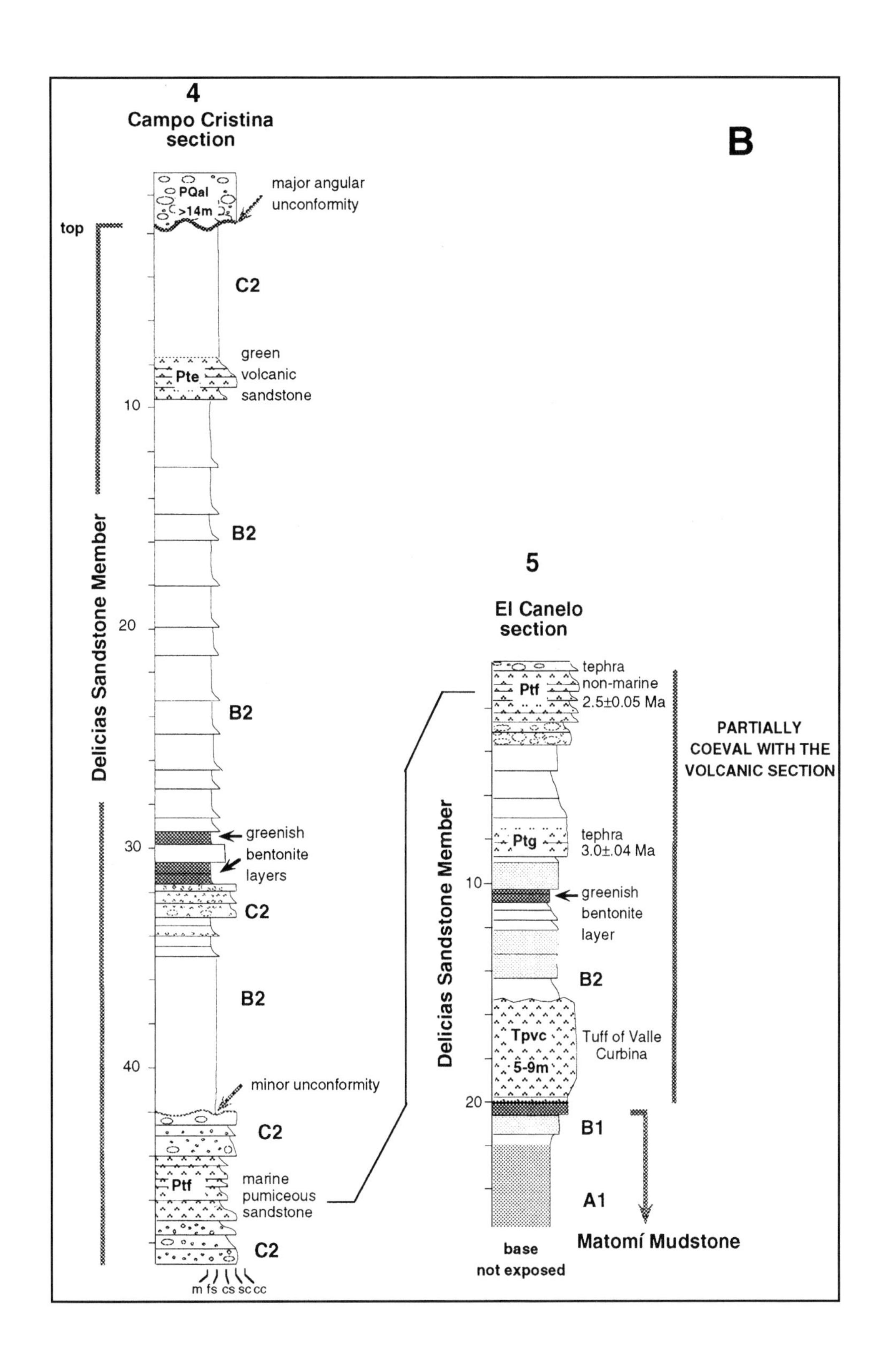
4
Campo Cristina section
B
PQal >14m
major angular unconformity
top
C2
green volcanic sandstone
Pte
10
B2
20
Delicias Sandstone Member
B2
greenish bentonite layers
30
C2
B2
40
minor unconformity
C2
Ptf
marine pumiceous sandstone
C2
m fs cs sc cc
5
El Canelo section
tephra non-marine 2.5±0.05 Ma
Ptf
PARTIALLY COEVAL WITH THE VOLCANIC SECTION
Ptg
tephra 3.0±.04 Ma
10
greenish bentonite layer
Delicias Sandstone Member
B2
Tpvc 5-9m
Tuff of Valle Curbina
20
B1
A1
base not exposed
Matomí Mudstone

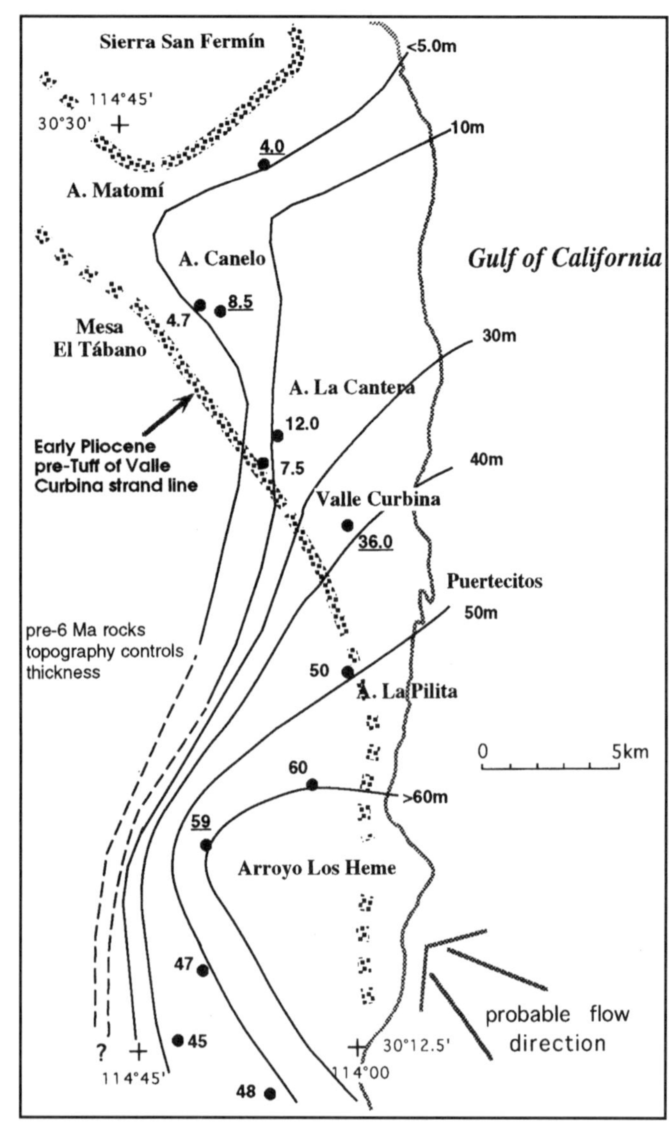

Figure 10. Isopach map of the Tuff of Valle Curbina (Tpvc). Thick patterned line shows probable strand line at the time of Tpvc deposition. Underlined thickness numbers indicate stratigraphic logs of Figure 11.

Thus deposition of the marine beds in the Arroyo Matomí area and in Sierra San Fermín began well after this period of tectonic extension. The presence of dominantly low-energy marine sediments close to the faulted range front in early Pliocene time suggests that these transgressive sequences may in general be attributed to isostatic changes causing westward motion of the strand line, rather than to strong tectonic subsidence and normal faulting near the shoreline. This observation is supported by two independent lines of evidence. First, the sedimentary sequence includes local unconformities. Tectonic subsidence was rapidly compensated by sedimentation and deposition of the large pyroclastic flow units, and the net results were low accumulation rates and condensed sequences during periods of subaerial exposure. During this time, small-scale synsedimentary faulting produced internal unconformities, and the hiatuses are marked by local angular unconformities and by the presence of reworked, reddish shell beds and conglomerates in the upper member (Facies A2).

Second, reliable isotopic ages of the Tuff of Valle Curbina, Ptg, and Ptf provide some constraints on the accumulation rate. The upper marine section in Arroyo Canelo is ~15 m thick and represents a time span of approximately 700 k.y. This yields an accumulation rate of between 5 and 9 mm/100 yr, which is consistent with accumulation rates estimated from magnetostratigraphy in the upper member east of the Sierra San Fermín. There, a sedimentation rate of <5 mm/100 yr was estimated for the Delicias Member (Rebolledo-Vieyra, 1994). These rates are two orders of magnitude less than reported sedimentation rates in the Vallecitos–Fish Creek section in southern California (Johnson et al., 1983) and in the Pliocene Loreto Basin (Umhoefer et al., 1994; Dorsey et al., this volume). This evidence suggests that here normal faulting and tectonic subsidence in Pliocene time were relatively small, compared with the large (>1 km) vertical offset in faults that bound the interbasinal mountain ranges to the north. Accumulation rates of subaerial volcanic deposits in some coastal areas were at least one order of magnitude greater than the accumulation rates of the marine sediments. For example, in Arroyo Los Heme, ~200 m of tuffs were deposited in ~0.5 m.y. (Martín-Barajas et al., 1995). Although speculative, we interpret differences in style of sedimentation to reflect higher heat flow and lower subsidence rates in the accommodation-zone setting.

DISCUSSION

Sedimentary facies and paleoenvironment

The Puertecitos Formation is largely composed of fine-grained volcaniclastic and epiclastic deposits, from subtidal to intertidal environments. Both the sedimentary facies and the fossil assemblages are consistent with a wide, tide-dominated shoreline, with some areas exposed to wave transport and other areas protected from the strong wave action. The Matomí Mudstone outcrops in the Sierra de San Fermín encircle a set of rhyolite hills at the southeast end of the range, suggesting deposition around an island in late Miocene–early Pliocene time.

In the Arroyo Canelo and El Coloradito areas, coarse alluvial deposits directly overlie and interfinger with the muddy sandstone facies. The beach facies that crop out in Valle Curbina–Arroyo La Cantera (lithofacies C1) are not present in Arroyo Canelo but are locally observed in the southeastern Sierra San Fermín. This facies variation suggests marine embayments or tide pools, probably protected by sand bars, as in the modern coast both north of the study area and south in the modern Puertecitos embayment. The transitional contacts among facies A1, B1, and C1 in Valle Curbina–Arroyo La Cantera indicate progressive shoaling and progradation of the coastline during deposition of the Matomí Mudstone Member.

The Puertecitos Formation includes depositional settings ranging from offshore, with a muddy substratum below the wave base (facies A1), to lower-upper shoreface to beach face (facies B1 and C1 respectively).

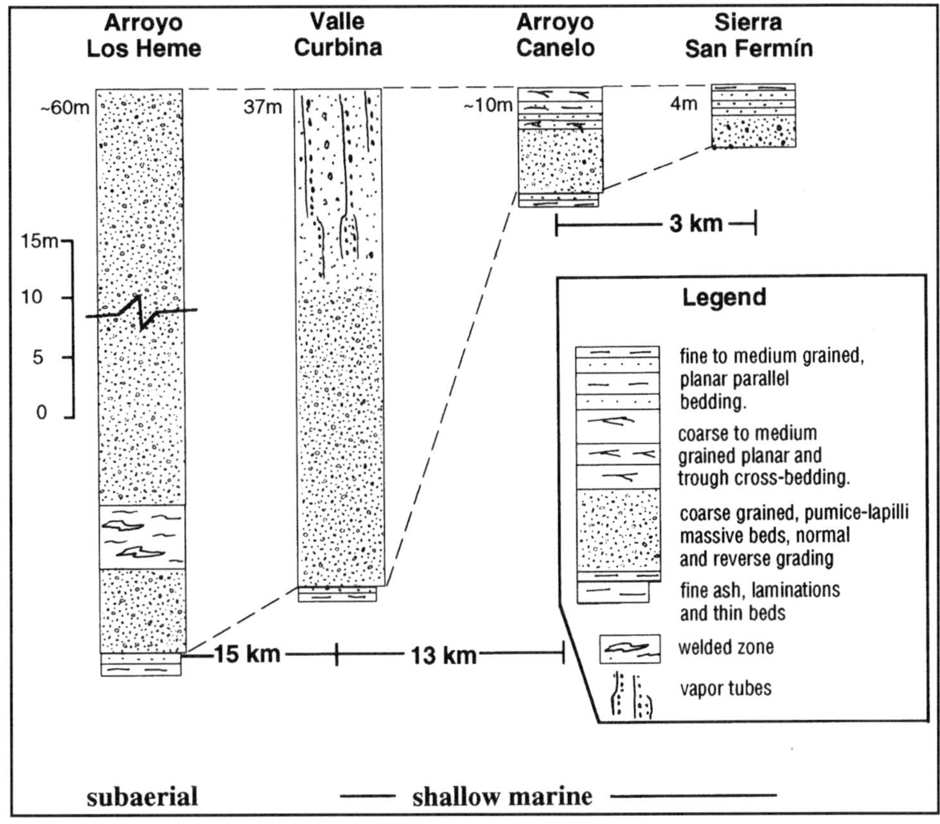

Figure 11. Depositional units of the Tuff of Valle Curbina along the eastern PVP. Locations of the stratigraphic logs are shown in Figure 10.

The fossil assemblage is relatively restricted compared with the modern mollusk communities living in the intertidal zone between Puertecitos and San Felipe. This less varied fossil assemblage cannot be explained by transport and destruction of the less resistant calcareous shells. The Pliocene calcareous shells and molds are in general well preserved, and in many cases the fossiliferous strata contain both reworked fossils and fossils in life position, principally *Chione* sp. cf. *jamainiana, Dosinia* sp. cf. *ponderosa,* and *Panopea* sp. Taphonomic studies in progress (M. Téllez-Duarte, unpublished data) indicate that in facies B1 up to 66% of the bivalves were found in a concave downward position, which is hydrodynamically less stable and normally occurs when the flow regime is low; 34% of the bivalves, on the other hand, were buried in a concave upward position, which is the more stable position during transport. Pronounced transport or strong currents tend to turn the bivalves concave upward to a more stable position. The lack of bedding and current sedimentary structures in facies A1 and B1 is due to intense bioturbation in the intertidal and subtidal zones. Bioturbation may have also contributed to the large proportion of the unstably positioned, disarticulated shells. These observations indicate moderate to weak transport and reworking in the intertidal zone.

Most of the macrofossil assemblage is Panamictic, with the exception of *Architectonica nobilis*, which has Caribbean affinities and has a stratigraphic range from Miocene to Recent. This species might be one of the first Caribbean species that reached the northern Gulf of California in late Miocene time. The microfossil association is also distinctively Panamictic (A. L. Carreño, written communication, 1994). No microfossils of the Californian Province and the Southern Province were identified. This is significantly different from the microfossil assemblage known from the Imperial Formation, where microfossils of apparently Californian and Caribbean affinities coexisted in early Pliocene time (Quinn and Cronin, 1984). The fossil assemblage in the Puertecitos Formation corresponds to warm waters and shallow conditions similar to the modern shoreline, with high tidal ranges. However, in Pliocene time the biodiversity was apparently less than today. This difference may be attributed to a more stressful environment produced by high pyroclastic and epiclastic input.

The marine sequence lacks well-developed carbonate deposits. This is probably because high epiclastic and pyroclastic input prevented high biogenic sedimentation and accumulation of calcareous debris and encrusting algae. Additionally, dissolution of the calcareous micro- and macrofossils is common in the mudstone facies. Most of the least robust calcareous shells (e.g., *Chione* sp. and turritelids) were dissolved. The microfossils were also probably dissolved during diagenesis within the mudstone facies.

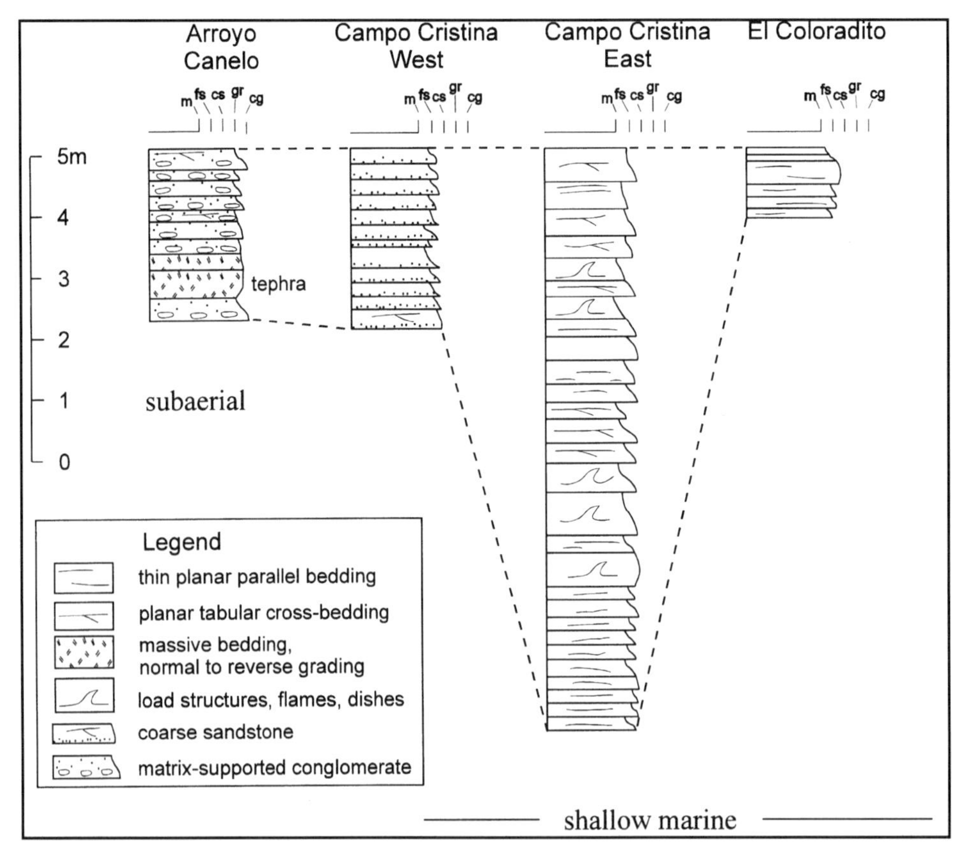

Figure 12. Depositional units of Ptf. Locations of the stratigraphic logs with unit Ptf deposited in a submarine setting are shown in Figure 8. Abbreviations: m = mud, fs = fine sandstone; cs = coarse sandstone; gr = gravel; cg = conglomerate.

Volcanism and sedimentation

The Puertecitos Formation represents marine sedimentation in a high-relief volcanic apron dominated by synsedimentary explosive (caldera type?) volcanism. Explosive volcanism and erosion of the pre ~6-Ma volcanic units supplied material to fill the basin. The location of the vents that produced the Pliocene pyroclastic deposits remains unknown. Isopach distribution and internal structure of the volcanic units suggest that the vent was located to the southeast and would be underwater at the present time. A reconstruction of the early Pliocene shoreline also suggests that the volcanic eruptions and the vent source may have been underwater in a shallow marine setting.

The Tuff of Valle Curbina is the first large pyroclastic flow of the Pliocene volcanic section deposited both in subaerial and shallow marine settings. This deposit contains a variety of accidental clasts including granitic rock fragments, andesite and rhyolite lava, and pyroclastic rocks. The clasts are both angular and subrounded, and it is possible that both sedimentary and country rock fragments from the vent were incorporated into this flow. No data permit us to estimate the influence of the seawater as a trigger for the volcanic eruption, but the heterolithologic composition of the accidental lithics in Tpvc and its poorly consolidated pumiceous ash and lapilli may suggest a phreatomagmatic eruption. Also, mineral contamination by biotite xenocrysts from the granitic basement is suggested by an argon date on biotite from the Tuff of Valle Curbina that yielded a Cretaceous age (P. Layer, written communication, 1993).

Outcrops of Tpvc in the southern Sierra San Fermín and in Arroyo Canelo show some bedding and lamination. In Arroyo Canelo, the flow was probably reworked in the upper intertidal zone. Thick pumiceous beds with internal lamination and thin beds with planar parallel cross-stratification probably indicate a low-energy beach setting. The top of the flow interfingers with thin beds of gravel and coarse sand lacking marine fossils, suggesting alluvial transport. In the Sierra San Fermín, Tpvc is massive and includes oyster shell fragments at the base. Upward, this unit shows laminations and grades into muddy sandstone deposits (facies B1 and B2) from the intertidal zone.

There is no strong evidence for water-magma interaction during the subsequent eruption of the dominantly densely welded, vitric tuffs of Mesa El Tábano. The mineral composition of Tpt and its welding features, even in distal facies, suggest high-temperature magmatic eruptions (Martín-Barajas et al., 1995) with little incorporation of accidental lithic fragments into the flow (see Martín-Barajas and Stock, 1993). The sharp decrease in thickness of unit b in preserved subaqueous sections may indicate that the flow moved above the air-water interface. Some authors maintain that

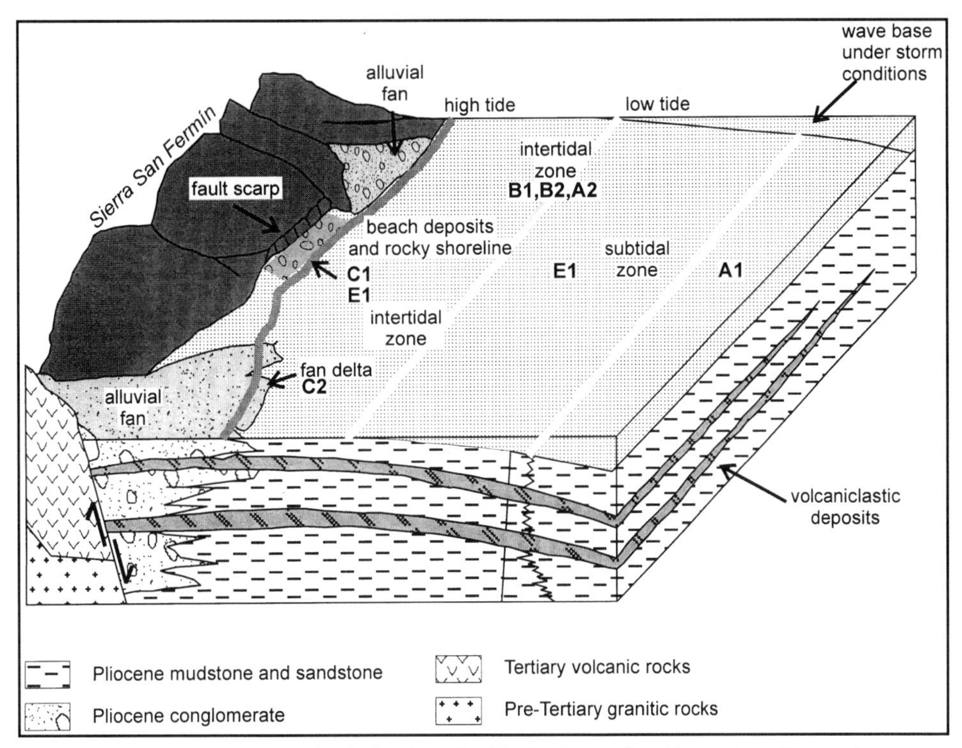

Figure 13. Depositional model for facies described in the Puertecitos Formation. Geometric proportion not to scale.

pyroclastic flows interact explosively with water and generate local secondary vents (Walker, 1979). Such interaction could prevent the flow from penetrating the air-water interface as a gaseous flow, which could explain the mass deficit in this tuff as it moved along the shore. The distal facies of the Tuff of Mesa El Tábano, however, maintained its welding texture in subaerial conditions beyond the marine embayment in the Sierra San Fermín (Lewis, 1995). The interaction between the pyroclastic flow and seawater may be invoked to explain the absence of the Tuff of Mesa El Tábano in a submarine environment. However, this is speculative, and a rigorous lithological analysis of the volcaniclastic facies is beyond the scope of this chapter.

The volcaniclastic units Ptf and Pte were probably deposited as fallout tephras. Unit Ptf in subaerial conditions is partially reworked but has a basal massive, well-sorted pumiceous layer (Fig. 11), which suggests a fallout deposit. The submarine equivalent of Ptf is totally reworked and no primary features were preserved. This deposit lacks the poorly sorted "massive unit" characteristic of subaqueous pyroclastic flows (cf. Fisher and Schmincke, 1984, p. 287–291), which is observed in the Tuff of Valle Curbina and in Ptg, even in the distal facies.

The distal facies of Pte in El Coloradito is enriched in phenocrysts (Cuevas-Jiménez, 1994), a condition that is likely the result of glass loss during transport and reworking as well as of mixing with epiclastic material from other sources. Reworking of Ptf and Pte in the intertidal to subtidal zone produced crystal concentrations at the base of both units (A. Martín-Barajas, work in progress).

Bedforms in Ptf and Pte indicate transport due to variable flow regimes. The flow regime was likely defined by interaction of tidal channels, tidal plain, and alluvial environments, the latter producing prograding coarse-grained deposits shoreward, probably in a fan-delta system similar to the modern Arroyo Matomí fan delta. In Campo Cristina, the thicker deposit of Ptf is interbedded with sandy and muddy marine deposits. Secondary structures (flames, dishes) are very common in the pumiceous ash and lapilli-rich layers. Planar and small-scale ripple cross-stratification suggest a lower flow regime. This site suggests reworking in the intertidal to subtidal zone. In contrast, Ptf is interbedded with marine conglomerate (facies C2) a few hundred meters to the southwest in Campo Cristina. In that part, bed forms are much larger, and include planar, low-angle, and trough cross-stratification. Reverse grading in pumiceous beds and lenses is common; Ptf also contains layers and lenses of gravelly lithics. The conglomerate (facies C2) at the base and on top of Ptf contains structures from the upper flow regime (trough and planar trough cross-stratification in thin sets) and contains intraclasts of the underlying muddy sandstone, strongly suggesting channel deposits in the intertidal zone. The coarse-grained alluvial deposits predominate to the west. Thus, the association of facies B2 and C2 may represent interaction between alluvial and shallow marine environments.

A change in the sediment provenance has been recorded between the lower and the upper member (Martín-Barajas et al., 1993; Cuevas-Jiménez, 1994). The clay mineral assemblage in the lower member, smectite >> illite and locally chlorite, suggests that meteoric and diagenetic alteration of volcanic material was an important contributor to the sediment input. Smectite is probably derived from the in situ alteration of volcanic ash, whereas chlorite probably results from hydrothermal alteration of andesitic rocks. Up section, a change in the sediment provenance has been reported on the basis of (1) an increase in the quartz and potassium feldspar content (Martín-Barajas et al., 1993), (2) the presence of kaolinite and a relative increase of illite over smectite in the upper member, and (3) the inclusion of both volcanic and granitic rock fragments in the upper member, whereas deposits in the lower member are exclusively composed of volcanic rock fragments (Martín-Barajas et al., 1993; Cuevas-Jiménez, 1994). This lithologic change is consistently observed throughout the area and suggests a major change in the source of sediments up section. The implication is that by this time, granitic detritus from batholithic rocks to the west had reached the Gulf of California in this region, presumably because coastal drainage at this latitude had extended westward to regions where the batholithic rocks had been tectonically denuded by normal faulting.

Global setting and correlation with other localities in the northwestern Gulf Extensional Province

On a regional scale, the Puertecitos Formation was deposited in a volcanic setting along an accommodation zone. The area at the latitude of Arroyo Matomí in the northern part of the Puertecitos volcanic province has been proposed to represent a structural transition zone between a highly extended area to the north and a less extended area to the south (Dokka and Merriam, 1982; Stock and Hodges, 1990), possibly related to structural segmentation along the Main Gulf Escarpment (Axen, 1995). The structural style in the south is characterized by north- to northwest-striking closely spaced normal faults that both predate and postdate the sedimentary sequence. The net effect of this style of deformation on the geometry of the basin and the sedimentary facies is a condensed sequence with rapid lateral variations of the sedimentary facies.

Northwest of the study area, long and relatively wide fault-bounded basins resemble the geomorphology of the Basin and Range Province. Although little is known about subsidence and accumulation rates in those basins, gravimetric studies (Slyker, 1970) in the Valle San Felipe suggest more than 2 km of sedimentary fill and rapid uplift of the San Pedro Mártir area during Plio-Quaternary time (Brown, 1978). Exploration wells in the Laguna Salada area indicate >2,000 m of post–Imperial Formation coarse-grained arkosic deposits (Alvarez-Rosales and González-López, 1995). The dominant facies in the eastern flank of the Laguna Salada area are alluvial fan/fan delta systems (Vázquez-Hernández, 1996). In the Fish Creek–Vallecitos area in southwestern Imperial Valley, California, paleontological data indicate an upper late Miocene age for the Fish Creek gypsum (Ingle, 1974; Dean, 1988). Paleomagnetic and fission track dating on the overlying sedimentary sequence indicates that up to 5 km of shallow marine and deltaic deposits from the Colorado River accumulated at rates of 5.5 to 0.5 mm/yr (Johnson et al., 1983). This is two orders of magnitude higher than accumulation rates in the Puertecitos Formation estimated with magnetostratigraphy (Rebolledo-Vieyra, 1994), isotopic ages, and stratigraphic considerations. We thus believe that slow subsidence rates and eustatic sea-level changes controlled the sedimentation and the facies distribution of marine and nonmarine deposits in the eastern PVP and the Sierra San Fermín.

Eustatic sea-level changes may have produced the two transgressive-regressive sequences that constitute the Puertecitos Formation. Facies A1 (internal platform facies), facies B1, and B2 (intertidal zone facies) are the most laterally continuous. Their distribution may have been controlled by a marine transgression-regression in early Pliocene time. The rapid transgressive onlap of the lower member over a ramplike volcanic basement may be related to a high sea-level stand. The Miocene-Pliocene boundary represents a time of sea-level high stand (Haq et al., 1987), and on the basis of the pre–3.2 Ma age for the lower member, we speculate that this marine transgression may be regionally recorded. Subsequent eustatic sea level changes, together with synsedimentary faulting, may have controlled deposition of the upper member.

The time of initiation of marine sedimentation along the northeastern margin of the peninsula may be synchronous in the Fish Creek–Vallecitos, Laguna Salada, and Puertecitos areas and probably in the area northwest of San Felipe. The time period between 6 to 5 Ma is thought to encompass the development of the plate boundary within the Gulf and initiation of the northward translation of Baja California (Curray and Moore, 1984; Lonsdale, 1989). Significant normal faulting had occurred prior to 6 Ma in some regions along the Main Gulf Escarpment (Stock and Hodges, 1990). Proto-Gulf extension occurred since middle Miocene time and produced marine sedimentation in Isla Tiburón (Smith et al., 1985) and extensive alluvial deposits in the Sierra Santa Rosa (Bryant, 1986). But, as we show here, the oldest known marine deposits to the east are 6 Ma and younger. In the Imperial Valley the onset of transtensional tectonism likely began in late Miocene time (Winker and Kidwell, 1986).

CONCLUSIONS

The early Pliocene Puertecitos Formation was deposited adjacent to an active Miocene-Pliocene silicic volcanic field (the Puertecitos Volcanic Province). Isopach distributions of Pliocene pyroclastic flows suggest a submarine source located east of the present coastline. Lithological and stratigraphic correlation among several localities bordering the Matomí alluvial fan and the eastern flank of Sierra San Fermín indicate that during Pliocene time this area was a shallow, tide-dominated marine embayment that received high epiclastic and pyroclastic input from explosive silicic eruptions.

The geometry of the volcano-sedimentary sequence consists

of two westward-thinning, wedge-shaped marine sequences, separated by one large (>5km^3) pyroclastic flow dated 3.27 ± 0.04 Ma. In northern areas sections are dominated by the two marine sequences, each less than 100 m thick, whereas the pyroclastic flow dominates sections in southern exposures. Sediment lithofacies distribution forms subparallel narrow belts along the present range front to the east. The coarse-grained facies extends less than 200 m from the range front and rapidly interfingers with the upper fine-grained intertidal to subtidal deposits.

Isotopic dating and facies geometry indicate that tectonic subsidence in Pliocene time was relatively small and accumulation rates were slow (<9 mm/100 yr), which is consistent with magnetostratigraphic data (e.g., Rebolledo-Vieyra, 1994). Subsidence could have been produced by an evenly distributed array of normal faults with small individual offset, which is common in the study area. This style of sedimentation contrasts with the coarse-grained alluvial fan/fan delta deposits of interbasinal mountain ranges to the north where rapid subsidence is associated with large (>1 km) vertical separation on basin-bounding faults. We interpret differences in style of sedimentation to reflect higher heat flow and lower subsidence rates in the accommodation-zone setting.

ACKNOWLEDGMENTS

Martín-Barajas was supported by Consejo Nacional de Ciencia y Tecnología, México, via grant 1224-T9203. Stock was supported by the Petroleum Research Fund of the American Chemical Society via grant 21291-G2 and by National Science Foundation grant EAR-89-04022. Special thanks to V. Frias for drafting and L. Skerll for improvement of the English usage. We thank G. Gastil and R. Sedlock for their constructive reviews.

REFERENCES CITED

Alvarez-Rosales, J., and González-López, M., 1995, Resultados de los pozos exploratorios en Laguna Salada, Baja California: Third International Meeting on the Geology of Baja California Peninsula, La Paz, Baja California Sur, Abstracts, p. 4.

Andersen, R. L., 1973, Geology of the Playa San Felipe Quadrangle, Baja California, Mexico [M.S. thesis]: San Diego, California, San Diego State University, 214 p.

Axen, G. J., 1995, Extensional segmentation of the Main Gulf Escarpment, Mexico and United States: Geology, v. 23, p. 515–518.

Bell-Countryman, P., 1984, Environments of deposition, Pliocene Imperial Formation, southern Coyote Mountains, Imperial County, California, *in* Rigsby, C. A., ed., The Imperial Basin—Tectonics, sedimentation and thermal aspects: Los Angeles, California, Pacific Section, Society of Economic Paleontologists and Mineralogists, p. 45–70.

Boehm, M. C., 1984, An overview of lithostratigraphy, biostratigraphy, and paleoenvironments of the late Neogene San Felipe marine sequence, Baja California, Mexico, *in* Frizzell, V., Jr., ed., Geology of the Baja California Peninsula: Pacific Section, Society of Economic Paleontologists and Mineralogists, v. 39, p. 253–265.

Brown, L. G., 1978, Recent fault scarps along the eastern scarpment of the Sierra San Pedro Mártir, Baja California [M.S. thesis]: San Diego State University, San Diego, California, 108 p.

Bryant, B. A., 1986, Geology of the Sierra Santa Rosa Basin, Baja California, Mexico [M.S. thesis]: San Diego State University, San Diego, California, 75 p.

Buising, A.V., 1990, The Bouse Formation, and bracketing units in southeastern California and western Arizona: Implications of the proto–Gulf of California and the lower Colorado River: Journal of Geophysical Research, v. 95, p. 20111–20132.

Carreño, A. L., 1982, Ostrácodos y foraminíferos planctónicos de la Loma del Tirabuzón, Santa Rosalía, Baja California Sur, e implicaciones bioestratigráficas y paleoecológicas: Revista del Instituto de Geología, Universidad Nacional Autónoma de México, v. 5, p. 55–64.

Cuevas-Jiménez, A., 1994, Estratigrafía de las terrazas marinas al este de la Sierra San Fermín, NE de Baja California, Mexico [Undergraduate Research Report]: Ensenada, Baja California, Facultad de Ciencias Marinas, Universidad Autónoma de Baja California, 80 p.

Curray, J. R., and Moore, D. G., 1984, Geologic history of the mouth of the Gulf of California, *in* Crouch, J. K., and Bachman, S. B., eds., Tectonics and sedimentation along the California margin: Society of Economic Paleontologists and Mineralogists, Pacific Section, v. 38, p. 17–36.

Damon, P. E., Shafiqullah, M., and Scarborough, R. B., 1978, Revised chronology for critical stages in the evolution of the Colorado River: Geological Society of America Abstracts with Programs, v. 10, p. 101–102.

Dean, M. A., 1988, Genesis, mineralogy, and stratigraphy of the Neogene Fish Creek gypsum, southwestern Salton Trough, California [M.S. thesis]: San Diego, California, San Diego State University, 150 p.

Dibblee, W. T., 1984, Stratigraphy and tectonics of the San Felipe Hills, Borrego Badlands, Superstition Hills, and vicinity, *in* Rigsby, C. A., ed., The Imperial Basin—Tectonics, sedimentation and thermal aspects: Los Angeles, California, Pacific Section Society of Economic Paleontologists and Mineralogists, p. 31–44.

Dokka, R. K., and Merriam, R. H., 1982, Late Cenozoic extension of northeastern Baja California, Mexico: Geological Society of America Bulletin, v. 93, p. 371–378.

Fisher, V. R., and Schmincke, U. H., 1984, Pyroclastic rocks: Berlin, Springer-Verlag, 472 p.

Gastil, R. G., Phillips, R. P., and Allison, E. C., 1975, Reconnaissance geology of the State of Baja California: Geological Society of America Memoir 140, 170 p.

Haq, B. U., Hardenbol, J., and Vail, P. R., 1987, Chronology of fluctuating sea level since the Triassic: Science, v. 235, p. 1156–1167.

Hertlein, L. G., 1968, Three late Cenozoic molluscan faunules from Baja California with a note on the west of San Felipe: California Academy of Sciences Proceedings, no. 30, p. 265–284.

Herzig, T. C., 1990, Geochemestry of igneous rocks from the Cerro Prieto geothermal field, northern Baja California, Mexico: Journal of Volcanology and Geothermal Research, v. 42, p. 261–271.

Herzig, T. C., and Elders, W. A., 1988, Nature and significance of igneous rocks cored in the State 2-14 research borehole: Salton Sea Scientific Drilling Project: Journal of Geophysical Research, v. 93, p. 13069–13080.

Ingle, J. C., Jr., 1974, Paleobathymetric history of Neogene marine sediments, northern Gulf of California, *in* Gastil, G., and Lillegraven, J., eds., A guidebook to the geology of peninsular California: Pacific Section, American Association of Petroleum Geologists, p. 209–217.

Johnson, N. M., Officer, C. B., Opdyke, N. D., Woodard, G. D., Zeitler, P. K., and Lindsay, E. H., 1983, Rates of late Cenozoic tectonism in the Vallecito–Fish Creek basin, western Imperial Valley, California: Geology, v. 11, p. 664–667.

Karig, D. E., and Jensky, W., 1972, The proto–Gulf of California: Earth and Planetary Science Letters, v. 17, p. 169–174.

Keen, A. M., and Coan, E., 1974, Marine mollusca genera of westhern North America (second edition): Stanford, California, Stanford University Press, 161 p.

Kerr, D. R., and Kidwell, S. M., 1991, Late Cenozoic sedimentation and tectonics, Western Salton Trough, California, *in* Walawender, M. J., and

Hanan, B. B., eds., Geological excursions in southern California and Mexico: Geological Society of America, San Diego, California, p. 397–416.

Lewis, C. J., 1994, Constraints in extension in the Gulf Extensional Province from the Sierra San Fermín, northeastern Baja California, Mexico [Ph. D. thesis]: Cambridge, Massachusetts, Harvard University, 361 p.

Lewis, C. J., 1996, Stratigraphy and geochronology of Miocene and Pliocene volcanic and marine rocks in the Sierra San Fermín and southern Sierra San Felipe, Baja California, Mexico: Geofisica Internacional, v. 35, p. 3–25.

Lewis, C. J., and Stock, J. M., 1997, Paleomagnetic evidence of localized rotation during Neogene extension in the eastern Sierra San Fermín, northeastern Baja California, Mexico: Journal of Geophysical Research (in press).

Londsdale, P., 1989, Geology and tectonic history of the Gulf of California, *in* Winterer, E. L., Hussong, D. M., and Decker, R. W. S., eds., The Eastern Pacific Ocean and Hawaii: Boulder, Colorado, Geological Society of America, Geology of North America, v. N., p. 499–521.

Martín-Barajas, A., and Stock, J. M., 1993, Estratigrafía volcánica y características petrológicas de la secuencia volcánica de Puertecitos, NE de Baja California: Transición del vulcanismo de arco a vulcanismo de rift en el golfo, *in* Delgado-Argote, L., and Martín-Barajas, A., eds., Contribuciones a la tectónica del occidente de México: Monografías de la Unión Geofísica Mexicana 1, p. 66–89.

Martín-Barajas, A., Téllez-Duarte, M., and Rendón-Márquez, G., 1993, Estratigrafía y ambientes de depósito de la secuencia marina de Puertecitos, NE de Baja California: Implicaciones sobre la evolución de la margen occidental de la depresión del golfo, *in* Delgado-Argote, L. and Martín-Barajas, A., eds., Contribuciones a la tectónica del occidente de México: Monografías de la Unión Geofísica Mexicana 1, p. 90–114.

Martín-Barajas, A., Stock, J. M., Layer, P., Renne, P., Hausback, M. B., and López-Martínez, M., 1995, Arc-rift transition volcanism in the Puertecitos Volcanic Province, northeastern Baja California, Mexico: Geological Society of America Bulletin, v. 107, p. 407–424.

McLean, H., 1988, Reconnaissance geologic map of the Loreto and part of the San Javier quadrangles, Baja California Sur, Mexico: U.S. Geological Survey Miscellaneous Field Studies Map MF-2000, scale 1:50,000.

Morris, R. H., Abbott, D. P., and Haderlie, E. C., 1980, Intertidal invertebrates of California: Stanford, California, Stanford University Press, 169 p.

North American Commission on Stratigraphic Nomenclature, 1983, North American Stratigraphic Code: American Association of Petroleum Geologists Bulletin, v. 67, p. 841–875.

Patchett, P. J., and Spencer, J. E., 1995, Sr isotope evidence for a lacustrine origin for the Upper Miocene to Pliocene Bouse Formation, Lower Colorado River Trough, and implications for timing of Colorado Plateau uplift: Eos (Transactions, American Geographical Union) (Fall Meeting Supplement), v. 76, p. F604.

Quinn, H. A., and Cronin, T. M., 1984, Micropaleontology and depositional environments of the Imperial and Palm Springs Formations, Imperial Valley, California, *in* Rigsby, C. A., ed., The Imperial Basin—Tectonics, sedimentation and thermal aspects: Los Angeles, California, Pacific Section, Society of Economic Paleontologists and Mineralogists, p. 71–86.

Rebolledo-Vieyra, M., 1994, Implicaciones tectónicas de la deformación en el Plio-Cuaternario de las terrazas marinas al norte de Puertecitos, NE de Baja California [M.S. thesis]: Ensenada, Baja California, Mexico, Centro de Investigación Científica y Educación Superior de Ensenada, 119 p.

Rendón-Márquez, G., 1992, Estratigrafía de los depósitos del Neógeno en la región de Puertecitos, NE de Baja California [Undergraduate research report]: Ensenada, Baja California, Facultad de Ciencias Marinas, Universidad Autónoma de Baja California, 59 p.

Slyker, R. G., Jr., 1970, Geological and geophysical reconaissance of the Valle San Felipe region, Baja California, Mexico [M. S. thesis]: San Diego, California, San Diego State College, 97 p.

Smith, J. T., 1991, Cenozoic marine mollusks and paleogeography of the Gulf of California, *in* Dauphin, J. P., and Simoneit, B. R. T., eds., The Gulf and Peninsular Province of the Californias: American Association of Petroleum Geologists Memoir 47, p. 637–666.

Smith, J. T., Smith, J. G., Ingle, J. G., Gastil, R. G., Boehm, M. C., Roldán, J. Q., and Casey, R. E., 1985, Fossil and K-Ar age constraints on upper middle Miocene conglomerate, SW Isla Tiburón, Gulf of California: Geological Society of America Abstracts with Programs, v. 17, p. 409.

Sommer, M. A., and García, J., 1970, Potassium-argon dates for Pliocene rhyolites sequences east of Puertocitos, Baja California: Geological Society of America Abstracts with Programs, v. 2, p. 146.

Stock, J. M., 1989, Sequence and geochronology of Miocene rocks adjacent to the Main Gulf Escarpment, southern Valle Chico, Baja California: Geofisica Internacional, v. 28, p. 851–896.

Stock, J. M., and Hodges, K. V., 1990, Miocene to Recent structural development of an extensional accommodation zone, NE Baja California, Mexico: Journal of Structural Geology, v. 12, p. 315–328.

Stock, J.M., Martín-Barajas, A., Suárez-Vidal, F., and Miller, M., 1991. Miocene to Holocene extensional tectonics and volcanic stratigraphy of NE Baja California, Mexico, *in* Walawender., M, and Hanan, B. B., eds., Geological excursions in southern California and Mexico: Geological Society of America, San Diego, California, p. 44–67.

Umhoefer, P. J., Dorsey, R. J., and Renne, P., 1994, Tectonics of the Pliocene Loreto basin, Baja California Sur, Mexico, and evolution of the Gulf of California: Geology, v. 22, p. 649–652.

Vázquez-Hernández, S., 1996, Estratigrafía y ambientes de depósito de la porción noroeste de la Sierra El Mayor, cuenca de la Laguna Salada, Baja California [Tesis de Maestría]: Ensenada, Baja California, Centro de Investigación Científica y Educación Superior de Ensenada, 156 p.

Vázquez-Hernández, S., Carreño, A. L., and Martín-Barajas, A., 1996, Stratigraphy and paleoenvironments of the Mio-Pliocene Imperial Formation in the eastern Laguna Salada area, Baja California, Mexico, *in* Abbott, P., and Cooper, J., eds., Field Conference Guide 1996, American Association of Petroleum Geologists and Pacific Section, Society of Economic Paleontologists and Mineralogists, v. 80, p. 373–380.

Walker, G. P. L., 1979, A volcanic ash generated by explosions where ignimbrite entered the sea: Nature, v. 281, p. 642–646.

Wilson, I. F., 1948, Buried topography, initial structures and sedimentation in Santa Rosalía area, Baja California, Mexico: American Association of Petroleum Geologists Bulletin, v. 32, p. 1762–1807.

Winker, C. D., and Kidwell, S. M., 1986, Paleocurrent evidence for lateral displacement of the Pliocene Colorado River delta by the San Andreas fault system, southeastern California: Geology, v. 14, p. 788–791.

Manuscript Accepted by the Society December 2, 1996

Geological Society of America
Special Paper 318
1997

Development and foundering of the Pliocene Santa Ines Archipelago in the Gulf of California: Baja California Sur, Mexico

Maximino E. Simian and Markes E. Johnson
Department of Geosciences, Williams College, Williamstown, Massachusetts 01267

ABSTRACT

Parts of an archipelago consisting of five Pliocene islands with high rocky shorelines are preserved in the Punta Chivato region of Baja California Sur, Mexico. The name for this old island group is taken from Islas Santa Ines (three small islands that originally formed one of the Pliocene islands) located 2 km southeast of the Punta Chivato promontory in the Gulf of California. With an elevation more than 100 m above early Pliocene sea level, the largest of the islands was 7 km^2. It now forms the Punta Chivato promontory. Island cores are composed of resistant Miocene volcanics (mostly andesite) belonging to the Comondú Group. They are skirted by carbonate ramps sloping at angles averaging 6.5° from present sea level up to elevations of approximately 80 m. Lithofacies representing intertidal conglomerates and siltstone to offshore limestones and siltstones are attributed to the lower Pliocene San Marcos and upper Pliocene Marquer Formations. Significant index fossils include the echinoid *Clypeaster bowersi*, the sand dollars *Encope sverdrupi* and *E. shepherdi*, the pectens *Aequipecten deserti* and *A. sverdrupi*, and the coral *Solenastrea fairbanksi*.

At their fullest development during the early Pliocene, the islands blocked and refracted waves driven by strong seasonal winds from the north. Windward biofacies found on the north and east sides of the Punta Chivato promontory include a diverse fauna of intertidal molluscs. Colonies of the coral *Solenastrea fairbanksi* also occur on the north side. Leeward biofacies occurring only on the sheltered south side of the Punta Chivato promontory include concentrations of oysters, the sand dollar *Encope sverdrupi*, the small echinoid *Agassizia scorbiculata*, and extensive horizonal burrows typical of "ghost shrimp." Fragments of fossil bone incorporated in basal conglomerate indicate that whales navigated the archipelago. A computer model is used to simulate Pliocene wave refraction around the main island at Punta Chivato. The archipelago completely sank below the surface of the Gulf of California by the start of late Pliocene time. Limestones rich in rhodoliths and siltstones bearing abundant pectens transgressed the onshore facies and buried the islands. Relative change in sea level was at least 100 m, based on the topography of the Pliocene-Miocene unconformity.

INTRODUCTION

Complete paleoislands with distinct rocky shorelines are rarely discussed in the geological literature (Johnson, 1992). A few reports provide examples under the rubric of *monadnocks* buried by marine strata. Elevations of basement gneiss in northern Norway are fully encircled by Eocambrian quartzites (Bjorlykke, 1967). Similar features are mentioned from the Precambrian-Cambrian boundary in the Grand Canyon (Sharp, 1940). Resistant Precambrian rocks are buried by Cambrian conglomerates and

Simian, M. E., and Johnson, M. E., 1997, Development and foundering of the Pliocene Santa Ines Archipelago in the Gulf of California: Baja California Sur, Mexico, *in* Johnson, M. E., and Ledesma-Vázquez, J., eds., Pliocene Carbonates and Related Facies Flanking the Gulf of California, Baja California, Mexico: Boulder, Colorado, Geological Society of America Special Paper 318.

sandstones in the Baraboo region of Wisconsin and by Silurian strata in the Longmynd region of Shropshire in England (Grabau, 1940; Raasch, 1958). Precambrian gneiss hills rising above the Prague peneplain in Bohemia are interpreted as islands flooded by a mid-Cretaceous transgression (Hercogová and Kříž, 1983).

Other reports refer to *inliers* of older resistant rock engulfed by younger strata. The Oscar Range of Western Australia is a Devonian reef complex that surrounds an island of Precambrian quartzite and schist (Playford and Lowry, 1966). Small inliers of granite-gneiss basement in western Orkney are described as islands inundated by Devonian Old Red sandstone (Fannin, 1969). More subtle inliers of Carboniferous limestone surrounded by Jurassic limestone along the Bristol Channel of South Wales were recognized as paleoislands by Trueman (1922) and subsequently named St. David's Archipelago by Ager (1974).

Much of the existing literature on monadnocks and inliers as paleoislands is anecdotal in nature. Present-day islands studied by botanists and marine biologists are self-contained field laboratories illustrating the division of habitats into wet and dry zones, windward and leeward shores, and their respective zones of high and low wave energy. Islands are conveniently discrete units of local geography, that interface ecologically on a larger regional scale with atmospheric and oceanographic patterns. Hayes et al. (1993), in their study of a small andesite island off the Gulf of California in peninsular Baja California's Bahía Concepción, attempted to assess the fossilization potential of Recent rocky-shore biotas occurring in distinct windward and leeward settings. Similar rocky shores with fossil associations around a complete paleoisland dating from the late Pleistocene were subsequently discovered and described at nearby Bahía Santa Ines (Libbey and Johnson, 1997). Another example of a complete paleoisland with well-documented rocky-shore biotas from contrasting windward and leeward settings is found in the Upper Cretaceous of northern Baja California (Lescinsky et al., 1991; Johnson and Hayes, 1993). One of the geologically older and most thoroughly studied paleoisland groups in terms of its original wind and wave exposure is that of the Baraboo district in southern Wisconsin (Raasch, 1958; Dott, 1974). Unfortunately, that Cambrian archipelago offers no adequate fossil record for incorporation into an ecological reconstruction.

This chapter describes the paleoecology of a 3- to 5- m.y.-old island group extraordinarily well preserved in the Punta Chivato region of Baja California Sur on the Gulf of California. An abbreviated report by us (Johnson and Simian, 1996) compares the Pliocene record of eustasy with the topography of the Pliocene-Miocene unconformity on the largest of five paleoislands in order to separate components of tectonic influence from the local mix of relative sea-level change. Here, we consider the entire island group in the context of variations in Pliocene lithofacies and biofacies associated with rocky shorelines. The resulting patterns of windward and leeward facies in the Pliocene record are compared with the contemporary system of atmospheric and oceanographic circulation in the Gulf of California.

LOCATION AND GEOLOGICAL SETTING

The Punta Chivato region is exposed to the Gulf of California on the east coast of peninsular Baja California, located between the principal towns of Santa Rosalia 45 km to the northwest and Loreto 130 km to the southeast (Fig. 1a). The region covers approximately 25 km^2 and includes a string of small islands and several mainland topographic features rising between 80 m and more than 100 m above sea level (Fig. 2b). Mesa Ensenada de Muerte occurs as a wide ridge to the northwest. The region's highest elevation occupies the broad back of the Punta Chivato promontory (Mesa Atravesada), projecting 4 km eastward into the Gulf of California. Islas Santa Ines are found in the southeast corner of the region. The northernmost of three islands is the site of a sea lion colony. South of the Punta Chivato promontory, the broad sheltered bay is called Bahía Santa Ines. Mesa El Coloradito is a narrow ridge near the region's center. To the southwest is Mesa Barracas, so called because its receding rows of eroded escarpments and dip slopes give the appearance of orderly barracks.

Geologically, each of these five areas is associated with a central core of volcanic rocks partly shrouded by sedimentary layers expressing very low attitudes of dip. Dip directions tend to be radial in pattern and symmetrical in slope, particularly around the perimeter of the Punta Chivato promontory. This relationship was used as evidence by Johnson and Simian (1996) to eliminate tectonic tilting as a postdepositional factor and apply the carbonate ramp model of Ahr (1973) as a syndepositional agent. The earliest geological assessments of the area, however, date back to a one-day reconnaissance by members of the 1940 Scripps cruise to the Gulf of California. Recording what he found in the vicinity of Punta Chivato, Anderson (1950, p. 34) detailed patches of "a pebbly limestone and siltstone" riding unconformably on basalt belonging to the Miocene Comondú Group. The pebbly material consists of eroded volcanic clasts in a carbonate matrix. Correlation of these strata with the lower Pliocene San Marcos Formation was supported by Durham (1950). The type locality, at Isla San Marcos, is only 15 km northwest of Punta Chivato. Durham (1950) recovered the lower Pliocene echinoid *Clypeaster bowersi* from limestone on the north shore of the Punta Chivato promontory as well as the lower Pliocene sand dollar *Encope sverdrupi* and lower Pliocene pectens *Aequipecten deserti* and *A. sverdrupi* from siltstone on the south shore. The only other part of the Punta Chivato area mentioned by Durham (1950) is Mesa Barracas, where the upper Pliocene sand dollar *Encope shepherdi* is recorded. Durham (1950) attributed strata at Mesa Barracas to the younger Marquer Formation.

Pliocene bivalves in the vicinity of Playa La Palmita are briefly mentioned in the report by Hertlein (1957). The only other published studies on the Punta Chivato area appear in the surveys of Ortlieb (1984, 1991), which focus on Pleistocene terraces. Some of the terraces delineated by Ortlieb (1984) appear to be confused with Pliocene ramps (Johnson and Simian, 1996). Additional evidence to this effect is presented herein.

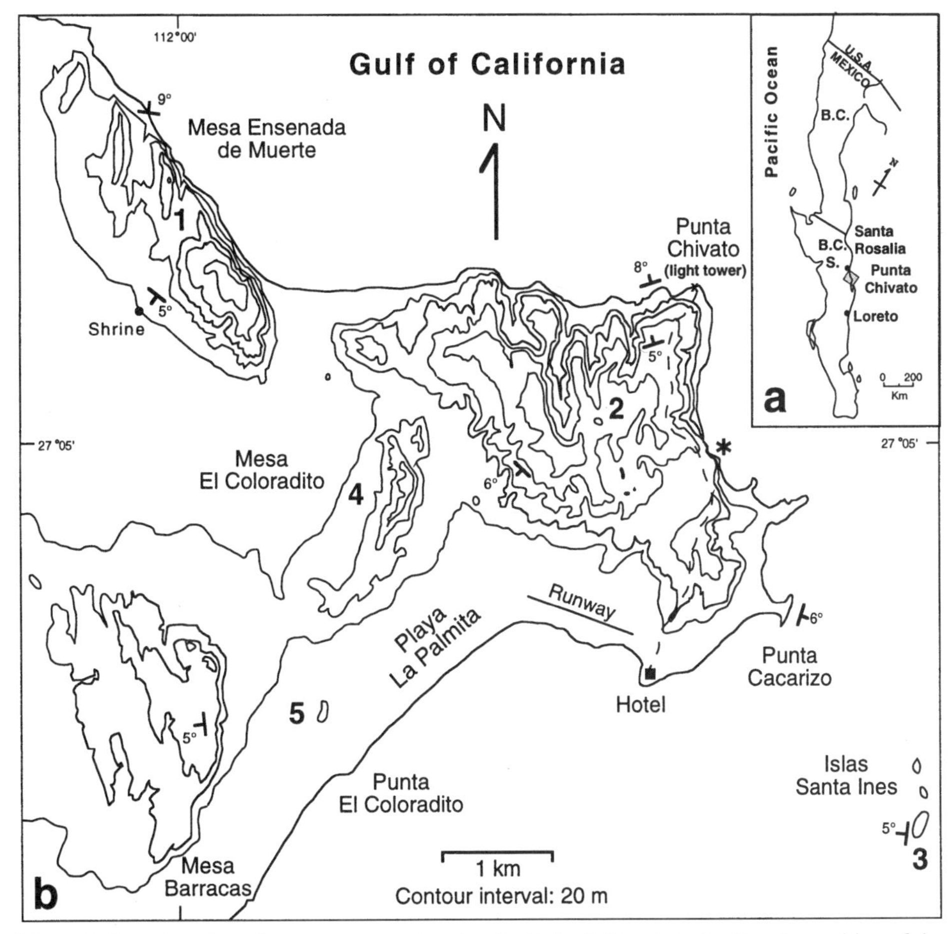

Figure 1. Location of study area. a, map of peninsular Baja California indicating the position of the Punta Chivato region between Santa Rosalia and Loreto in Baja California Sur on the Gulf of California. b, enlargement of the Punta Chivato region showing local topography clustered around five centers interpreted as Pliocene islands (numbered). Dashed line represents a road between the hotel and Punta Chivato.

METHODS AND MATERIALS

Geological mapping, on a topographic base map enlarged from the 1:50,000 San Jose de Magdalena sheet (G12A46) produced by the Instituto Nacional de Estadistica Geografia e Informatica, was the primary method followed in the deployment of this field study undertaken in January 1995. The distribution of Pliocene lithofacies and biofacies indicative of windward and leeward environments was tested against a computer model specific to the Punta Chivato region, using the program Wavemaker 3.0 within the parameters recommended by Fox (1989). The bathymetric data for this exercise were transposed from Chart V of Shepard (1950). Fossils illustrated in this chapter are reposited with the Colección Paleontologica de Referencia de la Universidad Autonoma de Baja California (UABC) in Ensenada, Mexico. For the most part, fossil identifications are keyed to the treatise by Durham (1950), and the life habits of extant species are checked against the guide by Brusca (1980).

PARADE OF ISLANDS

The most notable geological contact throughout the Punta Chivato region is the bold unconformity between Miocene volcanics and Pliocene strata, exhibiting a persistent ramp-shaped plane of erosion (Johnson and Simian, 1996). A superior example in cross section through its long axis is exposed on the eastern cliff face of the Punta Chivato promontary (Fig. 2a). The map position of this site is marked by an asterisk in Figure 1. Emplacement of a basal conglomerate derived from underlying andesite belonging to the Comondú Group may be traced directly over the unconformity surface, where it rises 41 m over a horizontal distance of 200 m. Thin stringers of conglomerate projecting into limestone show that marine onlap was not continuous (Fig. 2b) but was interrupted by a series of regressive progradations. Here, at one spot, nearly half the relative rise in sea level that affected the foundering of an entire island group is graphically represented. In this section we portray in succession the lithological and paleontological fabrics of ramp-associated

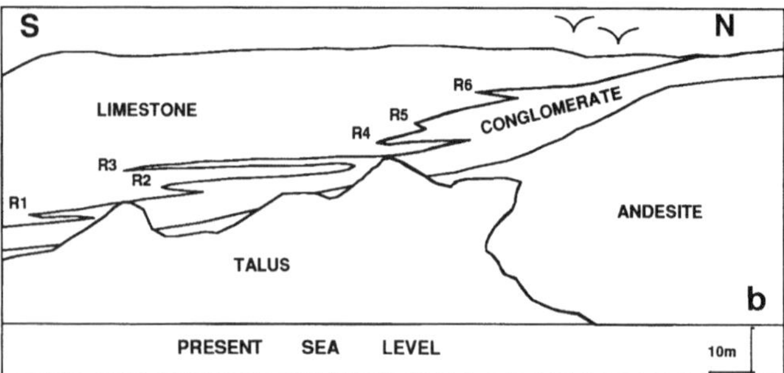

Figure 2. Pliocene-Miocene unconformity exposed in the high cliff face on the east side of the Punta Chivato promontory (location marked by asterisk in Fig. 1). a, photograph looking west from offshore. b, line drawing providing a slightly magnified overlay of the photograph's central field of view (after Johnson and Simian, 1996). Designations R1 through R6 refer to regressive conglomeratic wedges (R3 is 1 m thick and 55 m in length).

strata around each of the five topographic features enumerated in Figure 1. Summarization of these data will be by way of a final geological map.

Mesa Ensenada de Muerte

Rising steeply more than 100 m above sea level, the 3 km × 1 km ridge at Mesa Ensenada de Muerte is dominated by its dark Comondú andesites and red tuffs. Much of the Pliocene cover is now stripped away from the mesa, but sedimentary ramps dip 9° to the northeast off one side and 5° to the southwest off the other. Siltstone with abundant fossil pectens ramp against the west-central side of the ridge around the site of a religious shine (Fig. 1). Pockets of basal conglomerate are scattered across the southeast end at elevations between 70 and 90 m above sea level. Here, a mammalian bone fragment 1.6 m in length and 18 cm in diameter was found encased in the andesitic conglomerate. Given the size of this fragment, it must have belonged to a whale (Lawrence G. Barnes, 1995, personal commun.) that beached in the vicinity.

Punta Chivato promontory

Southeast of Mesa Ensenada de Muerte is the main Punta Chivato promontory (Mesa Atravesada), which terminates to the northeast and to the southeast at Punta Chivato and Punta Cacarizo, respectively. The promontory rises more than 100 m above sea level and occupies an area approximately 7 km^2 in size. Access across the east end of the promontory is provided by a 3.5-km-long road connecting the hotel with the light tower at Punta Chivato (Fig. 1). Only 0.25 km due west of the light tower is another example of a carbonate ramp resting on the Miocene-Pliocene unconformity (Fig. 3a). The ramp is well exposed across its long axis, and the unconformity surface dips north by northwest at an angle of 8° into the ocean. Cobble- to boulder-size andesite clasts (up to 70 cm in diameter) cemented within a fossil-rich carbonate matrix comprise the basal conglomerate of the San Marcos Formation at this locality (Fig. 3b). Immediately above this level is a 6-m-thick section of carbonate sand with well-developed laminations dipping 6° in the same direction as the ramp below. These strata of beach sand are beautifully eroded as karst pinnacles.

In contrast to the ramp near Punta Chivato, another ramp at Punta Cacarizo is exposed along strike, perpendicular to its long axis (Fig. 3c). This ramp supports a thinner basal conglomerate, only 3.75 m thick with andesite pebbles and cobbles up to 20 cm in diameter cemented in a fossil-rich carbonate matrix. Bedding planes of thick-bedded limestone succeed the basal conglomerate at Punta Cacarizo and dip 6° east by southeast into the ocean. Although the depression behind Punta Cacarizo is filled by a modern sand tombolo (Fig. 3d), the same Miocene-Pliocene unconformity occurs about 50 m above sea level in the bluffs due west. A trigonometric solution for the slope of the angle connecting the unconformity exposures between the bluffs and Punta Cacarizo yields a value of 6° for the intervening ramp removed by erosion.

Fossils from the basal conglomerates near Punta Chivato and Punta Cacarizo are dominantly internal and external molds belonging to molluscs. Rarer echinoids sometimes occur with their calcite tests largely intact. Some of the representative fossils from the ramp near Punta Chivato are illustrated in Figure 4. Presence of the robust echinoid *Clypeaster bowersi* (Fig. 5a) is confirmed. This is the only fossil previously reported by Durham (1950, p. 27) from the north side of the Punta Chivato promontory. The age range given by Durham (1950, p. 41) for this species is lower to middle Pliocene. Among the associated gastropods at this locality are *Conus* cf. *brunneus* (Figs. 4b and c), *Fasciolaria* cf. *princeps* (Figs. 4d and e), and *Mitra* cf. *tristis* (Fig. 4f). The largest of these, *F. princeps*, is reported from the lower Pliocene Imperial Formation of southern California (Hana, 1926), but its geological range is usually cited as Pleistocene to Recent. Durham (1950, p. 100) gives the age of *C. brunneus* as lower Pliocene, but Brusca (1980, p. 184) reports it as an extant species from the Gulf of California. The age range for *M. tristis* cited by Durham (1950, p. 105) is

Figure 3. Carbonate ramps on the north and east sides of the Punta Chivato promontory. a, view from Punta Chivato (see Fig. 1) looking west toward the eroded cross section of the north-dipping (8°), Pliocene-Miocene unconformity (middle ground). b, close-up view of the same unconformity showing conglomerate formed by andesite clasts in the basal Pliocene San Marcos Formation. c, oblique areal view of east-dipping (6°) San Marcos ramp at Punta Cacarizo (see Fig. 1). d, seaward view of Punta Cacarizo from an andesite basement ridge 50 m above present sea level.

upper Pliocene to Recent. According to Brusca (1980), live *C. brunneus* occupies habitats from the rocky, low intertidal to offshore; *F. princeps* thrives offshore in sand; and the allied species of *Mitra fultoni*, which is comparable in size to *M. tristis*, occurs under rocks in the low intertidal zone.

A variety of fossil bivalves co-occurs with echinoids and gastropods in the basal conglomerate of the San Marcos Formation near Punta Chivato. Among them are *Glycymeris* cf. *maculata* (Figs. 4g and h), *Crassatellites* cf. *digueti* (Fig. 4i), *Barbatia* cf. *reeveana* (Figs. 4j and k), and *Tellina* cf. *orhracea* (Figs. 4l and m). The age range of *G. maculata* is said to be upper Pliocene to Recent (Durham, 1950, p. 56); extant species in the Gulf of California live intertidally in sand (Brusca, 1980, p. 136). The range of *C. digueti* is reportedly Pleistocene to Recent (Durham, 1950, p. 70), but its life habits are not commented on by Brusca (1980). The range of *B. reeveana* is listed as upper Pliocene to Recent (Durham, 1959, p. 56); extant species live from the mid-intertidal to offshore in tidal flats, lagoons, and rock-mud interfacies (Brusca, 1980, p. 133). *T. ochracea* is given an age range of upper Pliocene to Recent by Durham (1950, p. 89); extant species thrive on tidal flats and soft bottoms extending offshore, according to Brusca (1980, p. 148).

Species identifications made from internal molds can be problematical, but the majority of these fossils are Pliocene if not early Pliocene in age. Likewise, the overall reflection of habitat is clearly the low intertidal zone, although there is some mixing of sand and rock-dwelling species. No encrustations of fossil barnacles, red coralline algae, or any other organic covering that might suggest an environment higher in the intertidal zone was observed on andesite clasts.

The spectacularly exposed carbonate ramp already discussed and illustrated in Figure 2a occurs roughly midway between Punta Chivato and Punta Cacarizo (marked by asterisk on Fig. 1). Talus accumulating below the east cliff face is rich in

a
d
b
c
e
f
g
j
i
k
h
l
m

Figure 4. Internal molds of lower Pliocene fossils from the San Marcos Formation at Punta Chivato (see Figs. 3a and 3b). All specimens are figured at 1×. a, top view of the echoinoid *Clypeaster bowersi* (UABCFCM 1938). b, side view of the gastropod *Conus* cf. *brunneus* (UABCFCM 1939). c, top view of the same specimen. d, side view of the gastropod *Fasciolaria* cf. *princeps* (UABCFCM 1940). e, top view of the same specimen. f, side view of the gastropod *Mitra* cf. *tristis* (UABCFCM 1941). g, hingeline of the bivalve *Glycymeris* cf. *maculata* (UABCFCM 1942). h, right valve of the same specimen. i, left valve of the bivalve *Crassatellites* cf. *digueti* (UABCFCM 1943). j, right valve of the bivalve *Barbatia* cf. *reeveana* (UABCFCM 1944). k, hingeline of the same specimen. l, left valve of the bivalve *Tellina* cf. *ochracea* (UABCFCM 1945). m, hingeline of the same specimen.

pectens preserved in a lime-rich siltstone. Fossil material is very abundant and easy to collect on the flats west of the road above this site, and it appears that much of the talus is derived from this level approximately 50 m above present sea level. The four most abundant species from this locality include *Argopecten circularis*, "*Aequipecten*" *corteziana*, *Nodipecten* cf. *nodosus*, and *Flabellipecten bosei* based on material collected by us and submitted for identification (Judith Terry Smith, personal communication, 1995). *A. circularis* has previously been found in the Infierno, Carmen, and Marquer Formations (all upper Pliocene); *A. corteziana*, *N. nodosus*, and *F. bosei* are known from both the San Marcos and Boleo Formations (ranging through the lower and upper Pliocene), according to the range chart published by Smith (1991, table 3). The paucity of fossils other than pectens and the total absence of reworked andesite clasts implies that these strata were deposited in an offshore setting, which is consistent with the facies relationship diagrammed in Figure 2b.

Other ramps occur on the south flank of the Punta Chivato promontory. A typical view looking across a deep arroyo incised on the middle of the southwest face shows a good exposure of the Pliocene-Miocene unconformity (Fig. 5a). Riding the unconformity above Comondú andesite is a carbonate ramp dipping 6° to the southwest (Fig. 1). The diverse fossil association characteristic of the basal conglomerate at Punta Cacarizo and Punta Chivato is absent here. Instead, the basal conglomerate is littered with the fragmented debris of large oysters (Fig. 5b). Carbonate ramps on this side of the Punta Chivato promontory are not exposed at the coast. Much of the wide shelf on which a dirt runway exists (Fig. 1) is underlain by siltstone and sandstone apparently unrelated to a ramp structure.

Bedding planes atop low-lying shore cliffs form the south extension of this shelf on the seaward side of the runway. These strata are exposed intermittently for almost a kilometer along the coast (Fig. 6). Some stretches expose local accumulations of pecten hash. It was from these strata that Durham (1950) collected the lower Pliocene pectens *Aequipecten deserti* and *A. sverdrupi* as well as the lower Pliocene sand dollar *Encope sverdrupi* (Fig. 5c). A population census of *E. sverdrupi* was undertaken over a surface area of approximately 1,300 m^2. A total of 173 complete to relatively whole individuals was counted. Among these, 111 are preserved in life position (dorsal surface up) and 62 are overturned (ventral surface up). Populations of small echinoids identified as *Agassizia scorbiculata* are concentrated over a short stretch of sandstone (Fig. 6). According to Durham (1950, p. 188), this species also has a geological range extending to the lower Pliocene. Other bedding planes in this area reveal extensive horizontal burrows characteristic of "ghost shrimp." At the far east end of the bay near the hotel, the ancient shoreline is marked by a small outcrop of siltstone with eroded andesite clasts abutting against Comondú cliffs (Figs. 1 and 6).

As for many of the fossils at Punta Cacarizo and Punta Chivato on the north and east sides of the promontory, an early Pliocene age is apparent for the constituents of these biofacies. In terms of lithofacies and biofacies composition, however, the depositional environment on the south side of the promontory strongly indicates a more sheltered setting in which finer clastics could accumulate and fragile organisms such as sand dollars could be preserved in life position.

Another ramp and biofacies are unique to the north side of the Punta Chivato promontory. A sequence consisting of 3.35 m of basal conglomerate and limestone perches directly on Comondú andesite at an elevation 75 m above sea level on the east edge of a deep gorge accessible west of the road to the Punta Chivato light tower. Here the beds dip 5° north by northwest (Fig. 1). The basal conglomerate consists of andesite clasts up to 18 cm in diameter and large dome-shaped colonies of the coral *Solenastrea fairbanksi* up to 22 cm in diameter (Fig. 5d). Comparable limestone rich in this coral also occurs farther to the west above the north shore of the promontory. The tumbled nature of these beds and their proximity to the Pliocene-Miocene unconformity indicates vigorous onshore wave activtity.

S. fairbanksi is known from few other localities outside Carrizo Creek in the Imperial Formation of southern California (Foster, 1979). That area was once the most northern extension of the Gulf of California, now called the Salton Sea. The geological range of the species is restricted to the Miocene and lower Pliocene.

Islas Santa Ines

Three low islands named Islas Santa Ines are situated southeast of the Punta Chivato promontory (Fig. 1). The Pliocene-Miocene unconformity is exposed at the southwest end of the chain on the largest island. Miocene basement rock here appears to be a volcaniclastic conglomerate composed of tuffaceous pebbles, but many small boulders as large as 40 cm in diameter in the overlying Pliocene conglomerate are distinctly composed of banded gneiss or schist. As no comparable rock is known to outcrop in the Punta Chivato region, an extraneous source is required. This conclusion is supported by the fact that the 3-m-thick sequence of conglomerate and limestone

Figure 5. Additional ramps and other facies from the Punta Chivato promontory. a, south-dipping (6°) carbonate ramp (light) resting on Miocene basement rocks (dark) on the south side of the promontory. Person for scale is standing on the Pliocene-Miocene unconformity (circle at center of the photograph). b, close-up view of oyster debris a few meters above the Pliocene-Miocene unconformity at the same locality (pocket knife for scale is 9 cm long). c, sand dollars, *Encope sverdrupi*, from siltstone layer exposed on the south shore of the Punta Chivato promontory (knife for scale). d, overturned coral colony of *Solenastrea fairbanksi* a few meters above the Pliocene-Miocene unconformity on the north side of the Punta Chivato promontory (knife for scale).

dips westward in opposition to the carbonate ramp at Punta Cacarizo on the other side of an open channel 20 m to 30 m deep. The Islas Santa Ines probably belong to a piece of a separate Pliocene island with a different kind of hard-rock core. *Clypeaster bowersi*, the same large echinoid known from the basal Pliocene conglomerate near Punta Chivato, also occurs together with abundant mollusc molds at this locality. Santa Ines is hereby adopted as the name for the Pliocene archipelago, because the present islands are the only ones remaining in the Punta Chivato region.

Mesa El Coloradito

Rising more than 80 m above sea level, the narrow 2 km × 0.5 km ridge at Mesa El Coloradito forms the central rim of Bahía Santa Ines (Fig. 1). Like Mesa Ensenada de Muerte to the northwest, this ridge of Miocene andesite is stripped of most of its Pliocene apron. A thin, discontinuous band of limestone abuts against the east, seaward face of the ridge above the 40-m-contour interval. The contact traces a former shoreline, but no appreciable conglomerate is recognized and no discernable dip is exhibited by the carbonates. Below the truncated front edge of Pliocene limestone lies Playa La Palmita with its extensive cover of upper Pleistocene carbonates. A few east-directed arroyos penetrate the 4 to 5 m of Pleistocene cover, revealing the presence of a brown siltstone identical to that exposed in low cliffs on the adjacent shore near the runway. No Pliocene fossils were observed in these limited arroyo exposures, but the contact is assumed to be the Pleistocene-Pliocene unconformity.

Low saddles about 30 m above sea level separate Mesa El

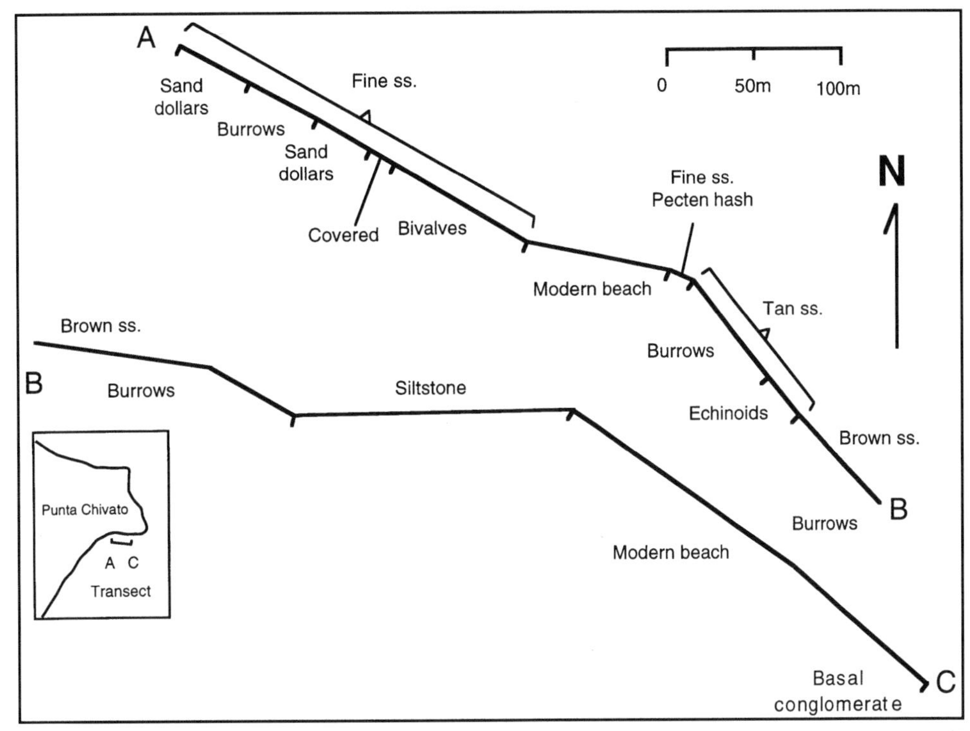

Figure 6. Transect in map view from the south shore of the Punta Chivato promontory showing the relationships of various lithofacies and biofacies in the lower Pliocene San Marcos Formation.

Coloradito from the Punta Chivato promontory to the northeast and Mesa Barracas to the southwest. Exposed around the two ends of the mesa are carbonate sandstones indicative of a paleobeach resting on Miocene andesite. These deposits together with the topography isolating Mesa El Coloradito indicate that it was a separate Pliocene island. No sedimentary rocks, however, are exposed along the mesa's steep west face.

Mesa Barracas

Two small horns of limestone that cap the northeast corner of Mesa Barracas barely surpass an elevation of 80 m above sea level (Fig. 1). The asymmetrical alignment of topography around the mesa conforms to an overall dip slope of about 5° toward the west by southwest. No Miocene basement rock is to be found on Mesa Barracas. Prominent dip slopes are protected by thin limestone cap rocks. Ongoing headward dissection of arroyos is undercutting the limestones to erode thick units of soft lime-rich siltstone. The two lithologies bear different fossil assemblages.

A 36-m-thick section of siltstone bearing abundant pectens is well exposed in the cleft below the limestone horns of Mesa Barracas. Based on material collected by us and submitted for identification (Judith Terry Smith, personal communication, 1995), the most common pectens are *Argopecten circularis*, *Nodipecten arthriticus*, and *Flabellipecten* sp. The first two are extant species also well known from the upper Pliocene Marquer and Carmen Formations elsewhere in Baja California (Smith, 1991, table 3).

Occurrence of the sand dollar *Encope shepherdi* elsewhere on the mesa led Durham (1950) to affiliate the rocks in this area with the upper Pliocene Marquer Formation. Distinctive fossils from limestone layers near the south end of the mesa are illustrated in Figure 7. The large oyster *Ostrea vespertina* (Fig. 7a) ranges through the entire Pliocene (Durham, 1950, p. 59). The arc shell *Anadara multicostata* (Figs. 7b and 7c) ranges from the upper Pliocene to the Recent (Durham, 1950, p. 54). According to Brusca (1980, p. 133), *A. multicostata* presently thrives throughout the Gulf of California on sandbars offshore. Locally abundant on Mesa Barracas are coralline red algal rhodoliths, an example of which is shown in Figure 7d.

Based on research by Foster et al. (this volume) regarding contemporary rhodolith beds in the Gulf of California, these unattached algae with their spherical growth forms are concentrated either in gently sloping, subtidal soft bottoms experiencing moderate wave action (2 to 12 m deep) or level-bottomed channels favored by tidal currents (12 to 25 m deep). Smaller, less robust fossil rhodoliths occur scattered throughout the Pleistocene deposits of nearby Playa La Palmita, where most of the carbonate debris is derived from crushed rhodoliths (Libbey and Johnson, 1977). The limestone beds on Mesa Barracas do not exhibit the same density of crushed algal material, nor is the concentration of relatively whole colonies anything like that of a live rhodolith bed. One level in the southern Barracas does expose a widespread intraformational conglomerate, but the carbonate host rocks on the mesa tend to be extensively tabular without signs of pinching and swelling or cross-bedding suggestive of transportation by

Figure 7. Some fossils from the upper Pliocene Marquer Formation at Mesa Barracas (see Fig. 1). All specimens are figured at 1×. a, right valve of the bivalve *Ostrea vespertina* (UABCFCM 1946). b, hingeline (internal mold) of the bivalve *Anadara multicostata* (UABCFCM 1947). c, right valve (internal mold) of the same specimen. d, top view of a rhodolith formed from coralline red algae (UABCFCM 1948).

strong currents. A direct correlation to the rhodolith habitat is tenuous, but it is likely that the depositional environment of these rhodolith-bearing limestones was between 2 and 12 m deep.

What is certain, however, is that none of the strata on Mesa Barracas were deposited immediately adjacent to rocky shorelines defined by Miocene basement rock. Furthermore, the geological position of these strata appears to be upper Pliocene, in contrast to all other deposits in the Punta Chivato region for which biostratigraphic evidence is available. We propose that Mesa Barracas represents a succession of consistently younger and more distal ramp deposits well off the topographic center of a Pliocene island core. Beginning 0.8 km east of Mesa Barracas is a small rise of Comondú andesite with an elevation above the 20-m-contour interval. This outcrop trails off to the coast at Punta El Coloradito (Fig. 1) before it disappears under water. We assume that lower Pliocene strata occur beneath Mesa Barracas, but they have been eroded away from the area surrounding Punta El Coloradito. Likewise, upper Pliocene strata once may have encircled the main Punta Chivato promontory. Remnants of such strata may still exist, but only offshore.

Geological map summary

Most of the above data are summarized in the accompanying geological map (Fig. 8), using a somewhat larger base map than offered in Figure 1. This scale necessarily excludes Islas Santa Ines but shows the principal lithofacies and biofacies interpreted for all the other Pliocene islands. The early development of the archipelago is lithologically represented by a conglomerate facies of the San Marcos Formation on the north and east sides of the Punta Chivato promontory in contrast with a siltstone facies of the same formation on the south side of the Punta Chivato promontory and the southwest side of Mesa Ensenada. Primary coastal biofacies include (1) intertidal molluscs (from basal conglomerate) restricted to the north and east sides of the Punta Chivato promontory as well as the west side of Islas Santa Ines, (2) robust corals (from massive limestone) limited to the north side of the Punta Chivato promontory, (3) oysters (from basal conglomerate and siltstone) found only on the south side of the Punta Chivato promontory, and (4) sand dollars (from siltstone) exclusively on the south side of the Punta Chivato promontory.

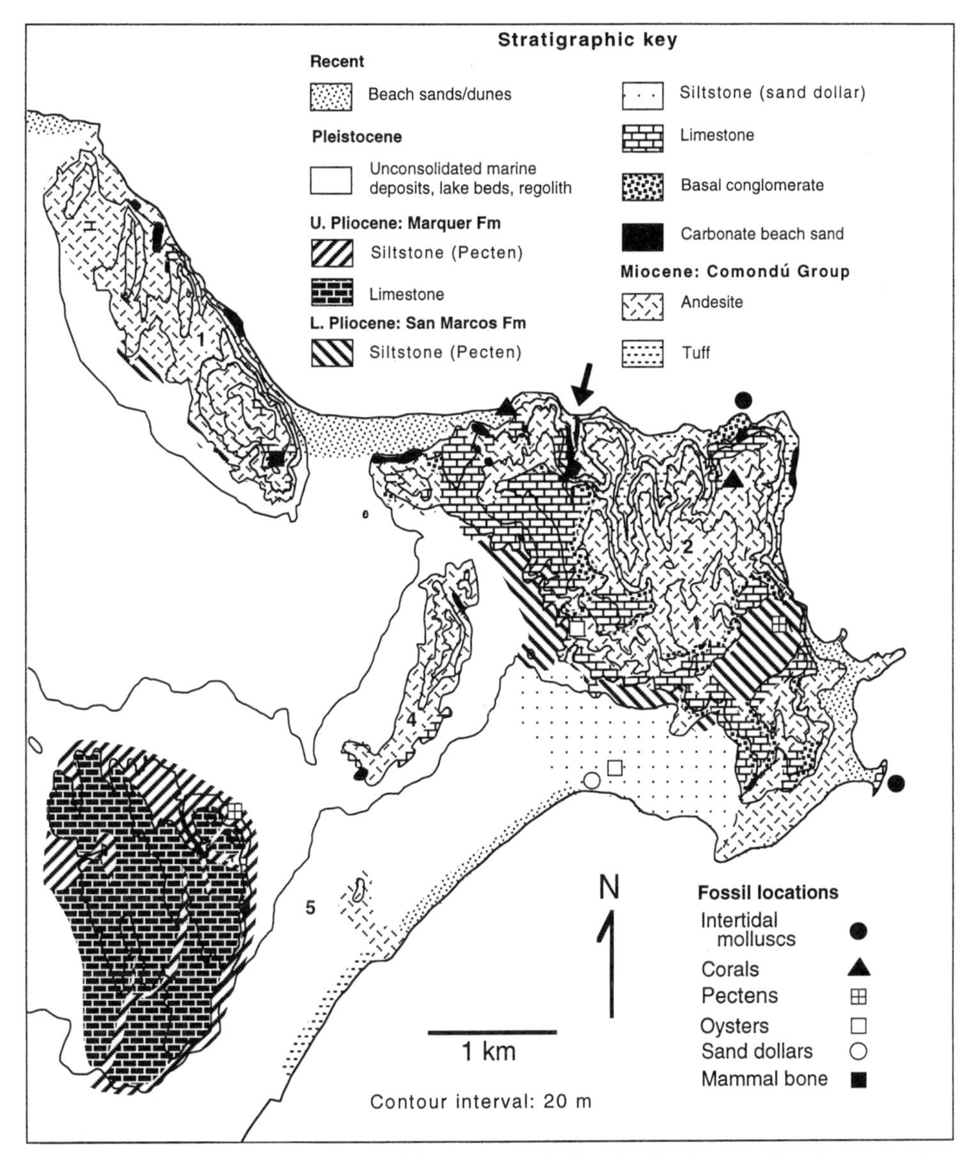

Figure 8. Geological map of the Punta Chivato area, including the distribution of biofacies sensitive to different levels of wave intensity. Corals are found exclusively on the north side of the Punta Chivato promontory, for example, whereas oysters and sand dollars occur only on the south side. Arrow (upper center) points to the location of a paleovalley filled with carbonate beach sand.

Offshore facies are represented by two comparable pecten-bearing siltstones belonging to the San Marcos and Marquer Formations as well as rhodolith-bearing limestones belonging to the Marquer Formation. The ages of these deposits and their map distribution with respect to elevations of Miocene basement rock indicate that the archipelago foundered by the start of late Pliocene time, if not somewhat earlier. Although about a 2 km^2 area of Comondú andesite is now exposed at the core of the Punta Chivato promontory, a thin limestone cap belonging to the San Marcos Formation reaches the promontory's highest elevation 100 m above sea level.

North-south valleys deeply penetrating the north side of the promontory have been enlarged, mainly as a result of Recent erosion, expediting removal of this limestone cap. One valley, however, is partially filled by stratified, carbonate-beach sand (Fig. 8, arrow) comparable to that now forming karst pinnacles near Punta Chivato. The disposition of these beds signifies that the valley must have existed as a Pliocene valley. A strong and persistent source of wind transport out of the north is required to account for the movement of these Pliocene sands 0.5 km inland.

COMPUTER MODEL FOR WAVE ASSAULT

Given a specific coastal configuration and the details of offshore bathymetry, wave assault may be modeled by computer simulation to illustrate a wide range of wave fronts moving from different compass directions, starting with different wave heights and wave periods. Normally, models are tested through a range of variabilities, particularly wave-front direction. In this case, the process was simplified because of the occurrence of a defining wind vector. Near the mouth of the paleovalley on the north side of Punta Chivato (arrow on Fig. 8), several examples of the leatherplant bush *Jatropha cuneata* are bent 90° at the base of the trunk from their normal skyward growth to assume a ground-hugging posture. Physical deformation of this sort in plants due to wind stress is termed a krummholz (Barbour et al., 1987). Such trees are organic wind socks pointing out the direction of prevailing winds. The leatherplant bushes are aligned on an axis of N 10°E to S 10°W, with their growing tips pointing south. This direction not only is consistent with the seasonal winter winds out of the north in the Gulf of California today but also conforms with the map orientation of the Pliocene valley and the distribution of its carbonate-sand deposit.

Waves attacking the Punta Chivato region from the north on a bearing of N 10°E to S 10°W are depicted in model form by the diagram in Figure 9. Constraints imposed by subsurface topography are implied by the location of the 50-fathom isobath. Maximum impact of waves is achieved on the windward north coast. As waves refract around the promontory they begin to lose some of their energy. South of the headland, waves are deflected through almost 90°, breaking at offshore locations where their orthogonal rays overlap. This process further reduces their energy, and minimum impact is achieved on the south (leeward) coast. The scenario modeled here is fully consistent with the geological data mapped in Figure 8, suggesting that the contemporary circulation pattern of seasonal winds in the Gulf of California has a long history reaching back at least to the early Pliocene.

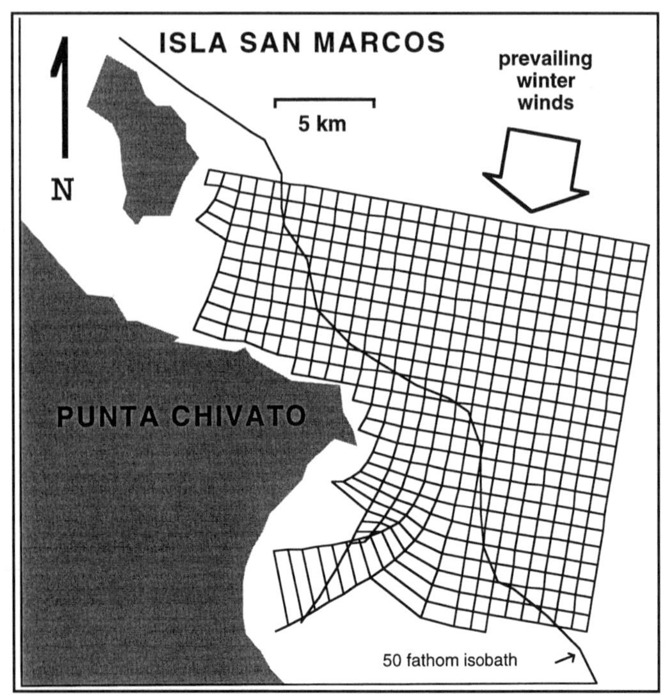

Figure 9. Computer-simulated wave action showing refraction around the Punta Chivato promontory, based on wave fronts pushed from the north to the south as typical of present atmospheric circulation during the winter season. The north shore occurs in a windward setting and the south shore in a leeward setting.

DISCUSSION

Winds, waves, and biotas

Wave movement in the present Gulf of California is strongly affected by atmospheric forcing. The wind field over the gulf is semimonsoonal in nature (Bray and Robles, 1991). During the colder winter months, winds emanating from high-pressure systems to the north are pulled down the axis of the gulf toward the equator. When the arid southwest deserts of North America heat up during the summer months, lighter southerly winds are pulled northward up the axis of the gulf by consuming low-pressure systems. Wind-amplitude and vector data collected from Isla Tortuga, located 35 km due north of Punta Chivato, confirm that the strongest winds blow out of the north from November through April; lighter winds blow out of the south from May through October (Bray and Robles, 1991, p. 514). Winter wind velocity may reach a peak of about 10 m/s on Isla Tortuga in January, whereas peak velocity during July may be only half as much. It is the winter winds, sometimes sustained over several days at a time, that are actively deforming the shape of the leatherplants on the north shore of the Punta Chivato promontory.

The same winter winds crossing open water from Isla Tortuga to assault the Punta Chivato promontory continue 20 km farther south to reach the north-facing mouth of Bahía Concepción. Located another 30 km south within this narrow bay is Isla El Requesón, where Hayes et al. (1993) studied modern rocky-shore biotas and their fossilization potential. The intertidal biotas encircling this small island are readily differentiated into a windward biofacies dominated by encrusting species of coralline red algae and the byssus-attached shell *Arca pacifica*, as opposed to a leeward biofacies dominated by barnacles and the oyster *Ostrea palmula* (Hayes et al., 1993; see their figs. 2 and 3). The outer shore of Isla El Requesón sustains harsh wave shock during the winter months. Lagoonal sediments are accumulating on the back side of the island, and a protected tombolo connects the island to the west shore of Bahía Concepción.

Some of the same species occurring at Isla El Requesón are found as fossils in deposits of late Pleistocene age associated with rocky shores at Playa La Palmita in the Punta Chivato region (Libbey and Johnson, 1997). One of several rocky-shore environments well preserved there is a complete andesite island 12,500 m^2 in area. It exhibits cobbles encrusted with fossil algae on its windward side and extensive populations of the fossil bivalves *Periglypta multicostata*, *Codakia distinguenda*, and *Chione californiensis* nestled in life position in sand and cobbles on its leeward side. This small paleoisland sits on the sheltered south side of the Punta Chivato promontory on Playa La

Palmita (Fig. 1b). Its seaward shore was affected less harshly by direct summer waves from the south and refracted winter waves from the north. Relationships at Isla El Requesón and Playa La Palmita underscore the continuity of atmospheric and oceanographic patterns along the gulf coast of Baja California Sur.

Few other examples of windward and leeward biofacies on ancient islands are well documented. Near Erendira on the Pacific coast of northern Baja California during late Cretaceous times, a small andesite island developed leeward facies, including encrusting oysters and bryozoans (Lescinsky et al., 1991), as opposed to windward facies dominated by encrusting red algae and rudist bivalves (Johnson and Hayes, 1993). Modern El Requesón, Pliocene Punta Chivato, and Cretaceous Erendira all show the development of islands with robust adaptations of ark shells, corals, or rudists in windward, rough-water settings in contrast to oysters in leeward, quieter water settings.

Sea level and tectonic relationships

Discrimination between Pliocene ramp and Pleistocene terrace structures makes a significance difference in the interpretation of the Punta Chivato region's relative sea-level history. Ortlieb (1984, pp. 125–130) describes an east-west transect across the Punta Chivato promontory, which attempts to establish a succession of Pleistocene terraces between 15 and 75 m above present sea level. From the highest of these, Ortlieb (1984, p. 129) cites "badly altered fossils" belonging to the Pleistocene Santa Rosalia Formation. It is argued by Johnson and Simian (1996) that these fossils actually are poorly preserved molds of Pliocene fossils from a carbonate ramp. Likewise, Ortlieb (1984) questions the occurrence of Pliocene pectens from an elevation 50 m above present sea level and implies the development of a Pleistocene terrace. Based on the location of Ortlieb's transect, the pecten beds in question are exposed on the flats directly above the east face of the Punta Chivato promontory with its exceptionally well developed ramp (Figs. 1 and 2). As detailed above, the fossil pectens from this locality are unquestionably Pliocene in age (Judith Terry Smith, personal communication, 1995). Parts of an extensive late Pleistocene terrace and other shorelines are confirmed throughout the Punta Chivato region at elevations between 6 and 12 m above present sea level (Johnson and Simian, 1996; Libbey and Johnson, 1997). No laterally consistent features resembling terraces at higher elevations can be justified in the Punta Chivato region. Instead, Pliocene ramps dominate the landscape.

The comparative tectonic calm experienced by this stretch of the Pliocene gulf along Baja California Sur during early Pliocene time makes it manageable to tease apart eustatic and tectonic components of relative sea-level change for the Santa Ines Archipelago (Johnson and Simian, 1996). An estimate of nearly 25 m of eustatic rise for the early Pliocene is based on adjustments by Kennett and Hodell (1995) to the work of Haq et al. (1988), taking into account variations of O^{16} as a proxy for ice-volume fluctuations off Antarctica. Subtracting this value from the approximate 100 m of relative sea-level rise recorded on the Punta Chivato promontory, it may be calculated that the remaining component was due to tectonic foundering at a very low rate of 0.05 mm/yr through the mid-Pliocene (Johnson and Simian, 1996). By comparison, complex faulting and very rapid subsidence of between 5 and 10 mm/yr characterize the mid-Pliocene Loreto basin (Umhoefer et al., 1994).

CONCLUSIONS

The study of island paleogeography and the paleoecological zonation of ancient rocky shores has moved well beyond the sparse and anecdotal literature compiled by geologists on monadnocks and inliers as paleoislands. Lithofacies and biofacies associated with unconformities can provide a wealth of information relating the geography of ancient islands to the atmospheric and oceanographic conditions in which they evolved both physically and biologically during the remote past. Research on the Baja California peninsula, with its wealth of contemporary ecosystems to apply as models and its varied and well-exposed geology, has played a stimulating role in these advancements (Hayes et al., 1993; Johnson and Hayes, 1993; Ledesma-Vázquez and Johnson, 1993; Libbey and Johnson, 1997; Johnson and Simian, 1996).

The Pliocene Santa Ines archipelago is the finest example of a paleoisland group yet described in Baja California. It includes parts of five islands with resistant andesite cores, ranging in size from 7 km^2 to 1 km^2 or smaller. The largest is reborn as the Punta Chivato promontory, which continues today to block and deflect seasonal wind-driven waves funneled down the Gulf of California's axis from the north much as it did as a Pliocene island 3 to 5 m.y. ago. Carbonate ramps developed around the circumference of the islands as coastal erosion attacked the topographic relief of Miocene andesites and tuffs belonging to the Comondú Group.

Reworked andesite clasts typically form a basal conglomerate in the lower Pliocene San Marcos Formation. Windward biofacies closely associated with the Pliocene-Miocene unconformity began with a diverse fauna of intertidal molluscs that was succeeded by corals forming massive limestone. *Solenastrea fairbanksi*, previously unknown from peninsular Baja California, is the dominant coral of this facies. Leeward biofacies probably started with dense populations of oysters closely associated with the Pliocene-Miocene unconformity, followed mainly by spatially monospecific populations of sand dollars (*Encope sverdrupi*) and small echinoids (*Agassizia scorbiculata*) living on a silty bottom near to shore.

Offshore facies consisting mainly of lime-rich siltstone with profuse pectens eventually overwhelmed the islands toward the close of the early Pliocene (upper San Marcos Formation) and beginning of the late Pliocene (lower Marquer Formation). A minor component of the burial was contributed by rhodoliths or unattached, coralline red algae adopting a spherical growth form. Complete foundering of the Santa Ines archipelago involved a relative sea-level change of at least 100 m. A rise of almost 25 m was caused by eustasy during the early Pliocene (Kennett and Hodell, 1995); the rest was due to gradual tectonic subsidence without appreciable tilting.

ACKNOWLEDGMENTS

Fieldwork was supported by a grant to M. E. Johnson from the National Science Foundation (INT-9313828). We are grateful for the interaction of several colleagues in the field, including Jorge Ledesma-Vázquez (Universidad Autonoma de Baja California, Ensenada), Keith H. Meldahl (Oberlin College), and Michael S. Foster (Moss Landing Marine Laboratories) as well as B. Gudveig Baarli and Laura K. Libbey (Williams College). Identification of the coral *Solenastrea fairbanksi* was confirmed by Nancy Bud (University of Iowa); pecten collections were examined by Judith Terry Smith (Stanford University). Lawrence G. Barnes (Natural History Museum of Los Angeles County) evaluated for us the bone fragment from Mesa Ensenada de Muerte. This contribution was reviewed by Robert H. Dott, Jr. (University of Wisconsin at Madison) and William N. Orr (University of Oregon).

REFERENCES CITED

Ager, D., 1974, The Jurassic Period in Wales, *in* Owen, T. R., ed., The Upper Palaeozoic and post-Palaeozoic rocks of Wales: Cardiff, University of Wales Press, p. 323–329.

Ahr, W. M., 1973, The carbonate ramp: An alternative to the shelf model: Transactions Gulf Coast Association of Geological Societies, v. 23, p. 221–225.

Anderson, C. A., 1950, 1940 E. W. Scripps cruise to the Gulf of California. Part I. Geology of the islands and neighboring land areas: Geological Society of America Memoir 43, 53 p.

Barbour, M. G. , Burk, J. H., and Pitts, W. D., 1987, Terrestrial plant ecology (second edition): Menlo Park, California, Benjamin/Cummings Publishing, 634 p.

Bjorlykke, K., 1967, The Eocambrian "Reusch Moraine" at Bigganjargga and the geology around Varanger fjord; northern Norway, *in* Studies on the latest Precambrian and Eocambrian rocks in Norway: Norges Geologiske Undersokelse, no. 251, p. 18–44.

Bray, N. A., and Robles, J. M., 1991, Physical oceanography of the Gulf of California, *in* Dauphin, J. P., and Simoneit, B. R. T., eds., The Gulf and Peninsular Province of the Californias: American Association of Petroleum Geologists Memoir 47, p. 511–553.

Brusca, R. C., 1980, Common intertidal invertebrates of the Gulf of California (second edition): Tucson, University of Arizona Press, 513 p.

Dott, R. H., Jr., 1974, Cambrian tropical storm waves in Wisconsin: Geology, v. 2, p. 243–246.

Durham, J. W., 1950, 1940 E. W. Scripps cruise to the Gulf of California. Part II. Megascopic paleontology and marine stratigraphy: Geological Society of America Memoir 43, 216 p.

Fannin, N. G. T., 1969, Stromatolites from the Middle Old Red Sandstone of Western Orkney: Geological Magazine, v. 106, p. 77–88.

Foster, A. B., 1979, Environmental variation in a fossil scleractinian coral: Lethaia, v. 12, p. 245–264.

Fox, W. T., 1989, Waves and tides, *in* Use of computers in geology (abstract section): Vermont Geological Society, Norwich University, Northfield, Vermont, February 1989, p. 10.

Grabau, A. W., 1940, The rhythm of the ages: Peking, Henri Vetch, 561 p.

Hanna, G. D., 1926, Paleontology of Coyote Mountain, Imperial County, California: Proceedings California Academy of Science, v. 14, p. 427–503.

Haq, B. U., Hardenbol, J., and Vail, P. R., 1988, Mesozoic and Cenozoic chronostratigraphy and eustatic cycles, *in* Wilgus, C. K., Hastings, B. K., Posamentier, H., Wagoner, J. V., Ross, C. A., and Kendall, C.G.St.C., eds., Sea-level changes: An integrated approach: Society of Economic Paleontologists and Mineralogists Special Publication no. 42, p. 71–108.

Hayes, M.L., Johnson, M. E., and Fox, W. T., 1993, Rocky-shore biotic associations and their fossilization potential: Isla Requeson (Baja California Sur, Mexico): Journal of Coastal Research, v. 9, p. 944–957.

Hercogová, J., and Kříž, J., 1983, New Hemisphaeramminae (Foraminifers) from the Bohemian Cretaceous Basin (Cenomanian): Vestnik Ústredniho ústavu geologického, v. 58, p. 205–216.

Hertlein, L. G., 1957, Pliocene and Pleistocene fossils from the southern portion of the Gulf of California: Bulletin Southern California Academy of Sciences, v. 56, p. 57–75.

Johnson, M. E., 1992, Studies on ancient rocky shores: A brief history and annotated bibliography: Journal of Coastal Research, v. 8, p. 797–812.

Johnson, M. E. and Hayes, M. L., 1993, Dichotomous facies on a Late Cretaceous rocky island as related to wind and wave patterns (Baja California, Mexico): Palaios, v. 8, p. 385–395.

Johnson, M. E. and Simian, M. E., 1996, Discrimination between coastal ramps and marine terraces at Punta Chivato on the Pliocene-Pleistocene Gulf of California: Journal of Geoscience Education, v. 44, p. 569–575.

Kennett, J. P., and Hodell, D. A., 1995, Stability or instability of Antarctic ice sheets during warm climates of the Pliocene?: GSA Today, v. 5, p. 1, 1–13, 22.

Ledesma-Vázquez, J. and Johnson, M. E., 1993, Neotectonica del area Loreto-Mulegé, *in* Delgado-Argote, L. A., and Martín-Barajas, A, eds., Contribuciones a la tectonica del Occidente de Mexico: Union Geofisica Mexicana Monografia 1, p. 115–122.

Lescinsky, H. L, Ledesma-Vázquez, J., and Johnson, M. E., 1991, Dynamics of late Cretaceous rocky shores (Rosario Formation) from Baja California, Mexico: Palaios, v. 6, p. 126–141.

Libbey, L. K., and Johnson, M. E., 1997, Upper Pleistocene rocky shores and intertidal biotas on the Gulf of California at Playa La Palmita (Baja California Sur, Mexico): Journal of Coastal Research, v. 13, p. 216–225.

Ortlieb, L., 1984, Neotectonics and sea level variations in the Gulf of California area—Field trip guidebook: Instituto de Geologia, Universidad Nacional Autonoma de Mexico, 152 p.

Ortlieb, L., 1991, Quaternary vertical movements along the coasts of Baja California and Sonora, *in* Dauphin, J. P., and Simoneit, B. R. T., eds., The Gulf and Peninsular Province of the Californias: American Association of Petroleum Geologists Memoir 47, p. 447–480.

Playford, P. E., and Lowry, D. C., 1966, Devonian reef complexes of the Canning Basin, Western Australia: Bulletin Geological Survey Western Australia, no. 118, 150 p.

Raasch, G. O., 1958, Baraboo monadnock and paleowind direction: Alberta Society of Petroleum Geologists Journal, v. 6, p. 183–187.

Sharp, R. P., 1940, Ep-archean and ep-algonkian erosion surfaces, Grand Canyon, Arizona: Geological Society of America Bulletin, v. 51, p. 1235–1270.

Shepard, F. P., 1950, 1940 E. W. Scripps cruise to the Gulf of California. Part III. Submarine topography of the Gulf of California: Geological Society of America Memoir 43, 32 p.

Smith, J. T., 1991, Cenozoic marine mollusks and paleogeography of the Gulf of California, *in* Dauphin, J. P., and Simoneit, B. R. T., eds., The Gulf and Peninsular Province of the Californias: American Association of Petroleum Geologists Memoir 47, p. 637–666.

Trueman, A. E., 1922, The Liassic rocks of Glamorgan: Proceedings of the Geologists' Association, v. 33, p. 245–284.

Umhoefer, P. J., Dorsey, R. J., and Renne, P., 1994, Tectonics of the Pliocene Loreto basin, Baja California sSur, Mexico, and evolution of the Gulf of California: Geology, v. 22, p. 649–652.

Manuscript Accepted by the Society December 2, 1996

Printed in U.S.A.

Geological Society of America
Special Paper 318
1997

Holocene sediments and molluscan faunas of Bahía Concepción: A modern analog to Neogene rift basins of the Gulf of California

Keith H. Meldahl
Department of Geology, Oberlin College, Oberlin, Ohio 44074
Oscar Gonzalez Yajimovich
Facultad de Ciencias Marinas, Universidad Autónoma de Baja California, Ap. Postal 453, Ensenada, Baja California, Mexico 22800
Christina D. Empedocles and Christofer S. Gustafson
Department of Geology, Oberlin College, Oberlin, Ohio 44074
Manuel Motolinia Hidalgo
Facultad de Ciencias Marinas, Universidad Autónoma de Baja California, Ap. Postal 453, Ensenada, Baja California, Mexico 22800
Timothy W. Reardon
Department of Geology, Oberlin College, Oberlin, Ohio 44074

ABSTRACT

Neogene rift basin deposits in the Gulf of California region record early phases in the tectonic evolution of the modern gulf. Potential Holocene analogs to these deposits occur today in Bahía Concepción, a rift basin forming a shallow marine bay on the eastern coast of Baja California Sur at ~26.5° North latitude. Bedrock geology and geomorphology suggest Bahía Concepción formed by drowning of an asymmetric graben during the Holocene transgression. A large normal fault zone bounds the bay's east side, associated with a 30-km-long shoreline bajada backed by small, steep drainages. A smaller fault zone probably bounds the bay's west side. The west side has larger drainages with gentler gradients and is characterized by rocky shorelines, pocket bays, mangrove swamps, and few exposed alluvial fans. Holocene sediments of the basin accumulate in alluvial fan, coastal interfan flat, mangrove swamp, fan-delta, pocket bay, nearshore shelf, and offshore shelf environments.

Three major types of marine sediment dominate the bay. *Green clastic mud* (with variable mollusc shell content) dominates offshore below 20 m. *Volcaniclastic sand* (with variable mollusc shell and calcareous algae content) dominates the shallow fan-deltas and nearshore shelf on the bay's east and south sides. *Carbonate sand* (both mollusc and calcalgal origin) dominates in shallow pocket bays and adjacent to rocky shorelines on the bay's west side. Comparison of drainage profiles from the east and west sides of the bay suggests that the restriction of carbonate sediments to the west side is due to trapping of terrigenous sediment in the large western drainages during Holocene sea-level rise.

Variation in species composition and taphonomy of molluscan assemblages in the bay is correlated with water depth. Cluster analysis on molluscan shell samples identifies four overlapping, depth-related biofacies: *mangrove swamp* (intertidal), *shallow nearshore* (1- to 5-m depth), *deeper nearshore* (5- to 12-m depth), and *offshore* (below 20 m). Taphonomic analysis demonstrates depth zonation in both the abundance and condition of mollusc shells. Shell abundance is greatest in intertidal mangrove

Meldahl, K. H., Gonzalez Yajimovich, O., Empedocles, C. D., Gustafson, C. S., Motolinia Hidalgo, M., and Reardon, T. W., 1997, Holocene sediments and molluscan faunas of Bahía Concepción: A modern analog to Neogene rift basins of the Gulf of California, *in* Johnson, M. E., and Ledesma-Vázquez, J., eds., Pliocene Carbonates and Related Facies Flanking the Gulf of California, Baja California, Mexico: Boulder, Colorado, Geological Society of America Special Paper 318.

swamps and the shallowest (less than 1 m deep) areas of pocket bays. Shells in less than 10-m depth exhibit variable taphonomic condition, whereas shells in greater than 20-m depth are in uniformly good condition.

The Holocene of Bahía Concepción provides useful insights into analogous ancient rift basin facies. In the nearby Pliocene Loreto Basin, the distribution of shallow-water carbonates may be controlled by same process operating at Bahía Concepción: upstream entrapment of terrigenous sediments in incised canyons during base-level rise. The Loreto Basin contains little biologic or taphonomic evidence of deposition in water depth greater than 10 m, indicating that, unlike Holocene Bahía Concepción, sedimentation kept pace with subsidence and sea-level changes throughout Loreto Basin evolution. This distinction likely reflects distinct tectonic or climatic differences between Pliocene and Holocene times.

INTRODUCTION

The Gulf of California is perhaps the world's premier example of a young ocean basin developed in a transform-rift plate margin setting. It provides excellent opportunities to integrate studies of modern and ancient sedimentary environments in order to better understand the geologic evolution of ocean basins. We report here on a study of the Holocene geology of Bahía Concepción, a rift basin forming an elongate northwest-southeast–oriented shallow marine bay at ~26.5° North latitude along the east coast of Baja California Sur, Mexico (Fig. 1). Bahía Concepción may be similar in many respects to Neogene shallow marine rift basins formed during the early evolution of the modern Gulf of California. Our objectives here are to describe the bay's geology and to document distributions of Holocene marine sediments, molluscan faunas and molluscan shell taphonomic features. Holocene sediments, faunas and taphonomy of Bahía Concepción could provide useful modern analogs for facies interpretation in Neogene rift basins of the Gulf of California.

Beal (1948) briefly discussed Bahía Concepción in his reconnaissance geologic study of Baja California, but the most comprehensive bedrock geologic study is that of McFall (1968), which includes a 1:70,000 geologic base map. More recently, Pliocene sediments in the region have been mapped and interpreted by Mayall et al. (1993), Johnson et al. (this volume), Ledesma-Vázquez et al. (this volume), and Simian and Johnson (this volume). The stratigraphy and structure of the Pliocene Loreto Basin, south of Bahía Concepción, have been investigated by Dorsey (this volume), Dorsey et al. (this volume) and Umhoefer et al. (1994). Ashby and Minch (1987) described the stratigraphy and paleoecology of Pleistocene marine sediments near Mulegé, several kilometers north of Bahía Concepción. Ortlieb (1987), as part of a comprehensive study of neotectonic movements in the Gulf of California, described and dated Pleistocene

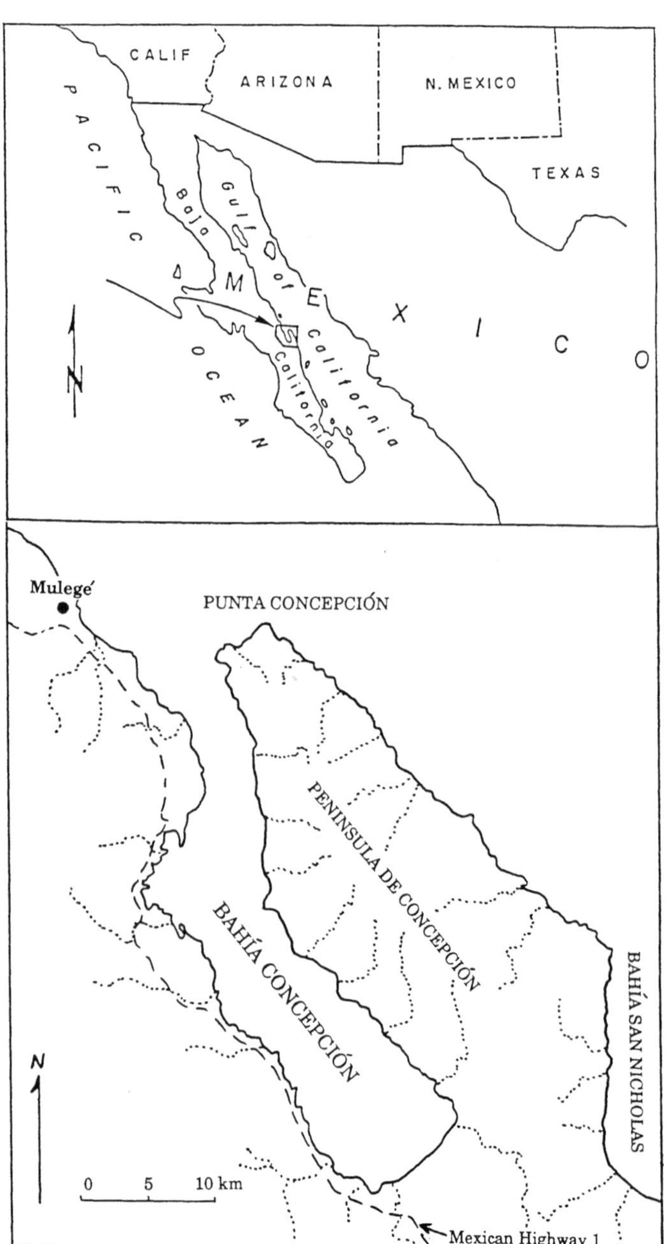

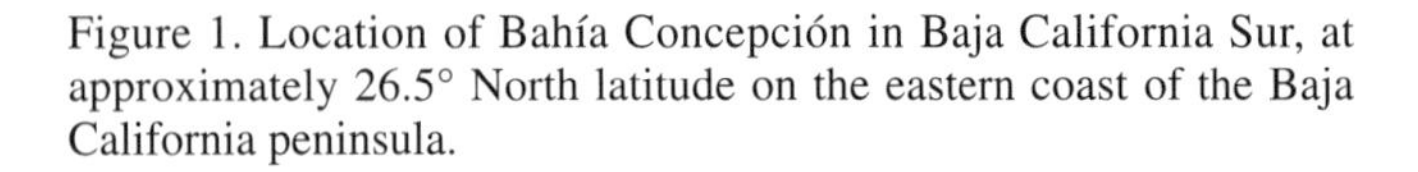

Figure 1. Location of Bahía Concepción in Baja California Sur, at approximately 26.5° North latitude on the eastern coast of the Baja California peninsula.

marine terraces in the Bahía Concepción area. Hayes et al. (1993) reported on the fossilization potential of rocky shore faunas near Isla Requesón within Bahía Concepción. The bay has been the focus of several marine biological studies, including rhodolith algae (Foster et al., this volume; Steller and Foster, 1995), phytoplankton (Romero-Ibarra and Garate-Lizarraga, 1991), bivalves (Singh-Cabanillas and Bojorquez-Verastica, 1987, 1990), gastropods (Baqueiro et al., 1983), polychaetes (Salazar-Vallejo, 1985; Salazar-Vallejo and Stock, 1987; Bastida-Zavala, 1990), shrimp (Rios-Gonzalez, 1989), and fish (Rodriguez-Romero et al., 1988, 1991).

GEOLOGIC SETTING

Bahía Concepción is about 40 km long and varies from 5 to 10 km wide. Figure 2 is a simplified geologic map of the bay area, based on McFall (1968). Figure 3 illustrates some of the major features of the bay area. The bay is open to the Gulf of California at its north end. The eastern side of the bay is formed by the 10- to 15-km wide Concepción Peninsula, which rises to elevations of over more than 700 m. Distinct geomorphic differences characterize the eastern and western shores of the bay. The eastern shore (western side of the Concepción Peninsula) has a well-developed, 30-km-long bajada backed by short, steep canyons that rarely extend more than 5-km back into the Concepción Peninsula (Fig. 3a, b). The western shore of the bay is dominated by steep rocky shorelines rising abruptly from the water. This rugged shoreline is interrupted in several places by long deep canyons, the largest of which extend inland 10 to 25 km. In distinct contrast to the extensive bajada on the east side, alluvial fans on the west side occur only at the mouths of these large canyons (Fig. 3c). In embayments along the western shore are several sheltered pocket bays rich in carbonate sand (Fig. 3d, e) and several mangrove swamps (Fig. 3f). The southern end of the bay is a low-lying alluvial plain that rises gradually away from the bay toward the southeast.

Bedrock geology and depositional environments

Rocks of the Bahía Concepción region consist of a basement complex of Cretaceous schistose and granitic rocks (K-Ar date on granite of 78.4 ± 2.9 Ma [McFall, 1968]), overlain unconformably by more than 4,000 m of Oligocene and Miocene volcanics of the Comondú Group. McFall (1968) recognizes six vertically stacked Comondú formations in the region. The distributions of these formations are shown on Figure 2. The oldest is the Salto Formation, consisting of over 300 m of red tuffaceous cross-bedded sandstone and interbedded tuffs (K-Ar date on the upper Salto Formation: 28.1 ± 0.9 Ma [McFall, 1968]). Two of the thickest Comondú formations dominate the area. The 23- to 28-Ma Pelones Formation dominates the Concepción Peninsula (Fig. 3a, b). It consists of over 2,000 m of volcanic agglomerates, basalt flows, and tuffs, with rare conglomerate and tuffaceous sandstone beds. The 17- to 22-Ma Ricasón Formation dominates the western side of the bay. It consists of more than 1,500 m of interbedded basalt flows, agglomerates, and tuffs. In some areas, the Pelones and Ricasón Formations are separated by one or more of the following formations (in ascending order): the Minitas Formation (30 to 150 m of coarse tuffaceous conglomerate), the Pilares Formation (100 m of aphanitic and porphyritic basalt), and the Hornillos Formation (150 m of coarse tuffaceous conglomerate). Where present, these intervening formations are grouped with the Pelones Formation in Figure 2.

Intruding the basement rocks and Comondú formations are two types of intrusions. Gabbro stocks and dikes intrude the Pelones Formation and older rocks and may represent conduits for magmas feeding the basalt flows of the Pelones Formation (McFall, 1968). Several tonalite intrusions also occur in the area. The largest is the Cerro Blanco stock (K-Ar date of 20.0 ± 2.0 Ma [McFall, 1968]), which underlies Cerro Blanco, the highest peak on the Concepción Peninsula.

Forming a thin veneer over the Comondú formations are conglomerates, sandstones, siltstones, and coquinas of the late Pliocene Infierno Formation (Mayall et al., 1993; Johnson et al., this volume; Ledesma-Vázquez et al., this volume). The Infierno Formation crops out widely in two areas: in the southeastern area of the bay and at the northern end of the Concepción Peninsula (Fig. 2). In both areas it is nearly flat. In the southeastern area of the bay, the Infierno Formation fills paleovalleys cut in the underlying Pelones and Ricasón Formations. Quaternary alluvium and fossiliferous marine terrace gravels cap the underlying units sporadically throughout the region.

Holocene sediments accumulate in seven distinct depositional environments at Bahía Concepción: alluvial fans, fan-deltas, coastal interfan flats, pocket bays, mangrove swamps, nearshore shelf, and offshore shelf. The major features of these environments are summarized in Table 1.

Structural setting

Bahía Concepción occupies a northwest-southeast–trending asymmetric graben of apparent Pliocene age (based on the occurrence of nearly flat-lying late Pliocene Infierno Formation in the basin). The main structural features of the area are northwest-southeast–trending faults. The most prominent of these faults make up the Bahía Concepción fault zone (McFall, 1968), which runs along the bajada on the eastern side of the bay (Fig. 2). The Concepción Peninsula forms part of the uplifted footwall block, and the bay itself occupies part of the down-dropped hanging-wall block (Fig. 4). Exposure of Cretaceous granites on the peninsula and offset of Comondú units indicate vertical displacements from 400 to 2,500 m along this fault zone (McFall, 1968).

The prominent, steep escarpment occurring along much of the bay's western shore suggests that this side is also bounded by a northwest-southeast–trending fault zone. However, McFall (1968) does not recognize major offset in the bedrock units on the bay's west side, suggesting that this fault zone has experienced less vertical throw than the Bahía Concepción fault zone (Fig. 4).

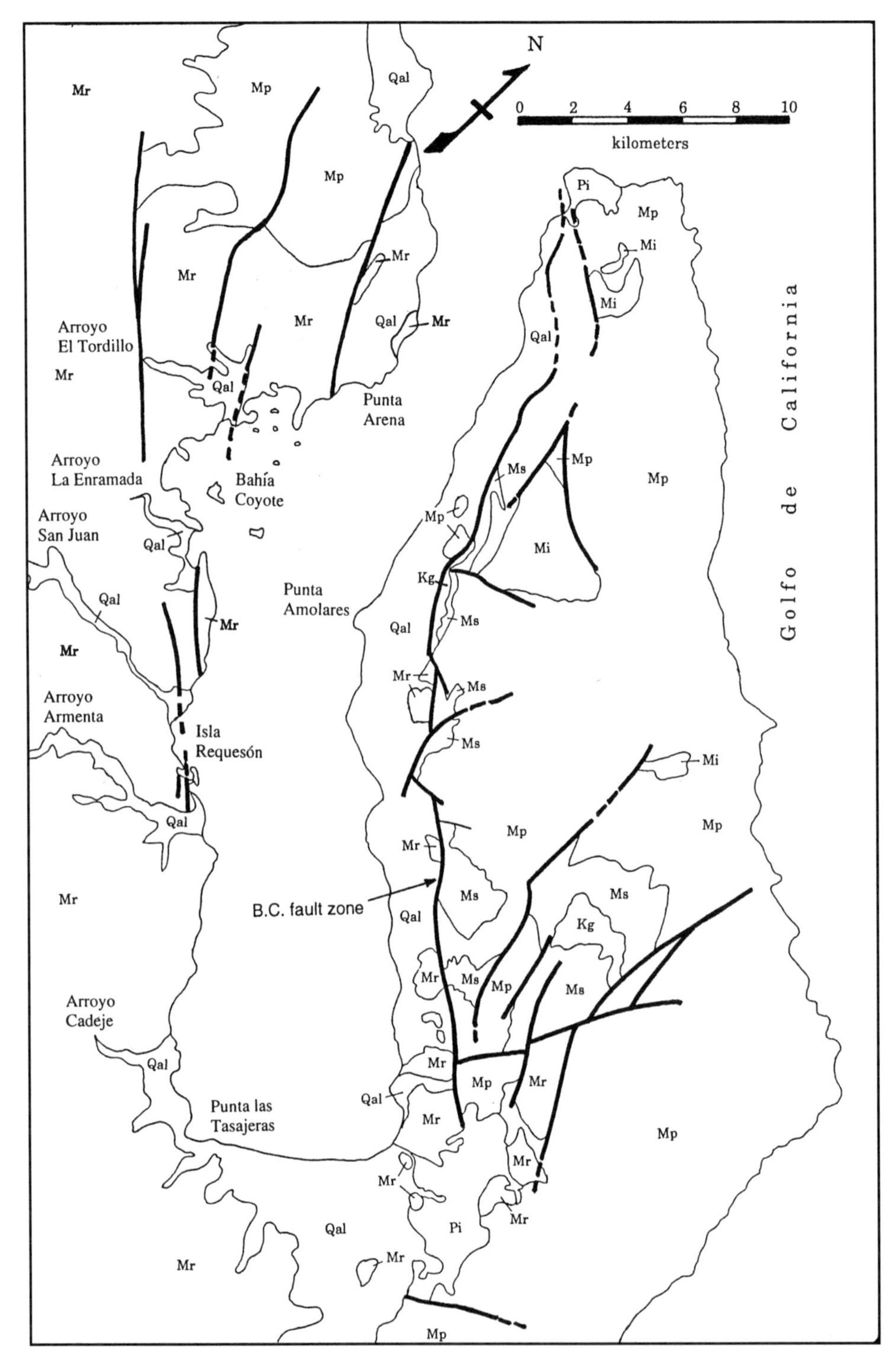

Figure 2 (on this and facing page). Geologic map of Bahía Concepción, simplified from McFall (1968). Heavy lines are faults; light lines are shoreline or contacts between major rock units. The Bahía Concepción (B.C.) fault zone runs subparallel to the shoreline along the eastern side of the bay (western side of the Concepción Peninsula). The peninsula is the uplifted footwall block; the bay occupies part of the down-dropped hanging-wall block. Note the prominent bajada along the eastern shore (broad band of Qal), derived from the uplifted footwall.

Qal	Quaternary Alluvium	volcaniclastic sand and conglomerate of alluvial fans, streambeds, and stream terraces
Pi	Pliocene Infierno Formation	marginal marine to shallow marine sandstones, siltstones, cherts, and coquinas
Mr	Miocene Ricasón Formation	andesitic and basaltic flows, agglomerates, and tuffs
Mi	Miocene intrusions (younger than Pelones Fm)	gabbro or tonalite
Mp	Miocene Pelones Formation (Here includes the overlying Hornillos and Pilares Formations at the base of Concepción Peninsula and the overlying Minitas Formation northwest of the bay mouth and on the east side of the Concepción Peninsula.)	andesitic agglomerates, flows, and tuffs
Ms	Miocene Salto Formation	cross-bedded sandstone and interbedded tuffs
Kg	Cretaceous basement	granodiorite and quartz monzonite with schist inclusions

In addition to vertical movement, McFall (1968) postulates up to 30 km of right-lateral displacement along the Bahía Concepción fault zone, based on the apparent truncation and offset of an anticline present in both the Mulegé area and on the Concepción Peninsula. However, this evidence appears equivocal, and only vertical offset is clearly demonstrated by bedrock mapping.

Effect of the Holocene transgression

Assuming minimal tectonic activity during Holocene time, marine waters must have flooded Bahía Concepción no earlier than about 8 ka. The bay averages 25 to 30 m deep, exceeding 30 m only at several stations in the southern third of the bay and at the northern mouth. Holocene sea-level curves indicate that eustatic sea level had risen to about 30 m below present by 8 ka (Fairbanks, 1989). Prior to this, and indeed throughout much of Pleistocene time, the bay was probably a nonmarine basin accumulating alluvial, fluvial, and possibly lacustrine/playa sediments. The low-lying alluvial plain extending south of the bay (along the axis of the graben) probably gives a reasonable view of the nonmarine appearance of the bay before Holocene drowning. Prior to Holocene flooding, there was probably a fluvial exit channel out the bay's northern mouth. Here the greatest measured depth (34 m) occurs very close to the eastern bajada. This possibly represents a drowned paleoriver channel. South of the mouth, the bay is generally flat floored, exhibiting no distinct east-west (cross-structural) bathymetric asymmetry.

The Holocene transgression may explain an initially puzzling aspect of alluvial fan development at Bahía Concepción. Studies of asymmetric extensional basins have shown that, where lithologies are similar, the following differences typically apply to catchments and fans developed on hanging-wall versus footwall blocks (Leeder, 1995; Leeder and Jackson, 1993; Leeder and Gawthorpe, 1987):

Hanging Wall	*Footwall*
• larger catchments	• smaller catchments
• gentler catchment gradients	• steeper catchment gradients
• larger alluvial fans/ fan-deltas with gentler gradients	• smaller alluvial fans/ fan-deltas with steeper gradients

Clearly the first two criteria apply to Bahía Concepción: the bay has small, high-gradient footwall catchments (east side) and large, low-gradient hanging-wall catchments (west side). Also, the footwall-sourced fans (east side) have markedly higher gradients than the hanging-wall–sourced fans (west side). However, contrary to expectation, the largest fans at Bahía Concepción are sourced out of the footwall, forming the 30-km bajada along the bay's east side, whereas the west side hanging wall has only a few isolated fans.

The rarity of large *exposed* fans on the bay's west side may be explained by their being drowned during the Holocene transgression. The bajada on the east side is currently exposed above sea level because it was deposited at higher elevations, having been sourced out of higher elevations and steeper gradients along the uplifted footwall block. There is probably a large, low-angle bajada on the bay's west side, but because it was deposited at lower elevations than the eastern bajada, it is

Figure 3. Major coastal environments and geomorphic features of Bahía Concepción. **a,** view northeast across the bay to the Concepción Peninsula. Note the continuous shoreline bajada and the uplifted, tilted Comondú volcanic strata beyond. The Bahía Concepción fault zone runs subparallel to the shoreline along the upper part of the bajada. Relief on the photo is about 300 m. **b,** view up the bajada from the shoreline, east side of Bahía Concepción. Note the short, steep nature of the drainages and the steep bajada gradient. Uplift of the footwall here has exposed Cretaceous granite (light colored rocks above the bajada), overlain by Comondú formations. **c,** view northwest across the Armenta fan-delta, one of the few isolated fans entering the west side of Bahía Concepción. **d,** view north across Playa Coyote, one of several carbonate sand pocket bays in Bahía Coyote, on the west side of Bahía Concepción. **e,** view north across Playa Requesón, a carbonate pocket bay formed behind Isla Requesón, on the west side of Bahía Concepción. The tombolo extending to the island is built of carbonate sand. **f,** view west up Arroyo El Tordillo on the west side of Bahía Concepción. In the foreground are the Los Cocos mangrove swamp and carbonate pocket bay. This canyon illustrates drainage aggradation in response to base level rise. Note the steep canyon walls separated by the wide, nearly flat canyon floor, indicating deep incision followed by backfilling in response to rising base level. Nearly all clastic sediments are trapped up the canyon, and carbonate sands dominate the pocket bay here.

TABLE 1. HOLOCENE DEPOSITIONAL ENVIRONMENTS OF BAHÍA CONCEPCIÓN

Environment	Location	Characteristics
Alluvial fans and fan-deltas (Fig. 3a, b, c)	On the eastern shore of the bay, multiple fans form a 30-km-long bajada along the Bahía Concepción fault zone. On the western shore of the bay, isolated fans occur at the mouths of the largest canyons. Fans meet the shoreline at cobble and boulder beaches. Large clasts extend offshore onto fan-deltas to 2–3 m depth, beyond which clastic sediments fine rapidly seaward, typically grading into mud-dominated sediments more than 1 km offshore.	Fans consist of moderately to very poorly sorted volcaniclastic sediments, ranging from sand to boulder size; granitic clasts may occur where granite is exposed along the Bahía Concepción fault zone. On fan-deltas, large clasts in 2-3 m depth commonly form substrates for *Porites* coral, algae, and various hard-substrate molluscs. Below 5 m, bottom sediments are typically heavily bioturbated by bottom-feeding elasmobranch rays and infaunal burrowers. The sandy parts of fan-deltas support diverse molluscan faunas dominated by *Pinna, Chione, Argopectin, Megapitaria, Glycymeris, Dosinia, Laevicardium elatum, Trigonocardia, Murex, Conus, Nassarius, Strombina,* and *Strombus.* Fan-deltas on the western shore of the bay commonly support dense beds of living rhodolith algae at 3–12 m depth (Foster et al., this volume).
Coastal inter-fan flats	Flat, low-lying areas along the coast between alluvial fans, best developed in the distal regions of the eastern shore bajada and near the shoreline on the alluvial plain at the south end of the bay.	Fine sand and silt, with common evaporitic crusts; surface commonly colonized by low bushy halophytic plants; sediments highly bioturbated by halophyte roots.
Pocket bays (Fig. 3d, e)	Occur only along the western shore of Bahía Concepción, throughout Bahía Coyote, and in several other embayments along the western shore. Pocket bays typically occur either at the mouths of back-filled canyons or in small, steep coastal embayments with minimal clastic input. Individual bays are separated by intervals of steep rocky shoreline.	Dominated by carbonate sands derived from breakdown of both calcareous algae (rhodoliths) and molluscs. Occurrence of carbonates is linked both to low clastic input in backfilled canyons and coastal embayments and to high carbonate production (especially by rhodoliths). Pocket bays support an abundant molluscan fauna similar to that of fan-deltas (see above). Bottom sediments in bays are typically heavily bioturbated by bottom-feeding elasmorbranch rays and infaunal burrowers.
Mangrove swamps (Fig. 3f)	Occur only along the western shore of Bahía Concepción in local embayments developed behind pocket bays or on the distal margins of alluvial fans.	Sediments range from coarse sand to mud, commonly very rich in decaying organic matter. The red mangrove *Rhizopora mangle* dominates the vegetation. The mangrove oyster *Ostrea palmula* is abundant, attached to mangrove roots. The gastropods *Theodoxus, Cerithidea,* and *Cerithium* and the bivalves *Anadara, Arca,* and *Chione* are common.
Nearshore shelf	Defined as the shallow shelf area at the south end of Bahía Concepción, not associated with fan-deltas, pocket bays, or rocky shorelines. Represents the submarine extension of the gently sloping alluvial plain at the south end of the bay.	Sediments range from coarse to fine sand. Abundant wave ripples form under high wave conditions during prevailing northerly winds, but under quiet conditions bottom sediments are heavily bioturbated by bottom-feeding elasmobranch rays and infaunal burrowers. Supports a molluscan fauna similar to that of fan-deltas and pocket bays (see above).
Offshore shelf	Defined as all areas offshore in more than 20 m water depth. Throughout the bay, water depths increase rapidly away from the shoreline and then typically level out between 25 and 30 m.	Dominated by green clastic mud containing a sparse mollusc fauna consisting of *Laevicardium elenense, Argopectin, Crucibulum,* and *Chione.* Some offshore areas have abundant hard-substrate fauna *(Anomia, Crucibulum),* probably indicating the proximity of submarine bedrock highs poking up through surrounding deep bay muds.

presently drowned and covered by a veneer of Holocene marine sediments, as shown schematically in Figure 4.

Drainages on the west side were deeply incised prior to the Holocene, and many appear to be presently aggrading upstream in response to Holocene transgression. This backfilling of western drainages means that little clastic sediment reaches the bay's western shore. This is important because clastic starvation due to upstream entrapment likely contributes to the accumulation of carbonate sediments along the western margin of the bay today. There are several excellent examples of drainage backfilling on the west side, most notably Arroyo El Tordillo and Arroyo La Enramada (Fig. 2, Fig. 3f).

METHODS

Our study focused on Holocene marine molluscan faunas and sediments. We collected faunal and sediment samples from

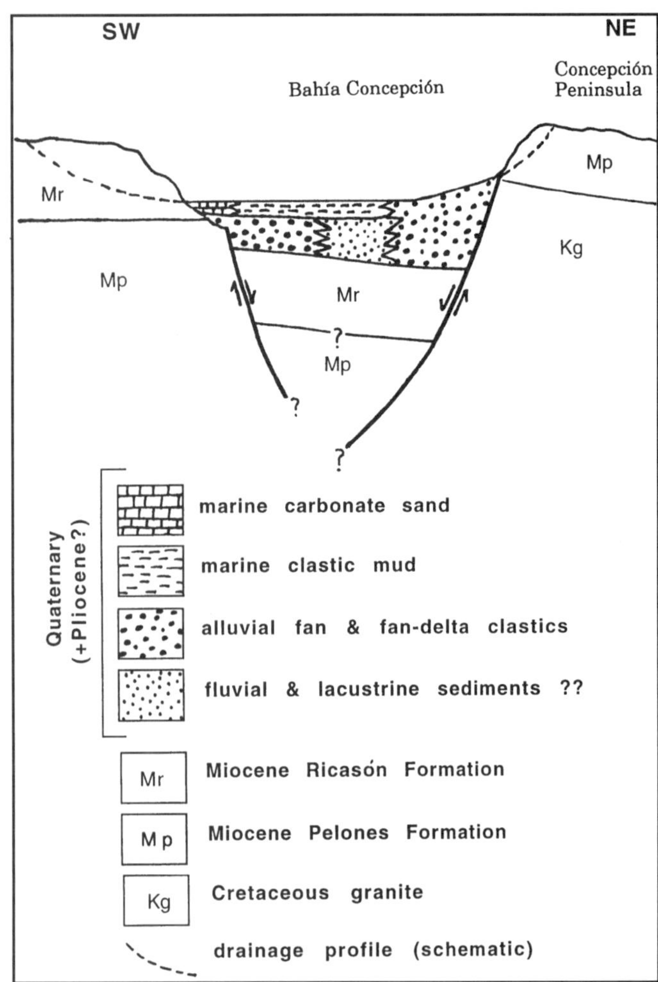

Figure 4. Schematic southwest-northeast cross section of Bahía Concepción, based on geologic map of McFall (1968) and personal observations. Not to scale. The basic structure of the basin appears to be an asymmetric graben with the main bounding fault on the east side. Cretaceous granites are exposed along parts of the uplifted footwall of the Concepción Peninsula (Fig. 3b). Drainages on the east side are short and steep and feed a continuous 30-km-long shoreline bajada (Fig. 3a). Drainages on the west side are long with gentle gradients, and many appear to have aggraded in response to Holocene sea-level rise (Fig. 3f). Only a few isolated alluvial fans are presently exposed on the west side (Fig. 3c). Carbonate sands dominate off rocky shores and in pocket bays along the west side (Fig. 3d, e). When sea level was low and the basin was nonmarine, the western drainages probably fed a large bajada, postulated here to be drowned and covered by marine sediments.

five environments: fan-deltas, pocket bays, mangrove swamps, nearshore shelf, and offshore shelf (see Table 1). We collected samples both by scuba diving along onshore-offshore transects and by grab-sampling (Ponar bottom grab) from a boat along east-west transects across the bay. At all stations we collected approximately 500 cm^3 of surface sediment for a sediment sample. At selected stations we sieved 1/64 m^3 of surface sediment through a 6-mm mesh screen, collecting remaining shells. In total we collected 101 sediment samples and 49 shell samples throughout the bay (Fig. 5). We shipped samples to either the Universidad Autónoma de Baja California (Ensenada, Baja California) or to Oberlin College (Oberlin, Ohio) for analysis.

Analysis of sediment samples consisted of grain size, sorting and composition, and visual estimates of carbonate content and composition.

Analysis of faunal samples included identification of molluscan species (using Keen, 1971), identification of biofacies using Q-mode cluster analysis, and identification of species assemblages using R-mode cluster analysis.

Taphonomic analysis included measurement of the abundance of shell carbonate in surface sediments (weight of shells and shell fragments greater than 6 mm in size) and assessment of the taphonomic condition of large (greater than 1.5 cm) disarticulated valves of infaunal bivalves. We focused on large valves because they are more likely to exhibit recognizable taphonomic alteration and carry that alteration to the stratigraphic record. We examined only the interiors of the disarticulated valves, since interior alteration is necessarily a postmortem development. To ensure adequate sample size, we examined only collections that had 15 or more large disarticulated valves.

We used taphonomic grades to quantify and compare the taphonomic condition of the valves in the samples. Taphonomic grades are semiquantitative categories of taphonomic alteration (Brandt, 1989; Flessa et al., 1993). We assigned individual valves to one of four taphonomic grades, following the system of Flessa, et al. (1993), summarized in Table 2. Grade 1 shells are in excellent condition (no or minimal alteration); Grade 2 shells are in good condition (slight alteration); Grade 3 shells are in fair condition (moderate alteration); and Grade 4 shells are in poor condition (extensive alteration).

We then calculated a Taphonomic Grade Index (TGI) for each sample. The TGI is a summary statistic that reflects the average taphonomic condition of a sample, based on the proportions of shells of different grades:

$$TGI = (g1 + (^2/_3)g2 + (^1/_3)g3) / n$$

where g1 = number of Grade 1 shells; g2 = number of Grade 2 shells; g3 = number of Grade 3 shells; and n = total number of shells in all four grades. The TGI varies between 0.0 and 1.0. A value of 0.0 means all shells in the sample are Grade 4 (poor condition, extensive alteration and degradation), whereas a value of 1.0 means that all shells in the sample are Grade 1 (excellent condition, minimal alteration). The TGI is a useful and consistent way to compare the taphonomic condition of samples from different environments.

RESULTS AND DISCUSSION: MARINE SEDIMENT DISTRIBUTIONS

Figure 6 illustrates the distribution of sediment types in the bay. Based on grain size, carbonate content, and carbonate composition in 101 sediment samples, we recognize seven distinct lithofacies that fall into three broad categories: (1) *clastic mud-*

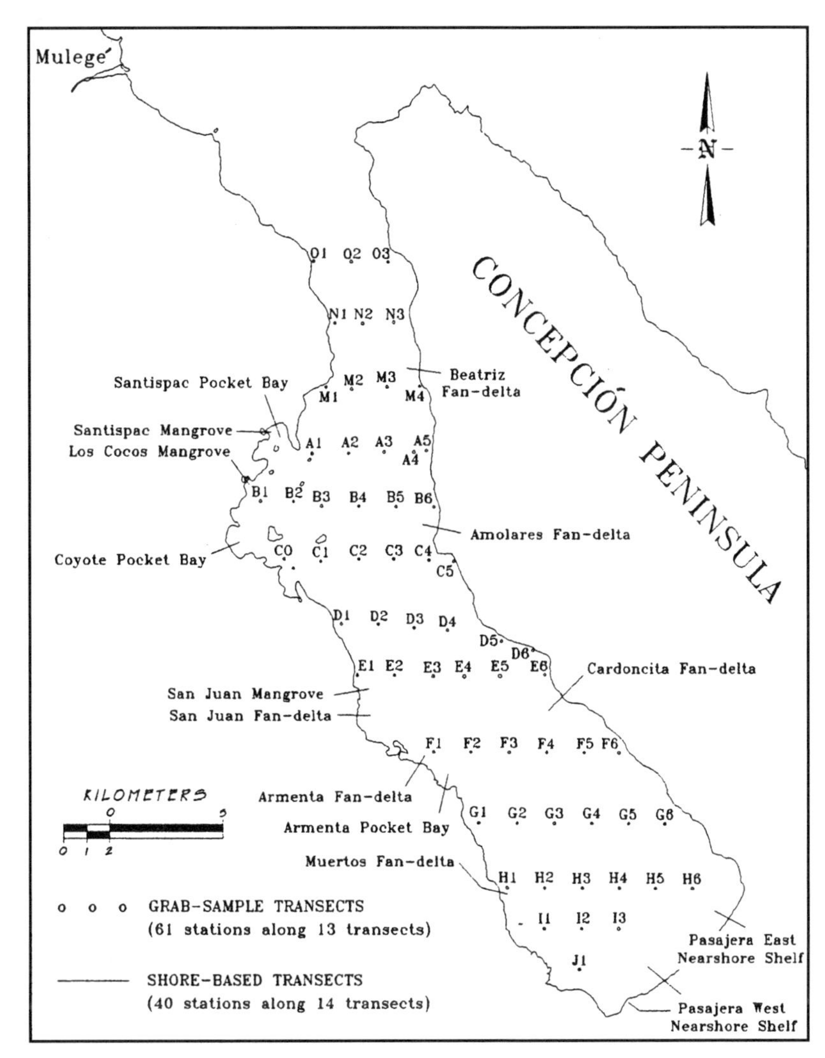

Figure 5. Location of sample stations for this study. Holocene sediments and mollusc shells were collected by grab-sampling from a boat along east-west transects (dots) and scuba diving along shore-based transects at selected locations (short lines).

TABLE 2. CRITERIA FOR ASSIGNING TAPHONOMIC GRADES TO BIVALVE SHELL SPECIMENS*

Grade 1 Excellent condition, no or minimal alteration	Grade 2 Good condition, slight alteration	Grade 3 Fair condition, moderate alteration	Grade 4 Poor condition, extensive alteration
Inner valve surface and muscle scars retain luster; pallial line clear; no visible chemical dissolution; all color present; no visible bioerosion or encrustation.	Inner valve surface slightly dull and rough and has lost most luster; muscle scars retain some luster; pallial line indistinct; significant loss of color; bioerosion or encrustation features cover <10% of interior surface.	Inner valve surface very dull and rough, all luster absent; muscle scars indistinct and lack luster; pallial line nearly obliterated or absent; dissolution pits may be present; color mostly absent; bioerosion or encrustation features cover 10 to 50% of interior surface.	Extensive loss of inner valve surface; dissolution pits common; muscle scars nearly obliterated or absent; pallial line absent; color usually absent; bioerosion or encrustation features cover >50% of interior surface.

*Modified from Flessa et al., 1993.

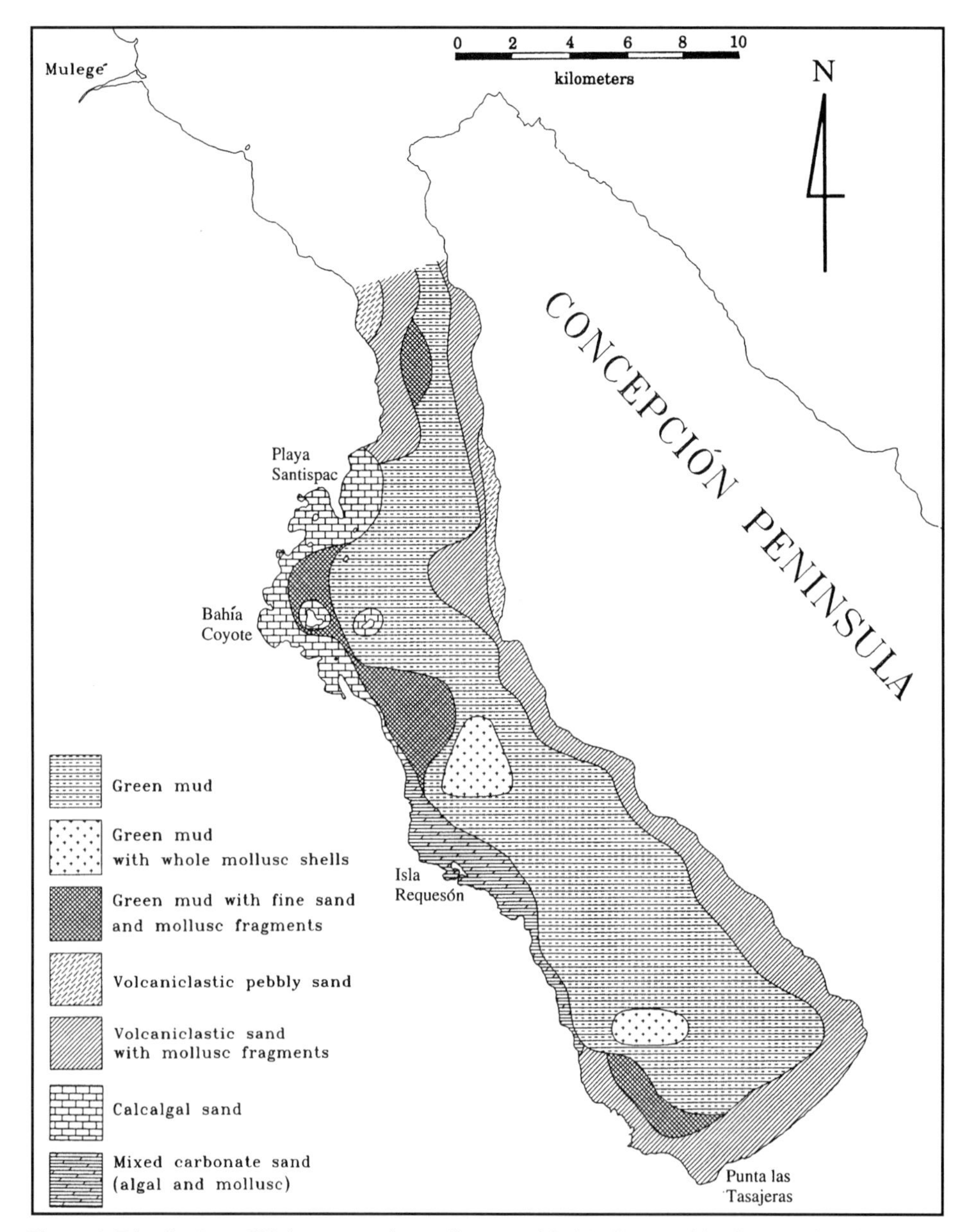

Figure 6. Distribution of Holocene marine sediments of Bahía Concepción. See text for discussion.

dominated facies: green mud, green mud with abundant whole mollusc shells, green mud with abundant mollusc fragments; (2) *clastic sand-dominated facies:* volcaniclastic pebbly sand with mollusc and calcalgal fragments and volcaniclastic sand with mollusc and calcalgal fragments; (3) *carbonate sand-dominated facies:* calcalgal (rhodolith) carbonate sand (with lesser amounts of mollusc fragments) and mixed carbonate sand (roughly equal amounts of rhodolith and mollusc framents).

Figure 6 shows that areas more than 1 km offshore are dominated by green mud. These areas are typically greater than 20 m deep. The mud is dominantly clastic, though diatom and foraminifera tests may be common. Most areas accumulating mud have very low shell content. However, there are several areas of shell enrichment within the green mud. In the southern mid-central areas of the bay, abundant whole mollusc shells occur in the mud, mostly *Crucibulum* and *Anomia*—taxa that attach to hard substrates. This suggests that extensive subsurface rocky outcrops may exist in these areas. Local fishermen report that several rocky shoal areas exist in the bay. In the area offshore and to the south of Bahía Coyote, abundant shell fragments occur in the mud. In this area, numerous small rocky islands and probable submarine rocky outcrops may also generate shell material. Given the rugged, highly faulted nature of the region, we expect that bedrock outcrops are common on the floor of Bahía Concepción.

Figure 6 also shows that sands rim the perimeter of Bahía Concepción, in water generally less than 20 m deep. There is a distinctive geographic separation between volcaniclastic sand

and carbonate sand. Volcaniclastic sand dominates the east and south margins of the bay and the northwest margin near the bay mouth. Carbonate sand dominates most of the western coast from Playa Santispac southward.

The occurrence of volcaniclastic sand on the east margin and the northwest margin of the bay is clearly associated with input from alluvial fans: the eastern bajada and the large coastal alluvial fans at Punta Arena (Fig. 2). The volcaniclastic sand along the southern margin of the bay is associated with the broad alluvial plain that slopes gently toward the bay from the southeast and with large fans in the southwest corner of the bay, such as the fan at Arroyo Cadeje (Fig. 2). Figure 6 shows a distinct band of pebble-enriched volcaniclastic sand occurring in shallow water along the northeastern margin of the bay. This enrichment of coarse material coincides with the steepest part of the eastern bajada and the highest elevations on the Concepción Peninsula.

Figure 6 shows that carbonate sands are associated with the pocket bays and steep rocky shores on the western margin of the bay. Carbonate sands are composed of both calcareous algae (rhodolith) fragments and mollusc fragments. Calcalgae-dominated sands are common in the Playa Santispac and Bahía Coyote areas (Fig. 3d). To the south, mixed carbonate sands occur in the areas north and south of Isla Requesón (Fig. 3e). Carbonate sands occur adjacent to even the most rugged areas of the western coast. Volcaniclastic sand may be enriched locally on some of the fan-deltas that enter the western margin of the bay, but even these areas are commonly carbonate dominated.

We think the restriction of carbonate sands to the west side of Bahía Concepción is due to the limited input of terrigenous sediments along this shoreline. The reason has to do with differences in stream gradients on the east and west sides of the bay. On the east side, short, steep canyons exit onto the steep eastern bajada (Fig. 3a, b). Rainstorms could easily transport clastic sediments down these high gradients and deposit them in the bay. In contrast, many west-side canyons have very gentle gradients in their lower reaches. Many western canyons, such as the one shown in Figure 3f, appear to have been previously deeply incised (during low sea level?) and have subsequently backfilled and aggraded to a gentle gradient in their lower reaches (in response to Holocene sea-level rise?).

This idea is supported by the data in Figure 7, which compares the gradients of two typical drainages on the east and west sides of Bahía Concepción. Stream gradients that are in equilibrium with base level generally exhibit smooth concave-up profiles that steepen upstream. Such equilibrium profiles plot as lines of constant slope when profile data are plotted on logarithmic axes. In contrast, stream gradients that have not adjusted by aggradation to a recent rise in base level often exhibit a sharp increase in gradient at some point along the profile, reflecting an upstream topographic "memory" of a previous time of lower base level (Hack, 1973). As Figure 7 shows, drainages on the east

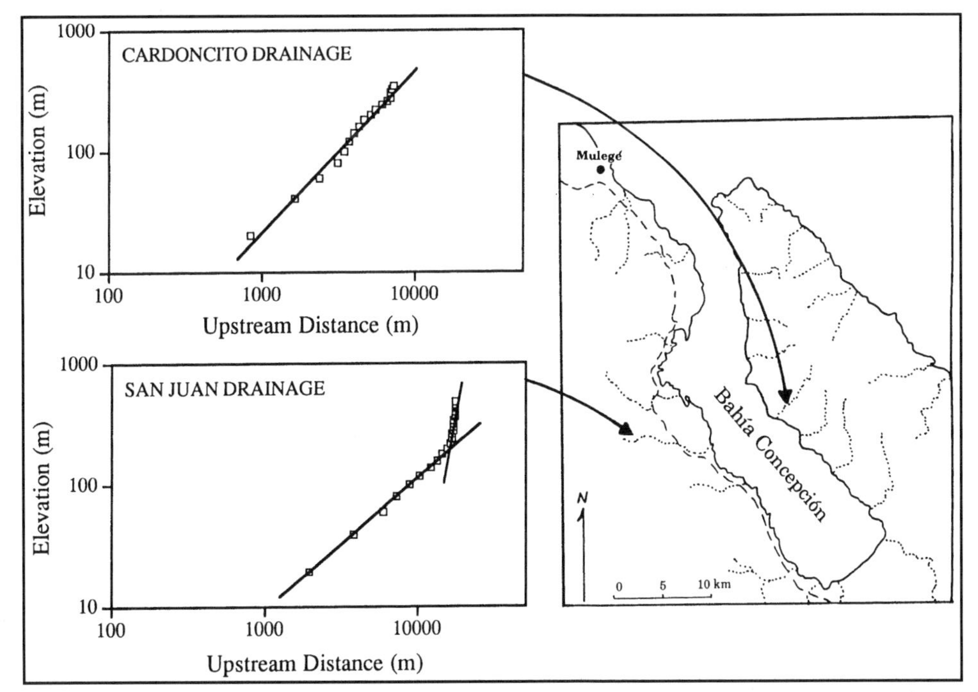

Figure 7. Comparison of drainage gradients on the east and west sides of Bahía Concepción. Drainages on the east side of the bay, as illustrated by the Cardoncito drainage, generally have profiles that plot as lines of constant slope (using logarithmic axes), suggesting that these drainages are in equilibrium with present base level (sea level). In contrast, drainages on the west side of the bay, as illustrated by the San Juan drainage, generally display a sharp increase in profile upstream. This suggests that western drainages are not in equilibrium with present base level (sea level) and are aggrading in their lower reaches in an effort to attain equilibrium. As a result, much terrigenous sediment shed into western drainages does not reach the shoreline, leading to clastic-starved conditions amenable to carbonate production along the west side of the bay.

side of Bahía Concepción, as illustrated by the Cardoncito drainage, appear to be in equilibrium with present sea level. In contrast, drainages on the west side, as illustrated by the San Juan drainage, exhibit a gentle gradient in their lower reaches and a distinct increase in gradient in their upper reaches. This profile indicates that many drainages on the west side of Bahía Concepción are not in equilibrium with present sea level but rather are aggrading in their lower reaches, probably in response to the Holocene transgression. Terrigenous sediments are trapped in these aggrading western canyons, creating clastic-starved conditions amenable to carbonate production along much of the western shoreline of Bahía Concepción.

Although we believe upstream terrigenous entrapment is the best explanation for the concentration of carbonate sediments along the western shoreline, we note that Steller and Foster (1995) and Foster et al. (this volume) suggest that prevailing wind patterns, not sediment input, are more important in controlling the distribution of living rhodolith beds (important sources of carbonate sand) in Bahía Concepción.

In summary, our results suggest a model for explaining the distribution of shallow-water carbonate facies in analogous Neogene rift basins. The distribution of shallow-water carbonates may be controlled, at least in part, by patterns of upstream terrigenous entrapment during sea-level rise.

RESULTS AND DISCUSSION: MOLLUSCAN MARINE FAUNA

Forty-nine molluscan shell samples collected throughout Bahía Concepción yielded 60 identifiable mollusc taxa (30 gastropods and 30 bivalves, identified using Keen, 1971). We tabulated species occurrences in each sample on a presence-absence basis. We performed both Q-mode and R-mode cluster analyses on these data using the clustering program in SYSTAT 5.1. Q-mode analysis groups samples that are similar in their species composition. This identifies biofacies—regions or habitats that support a similar fauna. R-mode analysis groups species that commonly co-occur in samples. This identifies species assemblages—groups of species that typically occur together.

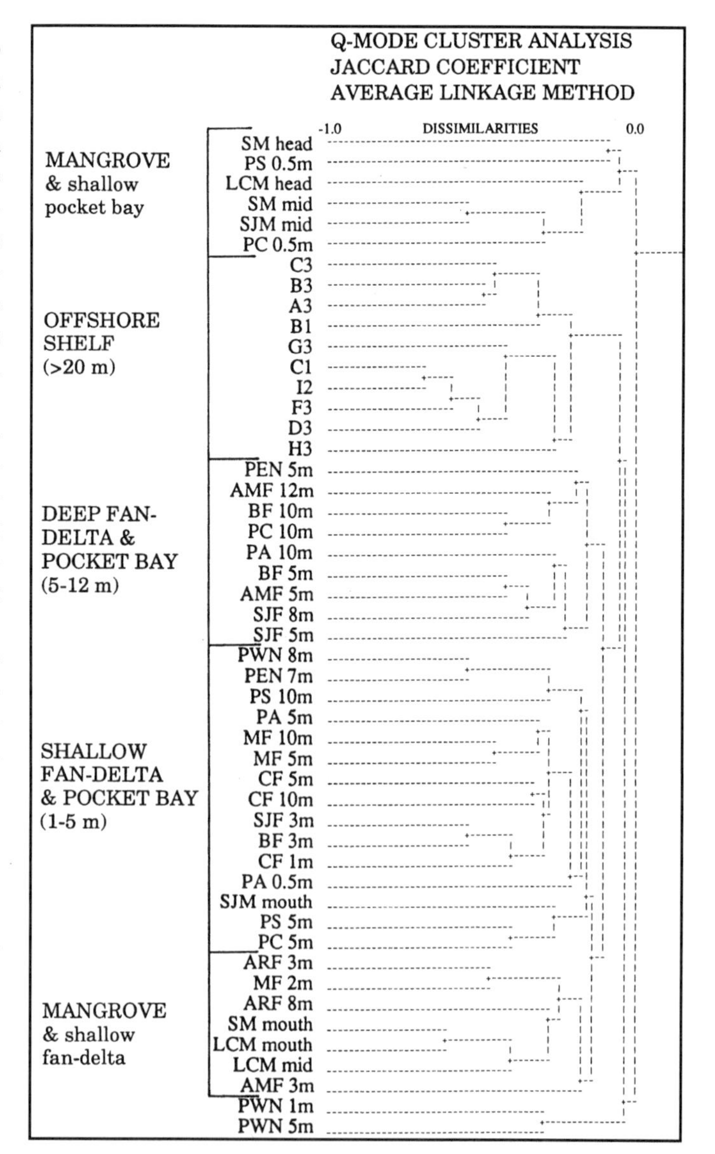

Figure 8. Q-mode cluster dendrogram based on species presence-absence data from 49 molluscan shell samples in Bahía Concepción (Jaccard coefficient, average linkage grouping method). The analysis demonstrates that variation in species composition is related to water depth. See text for further discussion.

Molluscan biofacies (Q-mode analysis)

We measured similarity in species composition between the 49 shell samples using the Jaccard coefficient, a common similarity index for presence-absence taxonomic data (Aldenderfer and Blashfield, 1984). Figure 8 is a Q-mode cluster dendrogram based on a matrix of Jaccard coefficients, using the average linkage method of grouping. The dendrogram separates samples into five loose clusters that primarily reflect depth zonation. Intertidal mangrove samples occur in two clusters, associated with samples from either shallow pocket bays or shallow fan-deltas. Most pocket bay, fan-delta and nearshore shelf samples occur in one of two depth-related clusters: samples mostly from 1- to 5-m depth, and samples mostly from 5- to 12-m depth. Offshore shelf samples (20-m depth or more) form their own distinct cluster. The looseness of the clusters and their low separation levels suggest that species assemblages among Bahía Concepción habitats are overlapping and gradational, with many species shared among habitats and high species variability within habitats.

Molluscan species assemblages (R-mode analysis)

At Bahía Concepción, only a few species make up the majority of individuals collected. To simplify our data and produce results more useful for paleoenvironmental interpretation, we limited our R-mode analysis to 20 species: the 10 most common bivalves and the 10 most common gastropods. We measured association among these 20 species using the simple matching coefficient (Aldenderfer and Blashfield, 1984). This coefficient counts both shared occurrences and shared absences of species when estimating similarity between samples (unlike the Jaccard coefficient, which counts only shared occurrences). We favor the simple matching coefficient in R-mode analysis, both because it yields a more interpretable dendrogram and because we believe that there is significance to both occurrences *and* absences of species when defining species assemblages, as discussed below. Figure 9 is an R-mode cluster dendrogram based on a matrix of simple matching coefficients, using the average linkage method of grouping.

Two distinct clusters on the dendrogram separate four species (top cluster of the dendrogram) that are common throughout nearshore and offshore habitats in Bahía Concepción from sixteen species that are common only in nearshore habitats (bottom cluster of the dendrogram). Within this bottom cluster is another cluster that represents species particularly common in mangrove swamps. We thus recognize three distinct species assemblages in Bahía Concepción: a mangrove assemblage, a nearshore assemblage, and an offshore assemblage (which consists of a subset of the nearshore assemblage).

The *mangrove assemblage* comprises the bivalves *Ostrea palmula* (the mangrove oyster, commonly attaching in dense accumulations to mangrove roots) and *Anadara* sp. (epibyssate filter feeder, attaching to rocks) and the gastropods *Cerithidea albonodosa* (deposit feeder, commonly reaching tremendous abundance in upper mangrove areas), *Theodoxus luteofasciatus* (algal grazer), and *Cerithium stercusmuscarum* (algal grazer).

The *nearshore assemblage* comprises a diverse assortment of species that are common in all the shallow subtidal habitats—fan-deltas, pocket bays, and the nearshore shelf. This assemblage includes the bivalves *Argopectin circularis* (epifaunal, motile scallop), *Chione* spp., *Laevicardium elenense, Megapitaria squalida, Dosinia ponderosa, Laevicardium elatum, Trigonicardia biangulata* (all shallow-burrowing filter feeders), and *Arca pacifica* (epibyssate filter feeder, attaching to shallow rocks). This assemblage also includes the gastropods *Crucibulum* sp. (sessile filter feeder attaching to hard substrates), *Nassarius* sp. (motile scavenger), *Modulus* sp., *Conus* sp., *Murex* sp., *Strombina* sp., and *Strombus gracilor* (all motile predators).

The *offshore assemblage* is a subset of the nearshore assemblage. In other words, species common offshore are also common nearshore, but most nearshore species are not common offshore. Since the offshore assemblage is distinguished by the absence of

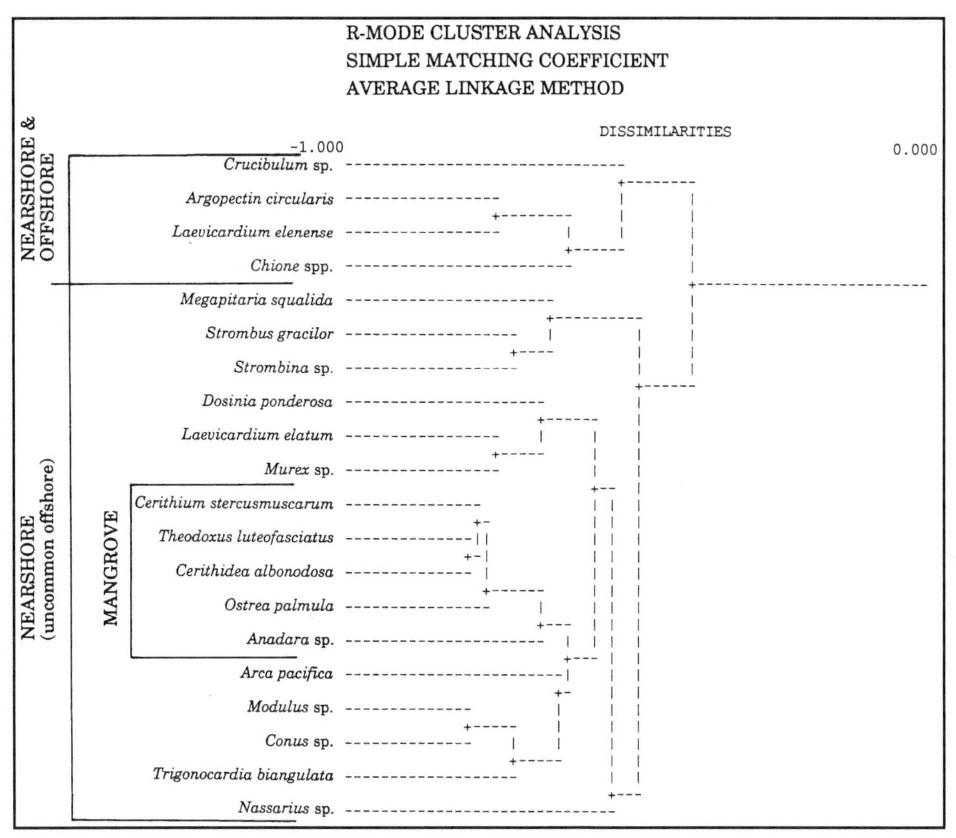

Figure 9. R-mode cluster dendrogram based on presence-absence data for the 20 most abundant Bahía Concepción mollusc species (simple matching coefficient, average linkage grouping method). The analysis identifies species commonly associated with the offshore shelf, nearshore habitats, and mangrove swamps. See text for further discussion.

certain nearshore species, we believe our use of the simple matching coefficient to delineate species assemblages is justified. The offshore assemblage is dominated by *Argopectin circularis, Laevicardium elenense, Chione* spp., and *Crucibulum* sp. (Fig. 9).

In summary, our results indicate that distributions of molluscan species at Bahía Concepción are related to water depth. In analogous Neogene rift basin facies, census of mollusc species on outcrop should yield useful bathymetric information.

RESULTS AND DISCUSSION: MOLLUSCAN TAPHONOMY

Variation among sedimentary environments in wave and current energy, light, sedimentation rate, and water chemistry often leaves diagnostic taphonomic "signatures" on organic remains, providing useful data for paleoenvironmental interpretation (e.g., Brett and Baird; 1986, Davies et al., 1990; Kidwell, 1991a; Meldahl and Flessa, 1990). Studies of taphonomic alteration of organic remains in modern sedimentary environments, where the relationship between environmental process and taphonomic result can be directly documented, provide valuable baseline data with which to make paleoenvironmental interpretations (e.g., Greenstein and Moffat, 1996; Meldahl, 1994; Parsons and Brett, 1991; Powell et al., 1989). At Bahía Concepción, two taphonomic parameters vary distinctly with water depth: shell abundance in the sediment and the taphonomic condition (taphonomic grade) of large bivalve shells.

Variation in shell abundance with depth

Figure 10 shows how shell abundance in samples varies with water depth. "Shell abundance" here refers to the dry weight of shells and shell fragments greater than 6 mm in size (the mesh size through which we sieved 1/64 m^3 of surface sediment to obtain shell samples). Figure 10 illustrates that the abundance of shells and fragments typically ranges from zero to about 80 g per 1,000 cm^3 of surface sediment. Values exceed 80 g only in areas that are less than 1 m deep, specifically the intertidal mangrove swamps and the shallowest parts of pocket bays.

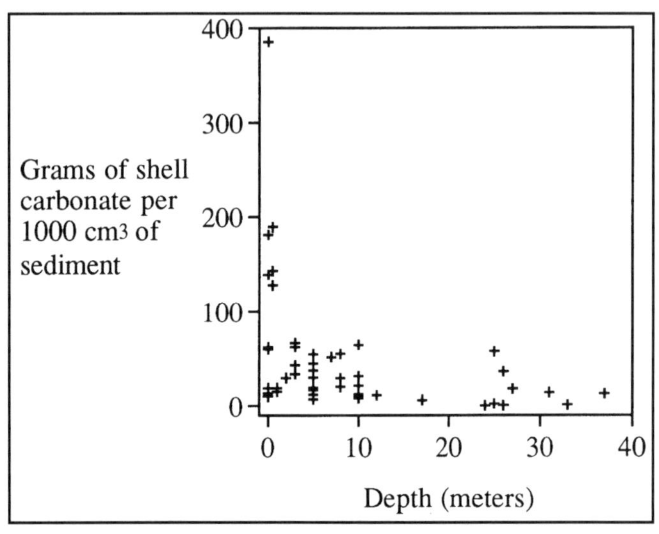

Figure 10. Depth variation in shell abundance (dry weight of shells and shell fragments greater than 6 mm in size), expressed as grams of shell carbonate per 1,000 cm^3 of surface sediment. Shell abundance is low throughout the bay, except in water depths of 1 m or less, as a result of increased wave and tidal current winnowing in very shallow water.

We think these results are due to increased wave and current winnowing in very shallow water. Wave reworking near the strandline in pocket bays and tidal current reworking in mangrove swamps winnow out finer sediment and concentrate shells. Furthermore, pocket bays and mangrove swamps develop best in coastal embayments with small catchments and thus little clastic input. In the absence of significant clastic input, shells and fragments accumulate in greater concentrations.

Variation in shell taphonomic grade with depth

Figure 11 illustrates variation in the taphonomic grade index (TGI) with water depth in Bahía Concepción (see "Methods" section for calculation of TGI). Samples from the offshore shelf (20-m depth or greater) have uniformly high TGI (i.e. offshore shells tend to be in good condition). In contrast, nearshore samples (less than 10-m depth) have highly variable TGI (i.e., nearshore shells range from good condition to highly altered condition).

These results are likely due to variation in either rates of taphonomic alteration or rates of net sedimentation. In other words, shallow-water shells become more altered than deep-water shells either because taphonomic processes occur more rapidly in shallow water or because shells are buried more slowly in shallow water and are thus exposed to the ravages of taphonomic processes for longer time spans. We suspect that interactions between rates of alteration and rates of burial explain why shallow-water shells have variable taphonomic grade, while deep-water shells have uniformly high taphonomic grade. Wave- and current-reworking in shallow water will increase rates of physical taphonomic alteration (shell abrasion) and also result in bypassing of much sediment to deeper water, burying deep-water shells while leaving shallow-water shells exposed for longer time spans. Furthermore, much taphonomic alteration is biogenic in origin (e.g., Cutler, 1994). Microboring sponges, polychaetes, and algae are ubiquitous on dead shells at Bahía Concepción, and microboring algae in particular flourish in shallow water (Cutler, 1994). However, in shallow water, migration of sedimentary bedforms and surface mixing of sediment by bioturbation (particularly bottom-feeding elasmobranch rays, whose feeding pits are common in shallow water throughout Bahía Concepción) could in some instances bury shells rapidly, protecting them from taphonomic processes. The net result is high taphonomic variability in shallow-water areas. In contrast, deep areas of the bay are likely to have lower rates of biogenic alteration due to less algal bioerosion (lower light levels), lower rates of physical alteration (quieter bottom conditions), and perhaps higher rates of net

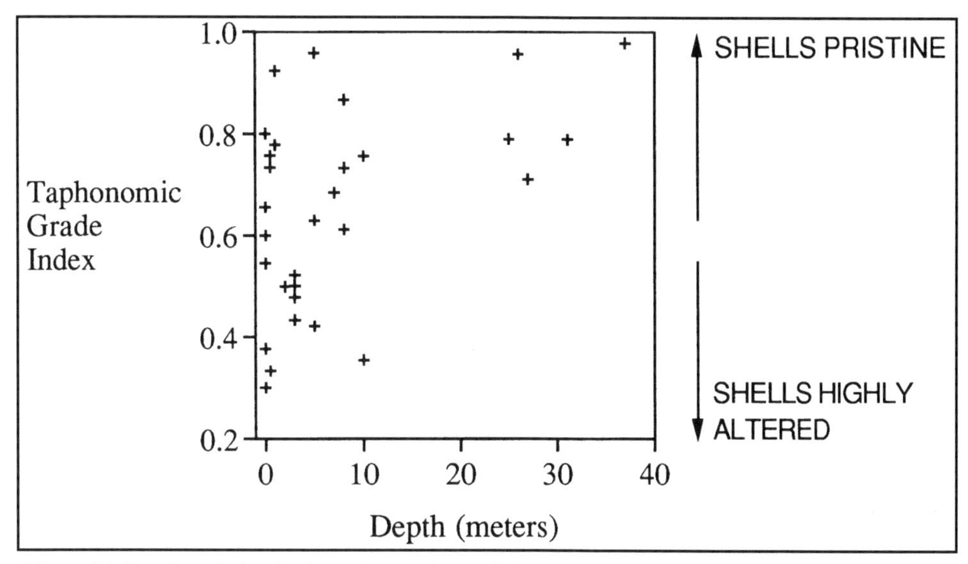

Figure 11. Depth variation in the taphonomic condition of assemblages of large (>1.5-cm diameter) disarticulated bivalve shells, expressed by the taphonomic grade index. Shells from the offshore shelf (>20 m deep) tend to be in good condition (high taphonomic grade index). Shells from nearshore habitats (<10 m deep) range from good condition to poor condition (taphonomic grade index varies from high to low). The trend is interpreted to reflect depth-related variation in rates of sedimentation and rates of taphonomic alteration. See text for further discussion.

sedimentation due to bypassing of sediment from shallow water to deep-water. The result is that deep water shell assemblages are in uniformly good condition.

In summary, our results illustrate depth-related variation in both shell abundance and shell taphonomic condition at Bahía Concepción. These data can be easily collected on outcrop and should provide useful bathymetric information in analogous Neogene rift basin facies.

SUMMARY AND IMPLICATIONS

The early evolution of the modern Gulf of California is recorded by the preserved remnants of Neogene rift basins in Baja California and southern Alta California. These basins formed in response to the separation of what is now the Baja California peninsula from mainland North America and its transfer to the Pacific plate (Lonsdale, 1989; Lyle and Ness, 1991; Stock and Hodges, 1989). The Neogene record of the nascent Gulf of California has been well studied in the Salton Trough region of Alta California (Kerr and Kidwell, 1991; Kidwell, 1988; 1991b; Winker, 1987). Studies of Neogene basins of Baja California are relatively new, and much work remains to be done (see Dorsey, this volume; Dorsey et al., this volume; McLean, 1988; Umhoefer et al., 1994). Holocene sediments and faunas of Bahía Concepción represent potentially useful modern analogs to lithofacies and biofacies in these Neogene basins.

Bahía Concepción formed by the drowning of a rifted alluvial basin during the Holocene transgression. The basin appears to be an asymmetric graben. A large normal fault zone bounds the east side, associated with a 30-km-long shoreline bajada backed by small, steep catchments. A smaller fault zone bounds the west side. The west side has larger catchments with gentler gradients and is characterized by rocky shorelines, pocket bays, mangrove swamps, and few exposed alluvial fans. Holocene sediments of the basin accumulate in alluvial fan, coastal interfan flat, mangrove swamp, fan-delta, pocket bay, nearshore shelf and offshore shelf, environments.

Three major types of marine sediment dominate Bahía Concepción. Green clastic mud (with variable mollusc shell content) dominates offshore below 20 m. Volcaniclastic sand (with variable mollusc shell and calcareous algae content) dominates the shallow fan-deltas and nearshore shelf on the bay's eastern and southern margins. Carbonate sand (both mollusc and calcalgal origin) dominates in shallow pocket bays and adjacent to rocky shorelines on the bay's western margin. We attribute the occurrence of carbonate sediments to clastic starvation along the bay's west side. It appears that many western drainages were deeply incised prior to the Holocene and then subsequently have aggraded to a gentle gradient, probably in response to Holocene sea-level rise. Terrigenous sediments are trapped upstream in these aggrading canyons, creating clastic-starved conditions amenable to carbonate production along much of the western shoreline. The occurrence of shallow-water carbonates in Neogene rift basins may be controlled by a similar process.

Variation in species composition and taphonomy of molluscan shell assemblages in Bahía Concepción is correlated with water depth. Cluster analysis on molluscan shell samples identifies four overlapping, depth-related biofacies: mangrove swamp (intertidal), shallow nearshore (1- to 5-m depth), deeper nearshore (5- to 12-m depth), and offshore (below 20 m). Taphonomic analysis demonstrates depth zonation in both the abundance and condition of mollusc shells. Shell abundance is greatest in intertidal mangrove swamps and the shallowest (less than 1 m deep) areas of pocket bays. Shells in less than 10-m depth exhibit variable taphonomic condition, while shells in greater than 20 m depth are in uniformly good condition. Data on molluscan species composition and taphonomy should provide useful bathymetric information in analogous Neogene rift basin facies.

A logical first step to test the applicability of the Concepción Basin as a modern analog for Neogene basins is to compare it with the Pliocene Loreto Basin, which lies along structural strike 60 km to the south of Bahía Concepción. The evolution of the Loreto Basin is related to the Pliocene tectonic development of the Gulf of California and is an important link in understanding the geologic evolution of this active plate margin (Dorsey et al., this volume; Umhoefer et al., 1994). The Concepción and Loreto basins have a similar asymmetric graben structure, though with opposite orientations: The Loreto Basin has its large bounding fault on the southwest side (Dorsey et al. [this volume]), whereas the Concepción Basin has its large bounding fault on the east side. The two basins may be structurally linked grabens separated by an accommodation zone (R. Dorsey and P. Umhoefer, personal communication, 1995). Both basins receive clastic sediment from alluvial fans and fan-deltas derived from highland sources produced by normal faulting; both are characterized by carbonate accumulation in areas of low clastic input; both have shoreline areas dominated by steep rocky coasts or alluvial fans; and both have an abundant and diverse molluscan fauna dominated by scallops.

A general comparison of the two basins (based on a reconnaissance study of the Loreto Basin) suggests that Holocene data from the Concepción Basin make two principal contributions to understanding the Pliocene Loreto Basin.

Origin of shallow water carbonates. Dorsey et al. (this volume) and Umhoefer et al. (1994) attribute the formation of shallow-water, shell-rich carbonates in the Loreto Basin to several possible mechanisms, including lateral lobe-switching on fan-deltas; episodic movements on basin-bounding faults; eustatic sea-level fluctuations; and high biological productivity. Data from Bahía Concepción suggest a specific, related mechanism for shallow-water carbonate formation: drainage incision during low sea level (creating potential accommodation space for clastic sediment), followed by drainage aggradation with rising sea level, trapping clastics upstream and creating starved conditions along the shoreline amenable to carbonate production. This mechanism could operate in any steep-sided basin surrounded by canyons when there is a rise in base level following a period of drainage incision. The origin of shallow-water carbonates in such basins is an interesting problem and a fruitful area for future research.

Sedimentation and sea level. Our reconnaissance study of strata deposited during Sequence 2 time in the Loreto Basin (a period of rapid subsidence [~8 mm/yr] during the most tectonically active phase of basin development; Dorsey et al., this volume; Umhoefer et al., 1994) indicates that nearly all deposition occurred in water less than 10 m deep. We found only molluscan assemblages similar in composition to mangrove and nearshore assemblages at Bahía Concepción; offshore assemblages are absent. We found that variable levels of shell preservation (typical of shallow-water areas of Bahía Concepción) dominated local sections, whereas uniformly good shell preservation (typical of deep water areas of Bahía Concepción) is uncommon. Local sections are dominated by clastic or carbonate sand, but mud (which dominates offshore at Bahía Concepción) is very rare.

These findings indicate that sedimentation rates nearly equaled rates of subsidence and sea-level change in the Loreto Basin, even during the most rapid phases of subsidence, so that water depths of greater than 10 m were rarely realized. In contrast, sedimentation in the Concepción Basin has not kept pace with Holocene subsidence or sea-level rise: Most of the bay is greater than 10 m deep, and much of it lies below 30 m. Assuming negligible Holocene subsidence, the rate of sea-level rise during the Holocene transgression in the Concepción Basin (~7 to 10 mm/yr; Fairbanks, 1989) is comparable to the maximum subsidence rates experienced in the Loreto Basin (~8 mm/yr). Why was sedimentation able to keep pace in the Loreto Basin but not in the Concepción Basin? One possibility is different tectonic settings of the Pliocene versus Holocene. During the Pliocene, much active rifting was concentrated in the continental crust of what is now Baja California. By the Holocene, active rifting had shifted eastward into oceanic crust of the central Gulf of California. Another possibility is different climatic conditions of the Pliocene versus Holocene. Perhaps wetter climates during Pliocene time enabled the Loreto Basin to accumulate sediment at rates not possible during arid Holocene time. Continued study of the similarities and differences between Holocene basins and their Neogene counterparts may resolve these questions, and further enrich understanding of Gulf of California basins.

ACKNOWLEDGMENTS

We thank Guillermo Avila Serrano for providing valuable assistance with the fieldwork and Jose Luis Pérez Soto for help with drafting figures. We thank Rebecca Dorsey, Paul Umhoefer, Mike Foster, Markes Johnson, Jorge Ledesma-Vázquez, and Larry Mayer for useful suggestions and discussions during various phases of the project. Comments by reviewers Benjamin Greenstein and Douglas Smith improved the manuscript. This research was supported by the National Geographic Society (5127-93), National Science Foundation (EAR-9316419), American Chemical Society (PRF 27801-B8), and the Universidad Autónoma de Baja California (0269). T. Reardon acknowledges support of a student research grant from Sigma Xi. We thank Markes Johnson and Jorge Ledesma-Vázquez for editing and producing this volume.

REFERENCES CITED

Aldenderfer, M. S., and Blashfield, R. K., 1984, Cluster analysis: Newbury Park, California, Sage Publications, 88 p.

Ashby, J. R., and Minch, J. A., 1987, Stratigraphy and paleoecology of the Mulege Embayment, Baja California Sur, Mexico: Ciencias Marinas, v. 13, p. 89–112.

Baqueiro, E., Masso-R., J. A., and Velez-B., A., 1983, Crecimiento y reproduccion de una poblacion de caracol chino *Hexaplex crythrostomus*, de Bahía Concepción, B.C.S.: Ciencia Pesquera, v. 4, p. 19–33.

Bastida-Zavala, J. R., 1990, Poliquetos (Annelida: Polychaeta) de Baja California Sur, *in* Programa y Resumenes, Sympisiom Internacional de Biologia Marina, 8th, Ensenada, Junio 4–8, 1990: Ensenada, Baja California, Mexico, Universidad Autónoma de Baja California, p. 62.

Beal, C. H., 1948, Reconnaissance of the geology and oil possibilities of Baja California, Mexico: Geological Society of America Memoir 31, 138 p.

Brandt, D. S., 1989, Taphonomic grades as a classification for fossiliferous assemblages and implications for paleoecology: Palaios, v. 4, p. 303–309.

Brett, C. E., and Baird, G. C., 1986, Comparative taphonomy: A key to paleoenvironmental interpretation based on fossil preservation: Palaios, v. 1, p. 207–277.

Cutler, A. H., 1994, Taphonomic implications of shell surface textures in Bahía la Choya, northern Gulf of California: Palaeogeography, Palaeoclimatology, Palaeoecology, v. 114, p. 219–240.

Davies, D. J., Staff, G. M., Callender, W. R., and Powell, E. N., 1990, Description of a quantitative approach to taphonomy and taphofacies analysis: All dead things are not created equal., *in* Miller, W., ed., Paleocommunity temporal dynamics: The long-term development of multispecies assemblages: Paleontological Society Special Publication 5, p. 328–350.

Fairbanks, R. G., 1989, A 17,000-year glacio-eustatic sea level record: Influence of glacial melting rates on the younger Dryas event and deep ocean circulation: Nature, v. 342, p. 637–642.

Flessa, K. W., Cutler, A. H., and Meldahl, K. H., 1993, Time and taphonomy: Quantitative estimates of time-averaging and stratigraphic disorder in a shallow marine habitat: Paleobiology, v. 19, p. 266–286.

Greenstein, B. J., and Moffat, H. A., 1996, Comparative taphonomy of modern and Pleistocene corals, San Salvador, Bahamas: Palaios, v. 11, p. 57–63.

Hack, J. T., 1973, Stream profile analysis and stream gradient index: U.S. Geological Survey Journal of Research, v. 1, p. 421–429.

Hayes, M. L., Johnson, M. E., and Fox, W. T., 1993, Rocky-shore biotic associations and their fossilization potential: Isla Requeson (Baja California Sur, Mexico): Journal of Coastal Research, v. 9, p. 944–957.

Keen, A. M., 1971, Sea shells of tropical west America (second edition): Stanford, California, Stanford University Press, 1064 p.

Kerr, D. R., and Kidwell, S. M., 1991, Late Cenozoic sedimentation and tectonics, western Salton Trough, California, *in* Walawender, M. J., and Hanan, B. B., eds., Geological excursions in southern California and Mexico: Fieldtrip Guidebook, Geological Society of America, 1991 Annual Meeting, San Diego, California, p. 397–416b.

Kidwell, S. M., 1988, Taphonomic comparison of passive and active continental margins: Neogene shell beds of the Atlantic coastal plain and northern Gulf of California: Palaeogeography, Palaeoclimatology, Palaeoecology, v. 63, p. 201–223.

Kidwell, S. M., 1991a, The stratigraphy of shell concentrations. *in* Allison, P. A., and Briggs, D. E. G., eds., Taphonomy: Releasing the data locked in the fossil record: New York, Plenum Press, p. 212–290.

Kidwell, S. M., 1991b, Condensed deposits in siliciclastic sequences: Expected and observed features: *in* Einsele, G., Ricken, W., and Seilacher, A., eds., Cycles and events in stratigraphy: Berlin, Springer-Verlag, p. 682–695.

Leeder, M. R., 1995, Continental rifts and proto-oceanic rift troughs, *in* Busby, C. R., and Ingersoll, R. V., eds., Tectonics of sedimentary basins: Cambridge, Massachusetts, Blackwell Science, p. 119–148.

Leeder, M. R., and Gawthorpe, R. L., 1987, Sedimentary models for extensional tilt-block/half-graben basins, *in* Coward, M. P., Dewey, J. F., and Handcock, P. L. eds., Continental extensional tectonics: Geological Society of London Special Publication 28, p. 139–152.

Leeder, M. R., and Jackson, J. A., 1993, The interaction between normal faulting and drainage in active extensional basins, with examples from the western United States and central Greece: Basin Research, v. 5, p. 79–102.

Lonsdale, P., 1989, Geology and tectonic history of the Gulf of California, *in* Winterer, E. L., Hussong, D. M., and Decker, R. W., eds., The Eastern Pacific Ocean and Hawaii: Boulder, Colorado, Geological Society of America, Geology of North America, v. N, p. 499–521.

Lyle, M., and Ness, G. E., 1991, The opening of the southern Gulf of California, *in* Dauphin, J. P., and Simoneit, B. R. T., eds., The Gulf and Peninsula Province of the Californias: American Association of Petroleum Geologists Memoir 47, p. 403–423.

Mayall, M., Gutierrez, S., Ledesma, J., Johnson, M., and Minch, J., 1993, Pliocene bedded cherts from Concepción Peninsula: Second International Meeting on the Geology of the Baja California Peninsula, Universidad Autonoma de Baja California, Ensenada, April 1993, Abstracts, p. 62–63.

McFall, C. C., 1968, Reconnaissance geology of the Concepción Bay area, Baja California, Mexico: Stanford University Publications in Geological Sciences, v. 10, no. 5 (25 p. text plus 1:70,000 scale geologic map).

McLean, H., 1988, Reconnaissance geologic map of the Loreto and part of the San Javier quadrangles, Baja California Sur, Mexico: U.S. Geological Survey Miscellaneous Field Studies Map MF-2000 (10 p. text plus 1:50,000 scale geologic map).

Meldahl, K. H., 1994, Biofacies and taphofacies of a Holocene macrotidal environment: Bahía la Cholla, Northern Gulf of California: Ciencias Marinas, v. 20, p. 555–583.

Meldahl, K. H., and Flessa, K. W., 1990, Taphonomic pathways and comparative biofacies and taphofacies in a Recent intertidal/shallow shelf environment: Lethaia, v. 23, p. 43-60.

Ortlieb, L., 1987, Neotectonique et variations du niveau marin au Quaternaire dans la region du Golfe de Californie, Mexique: Institut Francais de Recherche Scientifique pour le Developpement en Cooperation Collection Etudes et Theses (2 volumes: 779 p. and 257 p.).

Parsons, K. M., and Brett, C. E., 1991, Taphonomic processes and biases in modern marine environments: An actualistic perspective on fossil assemblage preservation, *in* Donovan, S. K., ed., The processes of fossilization: New York, Columbia University Press, p. 22–65.

Powell, E. N., Staff, G., Davies, D. J., and Callendar, W. R., 1989, Macrobenthic death assemblages in modern marine environments: Formation, interpretation and application: Critical Reviews in Aquatic Science, v. 1, p. 555–589.

Rios-Gonzalez, R., 1989, Un catalogo de camarones carideos de Mulege y Bahía Concepción, B.C.S., con anotaciones acerca de su biologia, ecologia, distribucion geographica, y taxonomia [Tesis]: Ensenada, Universidad Autónoma de Baja California, 208 p.

Rodriguez-Romero, J., Abita-Chardenas, L.A., Galvan-Magana, F., and Chavez-Ramos, H., 1988, Composicion y abundancia de la ictiofauna de Bahía Concepción, B.C.S., Mexico: California Cooperative Oceanic Fisheries Investigations Program Abstracts, Nov. 8–10, Lake Arrowhead, California, p. P-18.

Rodriguez-Romero, J., Abita-Cardenas., L. A., de la Cruz-Aguero, J., and Galvan-Magana, F., 1991, Composicion, diversidad y abundancia de peces de Bahía Concepción, B.C.S., Mexico, *in* Resumenes, Asociacion de investigadores del Mar de Cortes, Asociacion Civil Congreso, 3rd, Guaymas, Sonora, Abril 10–12, 1991: Hermosillo, Centro de Investigaciones, Cientificas y Tecnologicas de la Universidad Autonoma de Sonora, p. 13.

Romero-Ibarra, N., and Garate-Lizarraga, I., 1991, Dinamica del fitoplancton en Bahía Concepción, Baja California Sur, durante Julio 1990, *in* Resumenes, Asociacion de investigadores del Mar de Cortes, A.C. III Congreso, Guaymas, Sonora, Abril 10–12, 1991, p. 31–32.

Salazar-Vallejo, S. I., 1985, Contribucion al conocimiento de los poliquestos (Annelida: Polychaeta) de Bahía Concepción, B.C.S., Mexico [Tesis]: Ensenada, Centro Investigacion Cientifica Educacion Superior Ensenada, 311 p.

Salazar-Vallejo, S. I., and Stock, J. H., 1987, Apparent parasitism of *Sabella melanostigma* (Polychaete) by *Ammotella spinifera* (Pycnogonida) in the Gulf of California: Reviews of Tropical Biology, v. 35, p. 269–275.

Singh-Cabanillas, J., and Bojorquez-Verastica, G., 1987, Fifacion de moluscos bivalvos en colectores artificiales en Bahía Concepción, Baja California Sur, un avance: Resumenes, Segundo Congreso Nacional de Acuacultura, La Paz, B.C.S., Noviembre 1987: La Paz, Universidad de Autónoma de Baja California Sur, p. 3.

Singh-Cabanillas, J., and Bojorquez-Verastica, G., 1990, Captacion de juveniles de hacha larga (*Pinna rugosa*) en colectores artificiales en Bahía Concepción, B.C.S., *in* Resumenes, Congreso Nacional de Oceanografia, 8th, Hermosillo, Novembre: 21–23, 1990: Hermosillo, Escuela de Ciencias del Marinas, Universidad Autónoma de Sonora, p. 139.

Steller, D. L., and Foster, M. S., 1995, Environmental factors influencing the distribution and morphology of rhodoliths in Bahía Concepción, B.C.S., Mexico: Journal of Experimental Marine Biology and Ecology, v. 194, p. 201–212.

Stock, J. M., and Hodges, K. V., 1989, Pre-Pliocene extension around the Gulf of California and the transfer of Baja California to the Pacific plate: Tectonics, v. 8, p. 99–115.

Umhoefer, P. J., Dorsey, R. J., and Renne, P., 1994, Tectonics of the Pliocene Loreto basin, Baja California Sur, Mexico, and evolution of the Gulf of California: Geology, v. 22, p. 649–652.

Winker, C. D., 1987, Neogene stratigraphy of the Fish Creek–Vallecito section, southern California: Implications for the early history of the northern Gulf of California and Colorado delta [Ph.D. thesis]: Tucson, University of Arizona, 494 p.

Manuscript Accepted by the Society December 2, 1996

Printed in U.S.A.

Geological Society of America
Special Paper 318
1997

Upper Pliocene stratigraphy and depositional systems: The Peninsula Concepción basins in Baja California Sur, Mexico

Markes E. Johnson
Department of Geosciences, Williams College, Williamstown, Massachusetts 01267
Jorge Ledesma-Vázquez
Facultad de Ciencias Marinas, Universidad Autónoma de Baja California, Ap. Postal 453, Ensenada, Baja California, Mexico 22800
Mark A. Mayall
Department of Geosciences, Williams College, Williamstown, Massachusetts 01267
John Minch
26461 Crown Valley Parkway, Suite 200, Mission Viejo, California 92691

ABSTRACT

Geological mapping and stratigraphic differentiation of members belonging to the Upper Pliocene Infierno Formation at the base of Peninsula Concepción in Baja California Sur, Mexico, are the subjects of this chapter. Because of the regionally distinctive nature of chert deposits, names derived from local features are proposed for four new members. The stratigraphic order and lateral extent of these units indicate that at least two separate marine transgressions cover an area approximately 35 km^2. Three peninsulas and four islands up to 2 km^2 in size effectively subdivide the embayment into four basins interconnected with the Pliocene Bahía Concepción.

The islands and surrounding mainland are composed of andesite and basalt belonging to the Miocene Comondú Group. Included in the basins are some or all of the following units assigned to the upper Pliocene Infierno Formation (oldest to youngest): (1) Calabaza Member (mainly alluvial fans), (2) El Mono chert Member (includes fossil mangroves), (3) Bahía Concepción Member (lower and upper limestone units separated by an alluvial siltstone with a well-developed rhizolith), and (4) Cayuquitos chert Member. Buttress unconformities between the Miocene volcanics and Pliocene marine units delineate rocky shorelines of a well-sheltered nature. The extensive limestone beds of the Bahía Concepción Member are flat lying over most of the region, although they are cut by as many as four faults tending northwest by southeast. Two paleo hot springs are associated with faults within the study region, and another example is located nearby.

INTRODUCTION

Building on a series of review articles regarding ancient rocky shorelines (Johnson, 1988a, b, 1992), Johnson and Ledesma-Vázquez began a long-term survey of Baja Californian geology in search of buttress unconformities typical of abandoned coasts preserved in the geological record. Some of the results reviewed by Simian and Johnson (this volume) illustrate a diverse range of Cretaceous to Pleistocene rocky shores circumscribing entire islands with distinct windward and leeward lithofacies and biofacies. Nearly all other ancient rocky shorelines described in the geological literature (Johnson, 1992) pertain exclusively to open coasts that experienced intense wave shock resulting in the development of well-rounded and graded basal conglomerates. The purpose of this contribution is to characterize the late Pliocene shorelines defining a highly sheltered series of small marine

Johnson, M. E., Ledesma-Vázquez, J., Mayall, M. A., and Minch, J., 1997, Upper Pliocene stratigraphy and depositional systems: The Peninsula Concepción basins in Baja California Sur, Mexico, *in* Johnson, M. E., and Ledesma-Vázquez, J., eds., Pliocene Carbonates and Related Facies Flanking the Gulf of California, Baja California, Mexico: Boulder, Colorado, Geological Society of America Special Paper 318.

basins that cross onto Peninsula Concepción from the southeast corner of Bahía Concepción in Baja California Sur, Mexico (Fig. 1a).

Our original goal to map the Miocene-Pliocene unconformity at the base of Peninsula Concepción resulted in discovery of a spectrum of basin environments deserving careful description in their own right. In this chapter we establish a system of lithostratigraphic units with sufficient biostratigraphic control to enhance the paleogeographical context of our maps. A companion paper by Ledesma-Vázquez et al. (this volume) offers additional details on the shallow-water chert beds discovered in the Peninsula Concepción basins.

LOCATION AND PREVIOUS WORK

Situated southeast of the village of Mulegé on the gulf coast of Baja California Sur, Mexico, Bahía Concepción is a large southeast-northwest–oriented bay. Its elongated shape is related to a half-graben structure framed between the west side of Peninsula Concepción and the Baja California mainland (Fig. 1b). The bay is 40 km in length and varies between 5 and 10 km in width. Ending across the bay from Mulegé, Peninsula Concepción is 50 km in length and varies between 8 and 18 km in width. The Pliocene basins featured in this study are contained within a 7 km × 9 km area near the southern base of the peninsula (Fig. 1b). Access to the study area is by Mexican Federal Highway 1 south from Mulegé to the southwest corner of the bay, then by dirt road to the southeast corner of the bay in the vicinity of the abandoned Santa Rosaliita ranch. A rough road leading to the fishing camp at San Sebastián on the gulf coast of Peninsula Concepción traverses the field area from southwest to northeast.

The only previously published work of any detail from the same area is a reconnaissance report by McFall (1968), who

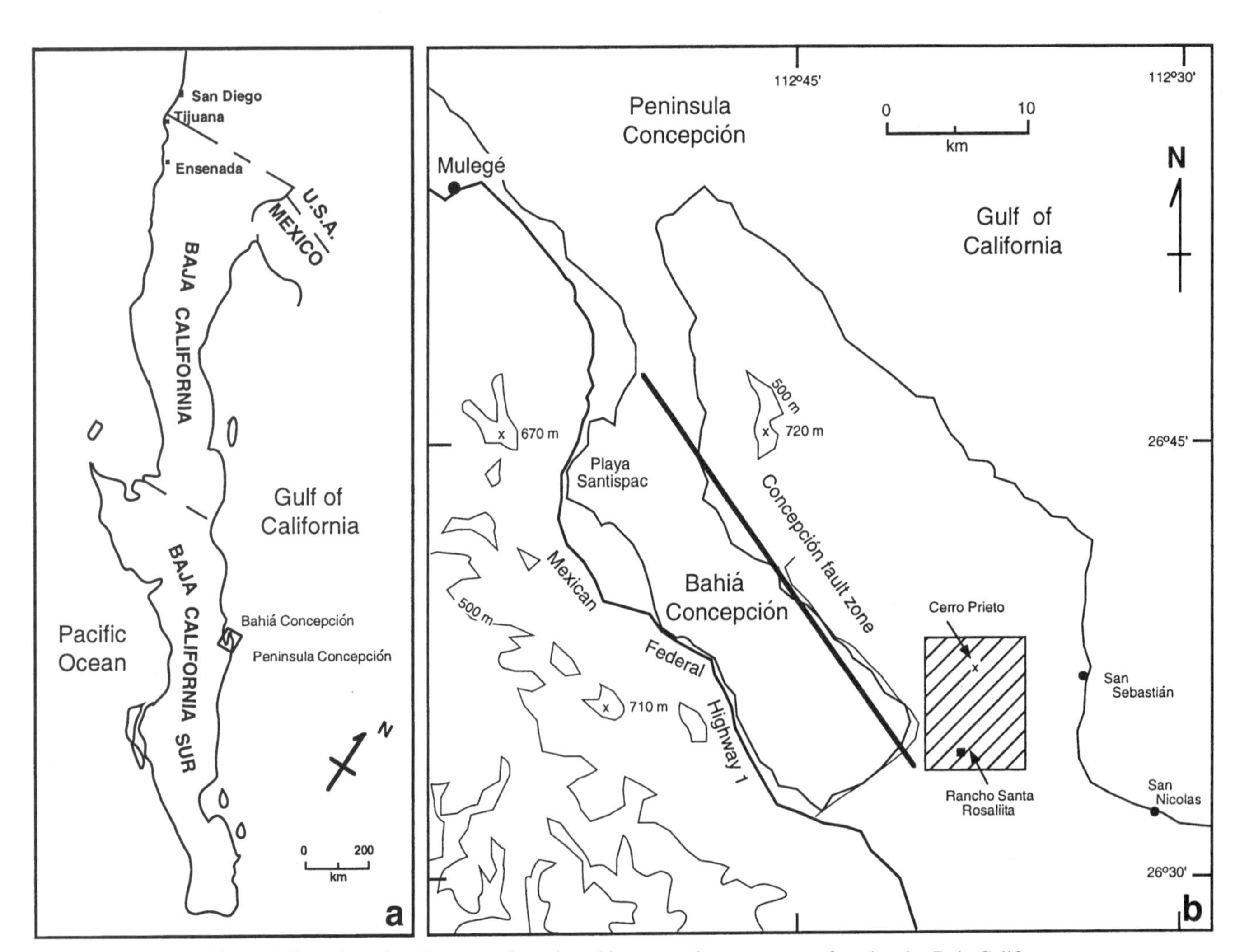

Figure 1. Location of study area and stratigraphic nomenclature. a, map of peninsular Baja California indicating the position of Bahía Concepción and Peninsula Concepción on the Gulf of California (open rectangle). b, the Bahía Concepción region (enlarged from a), showing the location of the study area at the base of Peninsula Concepción (diagonal-lined rectangle).

mapped a 1,165 km^2 region surrounding Bahía Concepción. He clearly recorded that a large basin filled with strata attributed to the upper Pliocene Infierno Formation sits within the topography of Miocene volcanics belonging to the Comondú Group at the base of Peninsula Concepción. Most of his report, however, is devoted to igneous geology. Only a single sedimentary sequence near Rancho Santa Rosaliita, described as 31.7 m of sandstones, conglomerate, coquina, and marl (McFall, 1968, p. 11), falls into the same area covered by this report. No fossils are identified therein to species level, and there is no mention of bedded chert.

METHOD OF STUDY

The stratigraphic scheme and geological map developed for this report were organized by standard field methods using teams of students from Williams College and the Universidad Autónoma de Baja California (Ensenada) to achieve dense ground coverage of a 63-km^2 area. A paleocoastline was mapped by walking out the Miocene-Pliocene unconformity and plotting its course on 1:50,000 topographic maps produced by the Instituto Nacional de Estadistica Geografia e Informatica for the El Coyote (G12A67) and San Nicolas (G12A68) regions. Aerial photos were employed to cross-check the surface geology. The treatise by Durham (1950) on Cenozoic paleontology of the gulf region in Baja California Sur was used to identify index fossils. The geographic coordinates of sites where stratigraphic sections were described were determined using the Global Positioning System and topographic base maps.

NEW STRATIGRAPHIC UNITS

Nomenclature developed by McFall (1968) for igneous formations within the Miocene Comondú Group in the Bahía Concepción region is retained in this report. McFall (1968, p. 11) noted that strata in the Peninsula Concepción basins resemble the Infierno Formation from the copper-mining district at Santa Rosalia to the north (Wilson, 1948; Wilson and Rocha, 1955). Based on biostratigraphic data summarized herein, a correlation with this upper Pliocene formation is appropriate except that the strata exposed at the base of Peninsula Concepción contain regionally unique bedded cherts (Ledesma-Vázquez et al., this volume). The stratigraphic scheme used in this report erects four new members of local significance under the regional cover of the Infierno Formation (Fig. 2). Names for these units are derived from arroyos within the study area, the formal names for which appear on the government topographic maps for the El Coyote (G12A67) and San Nicolas (G12A68) regions.

Stratigraphically, from youngest to oldest, the new members include the following: (1) Calabaza Member, dominated by andesitic conglomerates, sandstones, siltstones, and mudstones deposited in alluvial fans of variable thicknesses but sometimes broken by thin intercalations of in situ oyster beds; (2) El Mono Member, consisting of bedded cherts with occasional limestone lenses between 13 and 14 m in overall thickness; (3) Bahía Concepción Member, approximately 25 m in thickness and composed of lower and upper limestone beds separated by an alluvial siltstone with localized fossil rhizoliths and limestone lenses; and (4) Cayuquitos Member, another chert body up to 2 m in thickness.

Appendix 1 provides descriptions of 12 stratigraphic profiles used to determine physical correlation of Pliocene strata in the basins. Their locations are shown on the geological map (Fig. 2).

SUMMARY OF DEPOSITIONAL SYSTEMS

Containment

The shape and structure of Pliocene basins in the study area are controlled by the paleotopography of eroded Miocene andesites and basalts from the Comondú Group, regionally more than 4,000 m thick (McFall, 1968). The most significant unit is the Ricasón (Requeson) Formation consisting of dark-colored flows and agglomerates interbedded with light-colored tuffs. Perspective views across the field area from two different orientations capture a sense of topographic containment. Figure 3 offers a view looking west from the center of the largest basin to the south end of Bahía Concepción. Low andesite hills with elevations less than 50 m above sea level are planted to the far left (south) and right (north) in the middle distance fronting the bay. A thin, dark escarpment of Pliocene limestone stretches between the two hills in the center of the photograph, marking the egress of contemporary Arroyo El Mono to the bay. Figure 4a presents a view looking northeast over the same basin from a position on Arroyo El Mono toward dark andesitic hills formed by the Ricasón Formation on the horizon. The hill to the right is Cerro Prieto, a local landmark with an elevation 250 m above sea level. Mesas capped by light-colored limestone in the middle distance, before the Miocene hills, capture the nested effect of low-lying Pliocene strata.

The basement rock upon which the basins sit is discernible in a few places where contemporary stream erosion has cut through the entire Pliocene sequence. One such locality is the stream bed of Arroyo Cayuquitos 1 km south of Cerro Prieto (Fig. 4b), where angular basaltic agglomerates belonging to the Pelones Formation are exposed (Fig. 4c). The unconformably overlying Pliocene strata at this locality are about 55 m thick.

Subaerial fan deposits

Coalescing fan deposits eroded from surrounding Miocene hills form the basal beds of the Pliocene basins. The stratigraphic section at locality 7 (Fig. 2; Appendix 1) covers the sequence shown in the background of Figure 4b. Here the thickness of conglomeratic siltstone and mudstone amounts to more than 30 m, representing the thickest known profile of the Calabaza Member.

A 12-m-thick, red mudstone unit bracketed by limestone beds occurs higher in the same section at locality 7. Taken together, these three units make up the thickest and most complete profile of the Bahía Concepción Member. The mudstone notably lacks andesitic clasts and appears to represent the distal toe of subaerial fan

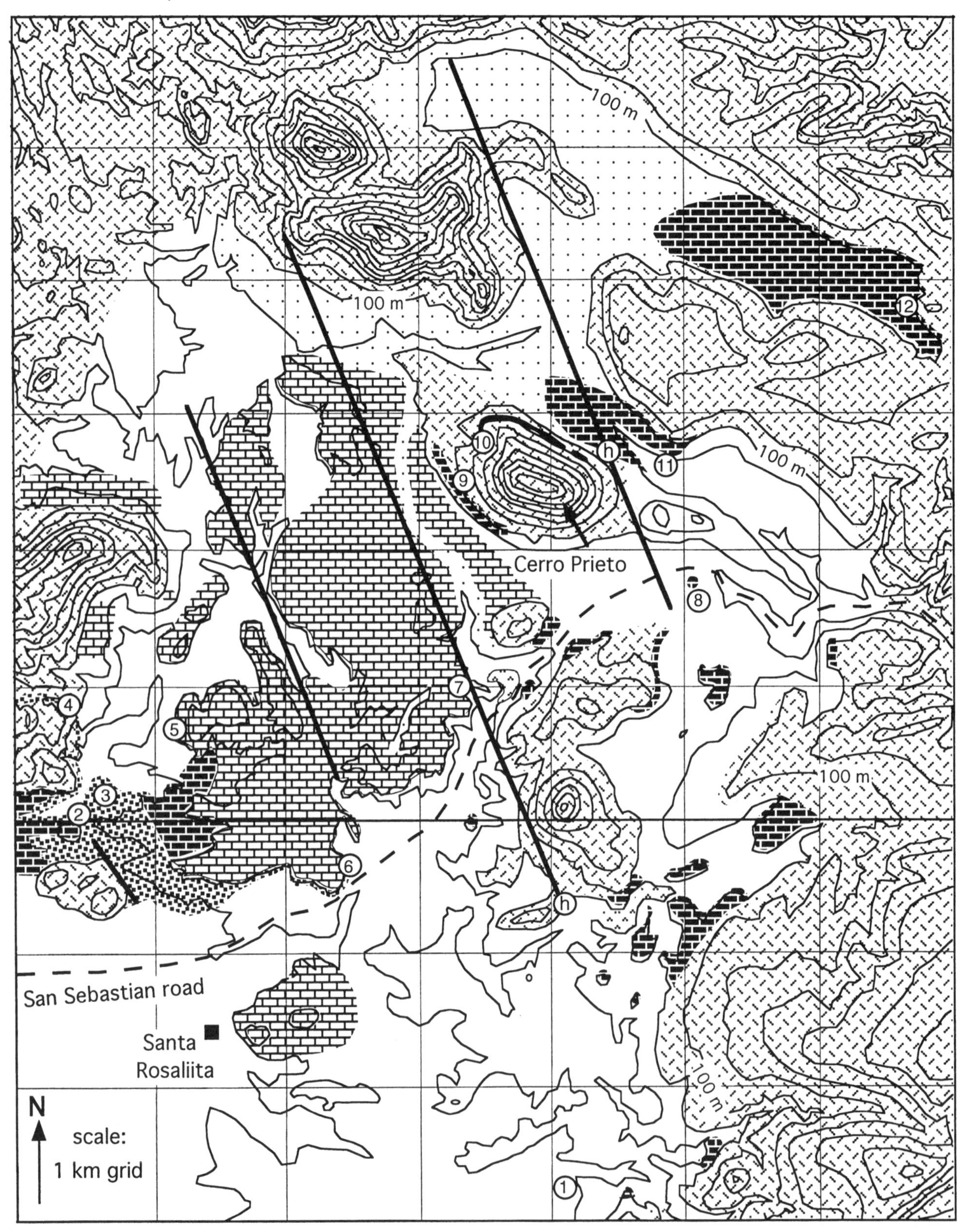

Figure 2 (on this and facing page). Geological map and key of the study area at the base of Peninsula Concepción (see Fig. 1b for geographic context).

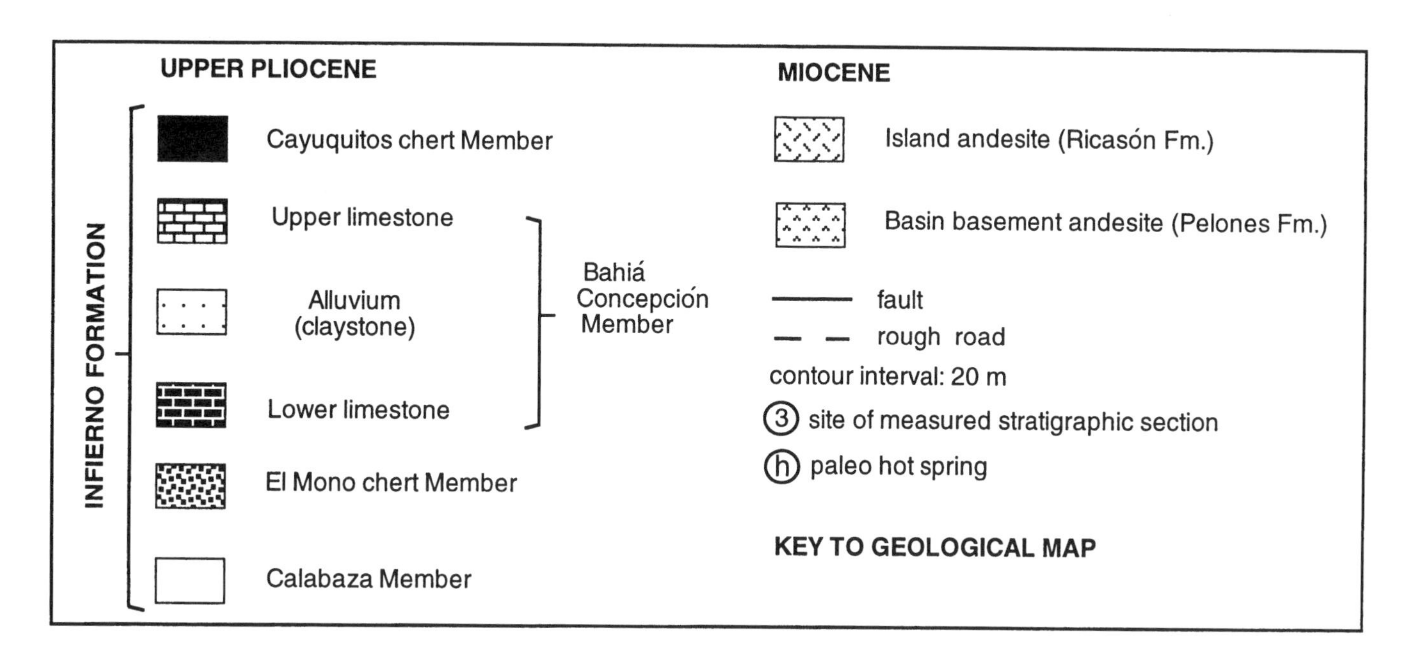

Figure 3. View west overlooking the south end of Bahía Concepción. Low hills to the left and right (fronting the bay) represent paleotopography of Miocene volcanics. The thin dark escarpment between the hills (center) is an upper Pliocene limestone in the Infierno Formation. Arroyo El Mono runs to Bahía Concepción along the base of this ridge.

deposits. At locality 6 (Fig. 2; Appendix 1), the equivalent mudstone includes a laterally extensive rhizolith. The lattice structure and root density at this locality (Figs. 5a and 5b) are similar to those previously attributed to fossil mangroves (Hoffmeister and Multer, 1965; Whybrow and McClure, 1981). A cross-sectional break across one of the roots (Fig. 5c) reveals the well-preserved concentric structure of the vascular and peridermal areas. The large diameter and downward-branching habit suggests that roots of the red mangrove (genus *Rhizophora*) might be represented. In contrast, smaller horizonal roots with upward-growing pneumatophores are typical of the black mangrove (genus *Avicennia*). A review of fossil mangroves by Plaziat (1995) emphasizes the unusual preservation of root systems and considers earlier published examples incorrect or unsubstantiated. Preservation of molluscs associated with the mangrove community is one test of validity. The rhizolith detailed in Figure 5 includes no marine invertebrates.

Chert deposits

The most unusual deposits in the Peninsula Concepción basins are the chert beds, named herein the Cayuquitos and El Mono Members. El Mono is the older of the two chert units,

Figure 4. Structure of the Peninsula Concepción basins. a, view looking northeast to Miocene volcanics (dark hills) on the horizon; the elevation to the right is landmark Cerro Prieto. Limestone-capped mesas (below the hills) belong to the upper Pliocene Infierno Formation. b, streambed of Arroyo Cayuquitos with figures standing on Miocene basement volcanics belonging to the Pelones Formation; nearly complete section of the Infierno Formation in background (Section 7, Appendix 1). c, closer view of the Pelones Formation, showing a tuff with angular pebble- to boulder-size basaltic fragments (pick for scale: 66 cm in length).

which is fully exposed near a sharp bend in Arroyo El Mono at a locality called the Notch (location 2 in Fig. 2; Appendix 1). The exposure features nearly 14 m of bedded cherts, including dense, competent layers of brown or white chert interbedded with soft, poorly consolidated layers of finely brecciated chert. One of the competent layers is a marker bed bearing *Ophiomorpha* ichnofossils (see Ledesma-Vázquez et al., this volume, their Fig. 4c). The poorly consolidated cherty breccia forming the top of the sequence includes abundant opalized root fragments (see Ledesma-Vázquez et al., this volume, their Fig. 4d). Excavation of this bed led to the discovery of in situ horizonal runners, 0.5 cm in diameter but up to 1 m in length, that may belong to a species of the black mangrove, *Avicennia* (Thomas Leslie, 1994, personal communication). No marine invertebrates, however, are associated. The origin of the chert beds is treated by Ledesma-Vázquez et al. (this volume).

Beds of El Mono chert abut directly against the sides of

Figure 5. Fossil rhizolith incorporated in the middle mudstone beds of the Bahía Concepción Member (Locality 6, Appendix 1). a, lateral extent of the rhizolith is marked by arrows (Palo Adam tree for scale). b, detail of fossil root lattice (centimeter scale). c, Cross-sectional break across a root fragment, showing growth rings (centimeter scale).

Miocene hills. A spectacular exposure of the unconformity between basal cherts and the Ricasón Formation may be examined at Locality 4 (Fig. 2; Appendix 1). Here, in a minor tributary of Arroyo El Mono on the east flank of El Mono ridge, is a 5-m-thick profile of andesitic conglomerate (Fig. 6a). The clasts are cobble to boulder in size, highly angular, and set in a matrix of white chert. Massive andesite of the Ricasón Formation forms both sides of the narrow arroyo, suggesting that the conglomeratic deposit represents the fill of a paleoarroyo. As the chert body is traced away

Figure 6. El Mono chert Member of the upper Pliocene Infierno Formation banked against andesite hills of the Miocene Ricasón Formation. a, exposure in a tributary of Arroyo El Mono showing an angular andesite conglomerate in a matrix of white chert (Section 4, Appendix 1). b, El Mono chert on the west side of Arroyo El Mono showing an abutment unconformity with Miocene andesite. c, closer view of chert at the same locality, with scattered cobbles (dark) of andesite eroded from Miocene hills (hammer for scale).

from El Mono ridge toward the basin center, more distal deposits contain fewer andesite clasts; these are smaller and better rounded (Figs. 6b and c). No andesite clasts are present in the chert beds at the Notch.

The younger Cayuquitos chert is known exclusively from a narrow, 2-m-thick belt of outcrop that extends intermittently around the north and northeast faces of Cerro Prieto at an elevation approximately 35 m above the valley floor (Section 10 on Fig. 2; Appendix 1). It is a white, finely brecciated chert that is poorly consolidated, much like some of the thicker beds in the older El Mono chert. In places, the Cayuquitos chert is underlain conformably by red mudstone.

Carbonate deposits

Thin limestone lenses occasionally are represented in the Calabaza and El Mono Members, but the most prominent and laterally extensive carbonates occur at the top and bottom of the Bahía Concepción Member, separated from one another by a red mudstone interval (Fig. 2). Both limestone units are typically 4 to 5 m thick. The lower beds of this member are often very fossiliferous, as found in several profiles at localities 2, 3, 7, 8, 11, and 12 (Fig. 2; Appendix 1). Molluscs dominate the faunas. Bivalves such as *Dosinia ponderosa* and *Tagelus californianus*, together with the large conch *Strombus subgracilior*, are prevalent along the west margin of the main basin. Bivalves such as *Aquepectin* sp. and *Ostrea* sp. are prevalent along the east margin of the main basin and the smaller basins farther north and east. Some of the fossil oysters are prodigious in size. One incomplete oyster shell was found to have a maximum length of 45 cm.

Rocky-shore deposits commonly occur at the juncture of the lower limestone beds and Miocene Ricasón Formation. Few fossils are present in these deposits, however. Rare fossil barnacles, belonging to a species of *Balanus* 1 cm in diameter, were observed in situ on andesite boulders at the shoreline of the large paleoisland skirted by the San Sebastián road 1 km south of Cerro Prieto (Fig. 2).

In terms of biostratigraphy, the most significant fossil occurrence from the lower limestone of the main basin is the sand dollar *Clypeaster marquerensis* (Durham, 1950, pl. 43, figs. 2 and 3). According to Durham (1950, p. 41), this species is restricted to the upper Pliocene of Baja California Sur. *Strombus subgracilior* is considered to range through the middle and upper Pliocene (Durham, 1950, p. 118). *Tagelus californianus* is extant but first appears in the upper Pliocene. *Dosinia ponderosa* is also extant but first appears in the lower Pliocene. All the large shelly fossils from the lower limestone beds of the Bahía Concepción Member are consistent with a late Pliocene age.

In contrast, the upper limestone beds of the Bahía Concepción Member are conspicuously barren of macrofossils. Rare fossil oysters were discovered in loose debris on the central limestone plateau about 1 km east of locality 5.

RESULTS OF GEOLOGICAL MAPPING

The geological map compiled for this project (Fig. 2) shows that the Pliocene basins of Peninsula Concepción occupy an area of approximately 35 km^2. The enclosing margins of the basins are formed entirely by andesite and basalt belonging to the Miocene Ricasón Formation. These rocks also protrude through Pliocene strata as inliers or monadnocks. The main Pliocene basin is open to Bahía Concepción on the west, but none of the smaller basins extends through to the Gulf of California on the east side of Peninsula Concepción.

The upper limestone of the Bahía Concepción Member is the most widely exposed Pliocene lithology, covering much of the main basin to the west. Based on the distribution of key profiles exposed in deep arroyos on opposite sides of this poorly dissected plateau, the basal limestone beds belonging to the same member are equally well represented in the subsurface (Fig. 7). The upper limestone beds are entirely absent from the smaller eastern and northern basins, where erosion has left widely scattered outcrops of the lower limestone standing as small mesas, as at locality 8 (Fig. 8a; Appendix 1).

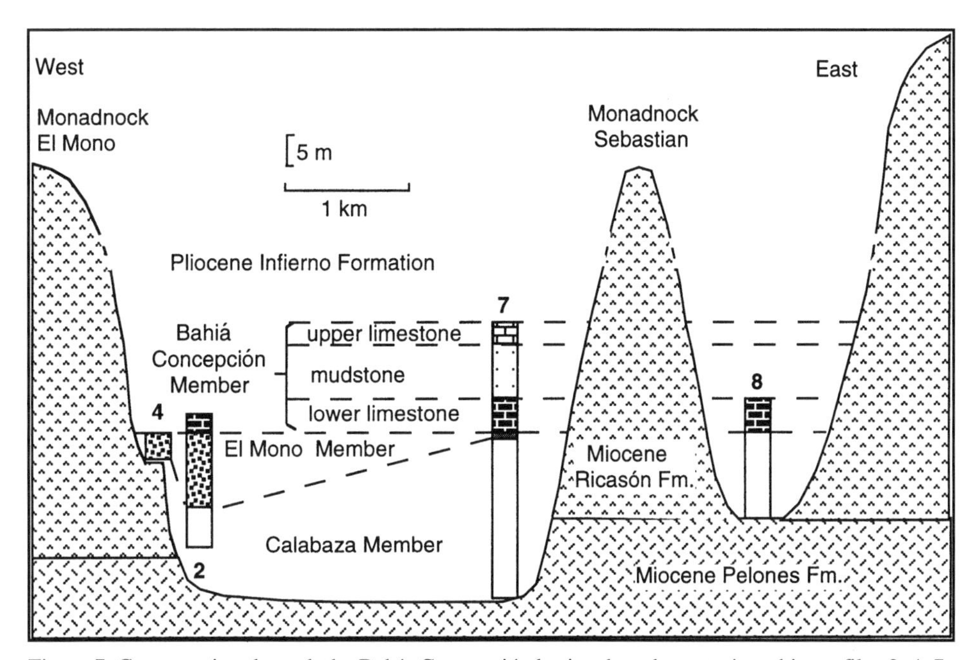

Figure 7. Cross section through the Bahía Concepción basins, based on stratigraphic profiles 2, 4, 7, and 8 (Appendix 1). Lithological patterns are the same as keyed in Figure 2.

In the western basin, the basal limestone beds of the Bahía Concepción Member rest conformably above El Mono chert. No outcrops of El Mono chert are to be found, however, in the smaller eastern basin. This is a certainty, because basement rock in the form of the Miocene Pelones Formation is exposed in both basins and the overlying Pliocene sequence is well known. Instead of El Mono chert, the basal limestone in the eastern basin rests conformably on mudstone of the Calabaza Member. This relationship implies that the smaller basins that are more distal from Bahía Concepción had a shelflike structure elevated topographically above the main western basin (Fig. 7). While the original sediments of the El Mono chert were deposited in the main basin under lagoonal conditions the smaller basins were still accumulating terrestrial sediments.

The limestone beds belonging to the Bahía Concepción Member are essentially flat lying. A dip of less than one degree occurs over a distance of 4 km from the top of the limestone-capped mesa at locality 8 to another mesa 0.5 km east of locality 1 (Fig. 2). This small variance is apparent with respect to the 100-m-contour level. The most northerly exposures of the lower limestone fall on Miocene andesite at or slightly above the 100-m-contour line, but the most southerly exposures reach the same contact at or below the 80-m-contour line.

The lower limestone beds flanking Arroyo El Mono immediately west of locality 2 (Fig. 2) dip perceptibly away from a small normal fault with its upthrown side on the west. At least three other faults cut through the field area along the same orientation from northwest to southeast, parallel to the great Concepción fault running along the east side of Bahía Concepción (Fig. 1b). Movement along these other faults, however, is not readily apparent. Tectonic tilting within the field area during Pliocene and post-Pliocene time was essentially minimal (Ledesma-Vázquez and Johnson, 1993).

Figure 8. Outcrops of the Bahía Concepción Member of the upper Pliocene Infierno Formation. a, part of a small mesa capped by the lower limestone beds of the Bahía Concepción Member with standing figure for scale (Section 8, Appendix 1). Cerro Prieto, in the background, is formed by andesite belonging to the Miocene Ricasón Formation. b, mudstone belonging to the middle beds of the Bahía Concepción Member (Infierno Formation) on the northeast side of Cerro Prieto showing disruptions due to rising gas bubbles from a Pliocene hot spring (hammer for scale).

Two paleo hot springs fall on major fault lines within the field area. One is located on the northeast side of Cerro Prieto (Fig. 2), where red mudstone belonging to the middle beds of the Bahía Concepción Member shows evidence of a peculiar disruption (Fig. 8b). The pattern of vertical partitions is reminiscent of the effect caused by gas bubbles rising through fine sediment in contemporary hot springs. Preservation of this feature in upper Pliocene strata signifies that a fault-fed hot spring was active in late Pliocene time. Another paleo hot spring is located at the terminus of a long fault line 2.5 km south of Cerro Prieto (Fig. 2), where a thermal aureole with aggregates of secondary quartz crystals occurs. Because this site is located on a Miocene inlier, it is difficult to know if the spring was active during Pliocene time. During exploration to check the local extent of Pliocene strata, a third paleo hot spring was discovered 2.25 km north of the field area (26°39′44″N; 111°38′43″W). The site is marked by a large mass of dogtooth spar in a thermal aureole surrounded by Miocene volcanics.

An unresolved problem evident in mapping the field area is the stratigraphic gap between the Bahía Concepción and Cayuquitos Members. The Cayuquitos chert is the most restricted lithological unit in the area, occurring only on the north and northeast faces of Cerro Prieto about 35 m above the surrounding valley floor. In view of the flat-lying strata typical of other units in the Infierno Formation, it is unlikely the Cayuquitos chert formed a deposit that sagged across the surrounding low topography. If the unit was originally more extensive in distribution, a considerable thickness of strata was eroded. Much of the upper limestone beds in the Bahía Concepción Member has been removed from the area, but the maximum known thickness of that unit is 5 m (Section 6, Appendix 1). Two m of red mudstone conformably underlie the Cayuquitos chert at locality 10 on the northeast side of Punta Prieto (Fig. 2; Appendix 1). The missing 35 m of Pliocene strata from the base of that section to the valley floor probably involves a transition between thick mudstone and the upper limestone of the Bahía Concepción Member.

PALEOGEOGRAPHIC INTERPRETATION

A series of paleogeographic maps for southeastern Bahía Concepción through late Pliocene time is interpreted in Figures 9, 10, and 11. In order to show the present position of Bahía Concepción's shoreline, the scope of these maps was enlarged from the field area represented in the geological map (Fig. 2) by adding a 2 km × 9 km strip to the west margin. The stratigraphic time frame is for deposition of the Calabaza Member through the lower limestone beds belonging to the Bahía Concepción Member (Infierno Formation). With open space indicating marine conditions, land is represented by the shaded area superimposed over present topography beginning at the 100-m-contour line. Although the match is not perfect, this line of elevation most closely approximates the late Pliocene shoreline indicated in Figure 11. Much of the area in the northwest corner of the maps is below the present 100-m-contour line, and apparently this area sustained significant post-Pliocene deflation. Considerable variation in the erosiveness of the Miocene Ricasón Formation is evident on the basis of present slope variability. Cerro Prieto in the middle of the field area has much steeper slopes than the range of hills to the east. In any case, no Pliocene strata are found today within the shaded areas.

From the geographic reconstructions (Figs. 9, 10, and 11), it is clear that the approximately 35-km^2 area onlapped by a late Pliocene transgression was divided into four separate basins. The largest is the west basin (1—numbered basins shown on Figs. 9, 10, and 11), situated between the peninsula terminating at Punta Los Monos and Islas Prieto, San Juan, and San Sebastián (Fig. 11). This basin supported the largest diversity of marine invertebrates, including large conchs (*Strombus subgracilior*) and sand dollars (*Clypeaster marquerensis*). The fact that diversity begins to fall off less than 1 km north of Punta Los Monos suggests that the interpretation of a peninsular barrier east of Isla Las Cuevas is correct.

The east basin (2) held a more restricted body of water located between the peninsular mainland to the east and Islas Prieto, San Juan, and San Sebastián (Fig. 11). Two narrow

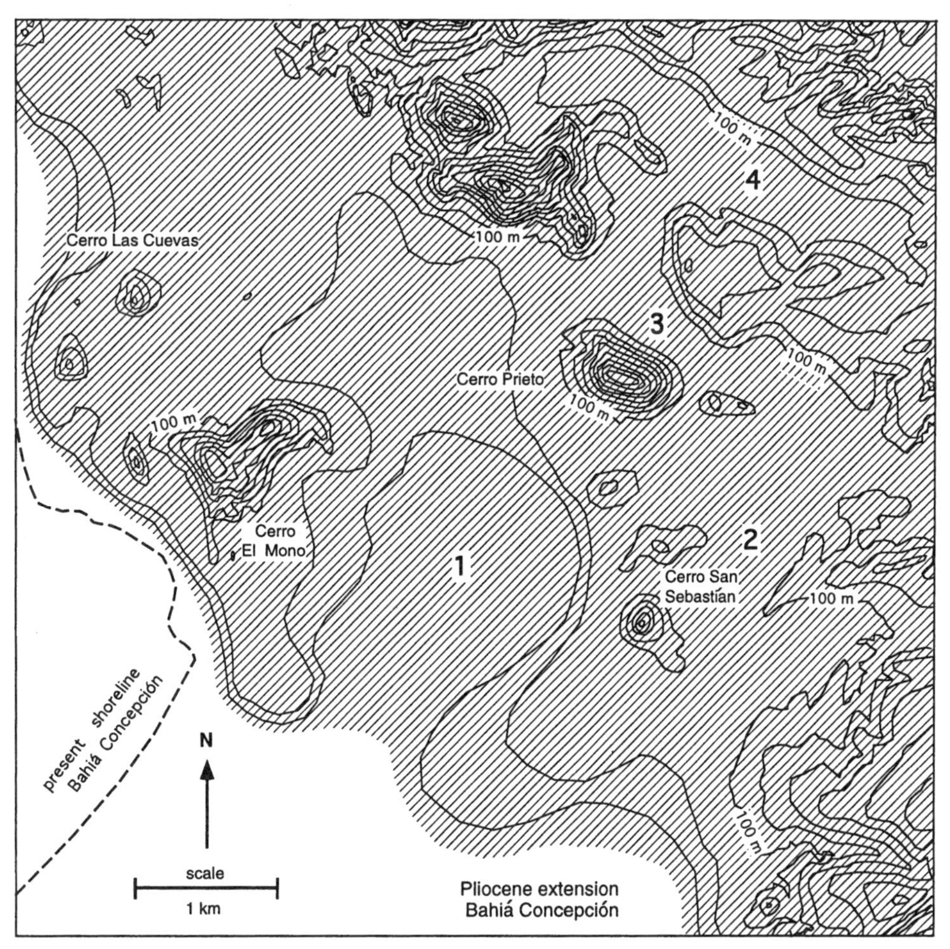

Figure 9. Paleogeography of southeastern Bahía Concepción during early late Pliocene time and development of the Calabaza Member (Infierno Formation). Land is indicated by diagonal shading with Recent topography above the present 100-m elevation retained. Contour interval is 20 m. Slight extent of Pliocene flooding is apparent by comparison with the present shoreline of Bahía Concepción (dashed line).

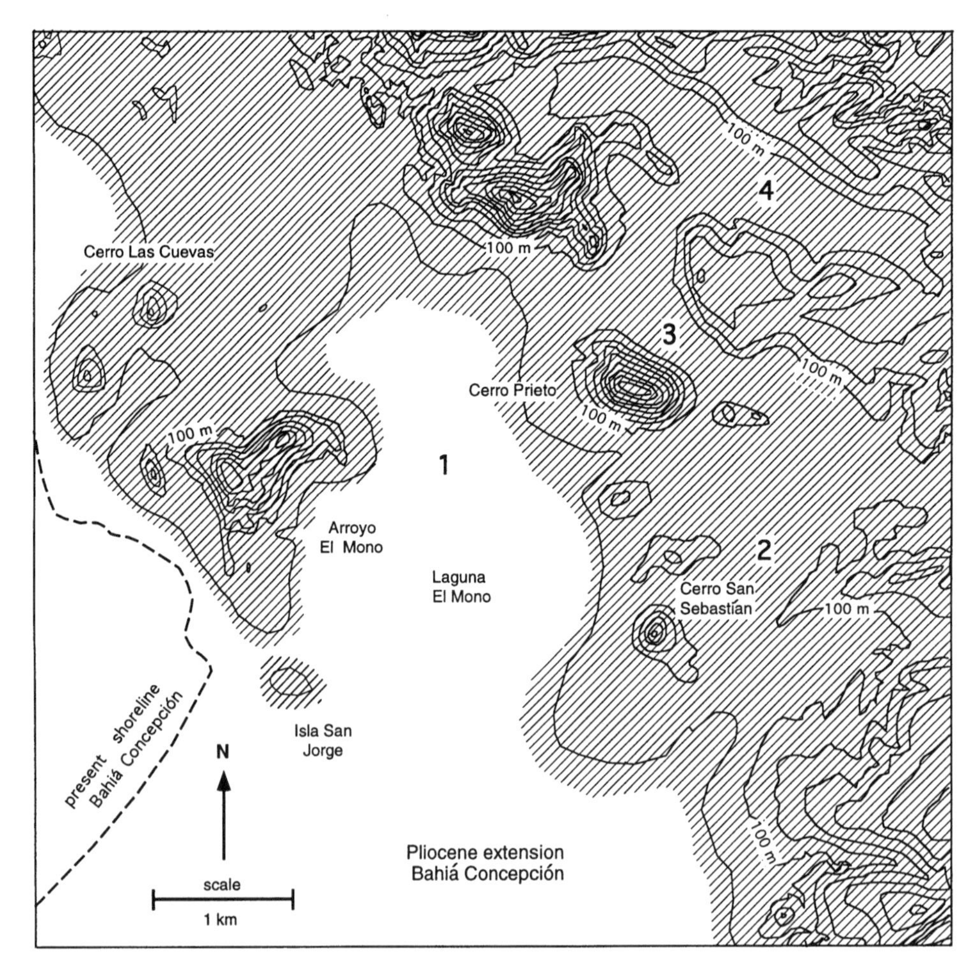

Figure 10. Paleogeography of southeastern Bahía Concepción during middle late Pliocene time. Flooding responsible for deposition of bedded cherts (El Mono Member) is apparent by comparison with the present shoreline of Bahía Concepción.

basins (3 and 4) were barely more than 0.5 km in width but 2 to 3 km in length. Laguna Cayuquitos (4) was the most restricted basin, with an entrance only 0.25 km wide. The eastern and northern basins (2, 3, and 4) shared the same low-diversity fauna of oysters and pectens. Population density was highest in the east basin, however, and lowest in Laguna Cayuquitos. A diversity gradient must have extended from west to east across the main basin, because the same low-diversity fauna of oysters and pectens completely encircles Isla San Sebastián.

Unfortunately, the lower limestone of the Bahía Concepción Member is inaccessible at the center of the west basin because of the pervasiveness of the upper limestone as a cap rock. It would be interesting to map the faunal zonation of the lower limestone in greater detail. Even though a geographic map reconstruction for the interval of time during deposition of the upper limestone beds would be little different, it is evident on the basis of rare macrofossils that marine circulation was much degraded. A complete regression occurred during the interval of time between deposition of the lower and upper limestones in the Bahía Concepción Member, when dense plant cover was rooted in red silt approximately midway between Isla San Jorge and the south end of Isla San Sebastián.

DISCUSSION

It is seldom necessary to go very far in Baja California to find contemporary settings matching the depositional environments of Cenozoic strata. To begin with, the disposition of the Peninsula Concepción basins proves that Bahía Concepción already existed as an embayment during the late Pliocene. There is no possible marine connection between the Peninsula Concepción basins and the Gulf of California except for a path through contemporary Bahía Concepción. Peninsula Concepción always remained a peninsula during Pliocene time, despite significant changes in relative sea level.

The present geography and range of depositional settings within Bahía Concepción provide the best key to understanding the Peninsula Concepción basins. Prevailing winds from the north push waves into the bay during the winter season, but conditions are calm during the summer season when lighter

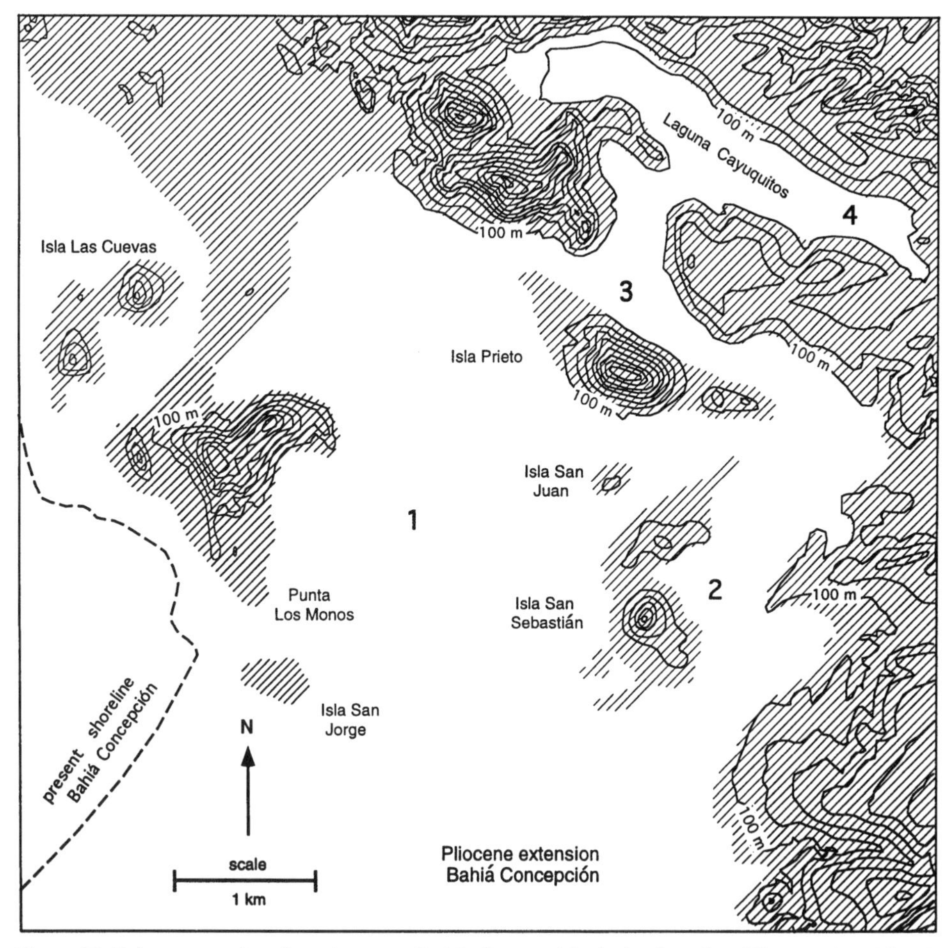

Figure 11. Paleogeography of southeastern Bahía Concepción during later late Pliocene time. Flooding responsible for deposition of the basal limestone unit (Bahía Concepción Member) is apparent by comparison with the present shoreline of Bahía Concepción.

winds are drawn north in a semimonsoonal effect brought on by the heating of the great American southwest deserts. The seasonal patterns of the wind field over the Gulf of California are well documented by Bray and Robles (1991). Wind-speed data from Isla Tortuga, only 65 km due north of the mouth of Bahía Concepción, are particularly relevant. Playa Santispac on the west shore of Concepción (Fig. 1b) is situated in a bay approximately 23 km^2 in area and well sheltered from prevailing winter winds. Ten small islands dot the bay. The largest is Isla Coyote, about 0.25 km^2 in area with an elevation more than 100 m above sea level. The beaches are composed of carbonate sand. Mangrove colonies thrive in most of the coves, often in close proximity to rocky shorelines. A coastal hot spring is active within the mangrove zone at Playa Santispac.

A low-lying basin about 3 km^2 in area is located on the west side of the highway at Playa Santispac. Alluvial fans ring the basin and it is easily flooded. Only a small rise in relative sea level would be necessary to create a small lagoon. The west side of Bahía Concepción is more rugged than Peninsula Concepción, however, and a major rise in sea level on the order of 100 m would not expand the size of the bay by more than 8 km^2. Rifting associated with the development of the Gulf of California clearly is related to the formation of the fault zones and hot springs in the Bahía Concepción region. One of the surprises of the local Pliocene geology, however, is that the strata are not as severely impacted by faulting as farther south in the Loreto basin (Umhoefer et al., 1994; Dorsey et al., 1995). The stacked Gilbert-type fan deltas characteristic of the Loreto basin did not have the same tectonic opportunity to develop in the smaller Peninsula Concepción basins. In this regard, the relative tectonic calm of the Bahía Concepción region is similar to the Punta Chivato region farther north (Simian and Johnson, this volume).

Two distinct transgressions are represented by the limestone beds of the Bahía Concepción Member (Infierno Formation). A complete regression between these two local events is indicated by the red mudstone and rhizolith in the middle of this unit. A third minor transgression, also coupled with a red mudstone facies, is suggested by the Cayuquitos chert. Haq et al.

(1988) proposed that three eustatic highstands in sea level occurred during the late Pliocene. None was capable of flooding to a level 100 m above present sea level without the assistance of regional tectonics. The three transgressive events recorded in the Peninsula Concepción basins may have been boosted by eustatic changes, but the available biostratigraphy is not sufficiently detailed to test this hypothesis. A more convincing match between Pliocene eustatic patterns and regional tectonics is made for the neighboring Punta Chivato region (Johnson and Simian (1996).

CONCLUSIONS

The Peninsula Concepción basins preserve a repetitive sequence of upper Pliocene strata reflecting three different depositional environments. Alluvial fan deposits, sometimes with fossil root casts, are represented by andesitic conglomerates, siltstones, and mudstones of the Calabaza Member; mudstones belonging to the middle beds of the Bahía Concepción Member; and mudstones conformably underlying the Cayuquitos Member. Chert deposits exhibited by the Cayuquitos and El Mono Members are associated with lagoonal sediments probably altered by hot-spring activity. Limestone deposits are well represented by the lower and upper beds of the Bahía Concepción Member. The lower beds, in particular, are rich in fossils, providing adequate biostratigraphic evidence that the conformable sequence is late Pliocene in age. All stratigraphic names are formally proposed herein as localized members of the upper Pliocene Infierno Formation.

The geography of four distinct but interconnected basins at the southeast end of Bahía Concepción lent itself to the development of sheltered rocky shorelines around several protective points and four paleoislands up to 2 km^2 in area. Although winter winds are capable of piling up waves at the south end of the present bay, the "fish-hook" layout of the Peninsula Concepción basins effectively shielded them from comparable prevailing winds and waves. If Pliocene Bahía Concepción were symbolized by the main shaft of the fish hook, then invertebrate faunas characteristic of a normal marine environment thrived at the bottom of the fish hook where communication with the larger bay was strong. A fauna limited to oysters and pectens dominated the more restricted basins that formed the distal, upturned hook and barb.

Despite the presence of major fault lines, very little tectonic tilting took place in the region during Pliocene and post-Pliocene times (Ledesma-Vázquez and Johnson, 1993). A relative rise in water depth on the order of 100 m above present sea level was necessary to fill the basins with Pliocene strata. Three eustatic highstands are defined for the upper Pliocene (Haq et al., 1988), but the available biostratigraphy is inadequate to determine whether or not the three transgressive phases represented by the Infierno Formation in this area are coeval with global events. In any case, regional tectonic-eustatic factors would have been required to flood the basins.

ACKNOWLEDGMENTS

Fieldwork on the Peninsula Concepción basins was carried out during three successive seasons in January 1993, 1994, and 1995. The first season was partially supported through a grant to M. E. Johnson from the Petroleum Research Fund (American Chemical Society). Mark Mayall (Williams College) was the principal student mapper. Research during January 1994 and 1995 was supported by joint grants to M. E. Johnson from the National Science Foundation (INT-9313828) and to J. Ledesma-Vázquez from CONACYT (3171-T9307). The following students participated in the 1994 mapping project: Cassity Bromley, Brett Dalke, Tim Farnham, Aengus Jeffers, Willard Morgan, Pete Taylor, and Maximino Simian (Williams College); Hugo Brodziak-Ochoa, Elizabeth Garcia-Galvan, Marissel Gonzalez-Bustos, and Hector Lazcano-Venegas (Universidad Autónoma de Baja California). M. Simian revisited the field area in 1995, and the authors are grateful to him for compiling the penultimate version of the geological map (Fig. 2).

APPENDIX 1. STRATIGRAPHIC SECTIONS FROM THE UPPER PLIOCENE BASINS OF PENINSULA CONCEPCIÓN

Section 1. La Calabaza stream bank, approximately 250 m west of ranch well (26°33′52″N; 111°37′38″W).

Upper Pliocene Infierno Formation

Calabaza Member

Mudstone, yellow-tan, abundant horizontal and vertical root casts . . . . . 3.00 m
Sandstone, tan, fine grained, with disarticulated oysters scattered at base . . . . . 0.50 m
Mudstone, yellow-tan, abundant horizontal and vertical root casts . . . . . 0.50 m
Oyster limestone, tan, in situ accumulation of articulated *Ostrea* sp. . . . . . 0.30 m
Mudstone, yellow-tan, abundant horizontal and vertical root casts . . . . . 0.40 m
Mudstone, red, with floating andesite clasts ranging from 1 to 6 cm in diameter . . . . . 0.50 m
Mudstone, red, with few root casts . . . . . 1.00 m

Section 2. The Notch near a bend in the stream bed of Arroyo El Mono (26°35′28″N; 111°40′10″W). See fig. 4 in Ledesma-Vázquez et al. (this volume).

Upper Pliocene Infierno Formation

Bahía Concepción Member

Lower limestone, gray to dark gray, massive, richly fossiliferous with *Clypeaster marquerensis* (sand dollar), *Strombus subgracilior* (gastropod), and *Tagelus californianus* (bivalve) . . . . . 2.25 m

El Mono Member

Chert, dark brown, well bedded . . . . . 1.52 m
Cherty breccia, white, poorly consolidated . . . . . 4.57 m
Chert, beige, well bedded, bearing *Ophiomorpha* ichnofossils . . . . . 1.34 m
Cherty breccia, white, poorly consolidated . . . . . 3.05 m
Chert, beige, well bedded . . . . . 0.20 m
Cherty breccia, white, well bedded . . . . . 3.05 m

Calabaza Member
Conglomeratic siltstone, reddish-brown with
andesitic pebble-size clasts . . . . 4.00 m

Section 3. Limestone-capped outlier, east side of El Mono stream bed (26°35′47″N; 111°39′94″W).

Upper Pliocene Infierno Formation

Bahía Concepción Member
Lower limestone, gray to dark gray, massive, richly fossiliferous with *Strombus subgracilior* (gastropod), *Dosinia ponderosa*, *Tagelus californianus* (bivalves), and rare opalized *Porites* sp. . . . . 3.50 m
El Mono Member
Chert, white . . . . 0.25 m
Cherty breccia, white-gray, well bedded . . . . 5.50 m
Chert, white-brown, massive . . . . 3.50 m
Calabaza Member
Conglomeratic siltstone, reddish-brown
with andesitic pebble-size clasts . . . . 6.50 m

Section 4. Tributary arroyo (and paleoarroyo) on east flank of El Mono ridge, west side of Arroyo El Mono (26°35′50″N; 111°40′05″W). See Figure 6a.

Upper Pliocene Infierno Formation

El Mono Member
Conglomeratic chert, boulder size, angular clasts of red-brown andesite set in a white chert matrix (Fig. 6a) . . . . 5.00 m

Section 5. Rattlesnake ridge (26°25′70″N; 111°39′66″W).

Upper Pliocene Infierno Formation

Bahía Concepción Member
Upper limestone, beige, massive, unfossiliferous,
upper 2 m contains fragments of white chert . . . . 4.50 m
Mudstone, red . . . . 4.00 m
Lower limestone, gray-white, thick bedded,
conspicuously unfossiliferous . . . . 2.50 m
Mudstone, red . . . . 2.50 m
El Mono Member
Cherty breccia, white, reworked aspect with limey matrix . . . 2.50 m
Mudstone, red, with calcite concretions . . . . 1.50 m
Cherty breccia, white, reworked aspect with limey matrix . . . 4.25 m
Calabaza Member
Conglomeratic siltstone, reddish-brown with andesitic
cobble to pebble-size clasts . . . . 11.50 m

Section 6. Fossil rhizolith, north side of road to San Sebastián (26°35′25″N; 111°38′91″W). See Figure 5.

Upper Pliocene Infierno Formation

Bahía Concepción Member
Upper limestone, beige weathering gray, thick bedded,
unfossiliferous, scattered chert fragments . . . . 5.00 m
Mudstone, red, with laterally extensive root casts . . . . 1.00 m
Lower limestone, beige weathering gray, massive,
unfossiliferous, with scattered pebble-size chert fragments . . . 6.25 m
El Mono Member
Chert, brown, thick bedded . . . . 0.50 m
Limestone, beige weathering gray, massive, reworked
pebble-size chert fragments . . . . 7.00 m

Section 7. West side of gorge on Arroyo Cayuquitos on the road to San Sebastián (26°35′82″N; 111°38′34″W). See Figure 4b.

Upper Pliocene Infierno Formation

Bahía Concepción Member
Upper limestone, beige weathering to a rose color,
thick bedded, unfossiliferous . . . . 3.25 m
Mudstone, red . . . . 12.00 m
Lower limestone, beige to tan, massive, fossiliferous
with *Aquepectin* sp. and *Ostrea* sp. (bivalves) . . . . 4.75 m
Mudstone, red . . . . 1.50 m
El Mono Member
Limestone, white, thick bedded, unfossiliferous
but with scattered chert fragments throughout . . . . 0.50 m
Calabaza Member
Mudstone, red . . . . 3.50 m
Conglomeratic siltstone, red with angular clasts
of pebble- to boulder-size andesite . . . . 27.00 m

Section 8. Limestone-capped outlier, south side of road to San Sebastián (26°36′20″N; 111°36′20″W). See Figure 8a.

Upper Pliocene Infierno Formation

Bahía Concepción Member
Lower limestone, beige weathering gray, massive,
richly fossilferous with abundant *Aquepectin* sp.
and *Ostrea* sp. (bivalves) . . . . 4.75 m
Calabaza Member
Mudstone, red with scattered pebble-size
clasts of red-brown andesite . . . . 18.50 m

Section 9. Southwest flank of Cerro Prieto (26°26′80″N; 111° 38′54″W).

Upper Pliocene Infierno Formation

Bahía Concepción Member
Lower limestone, white-gray, massive,
unfossiliferous, with scattered cobble- to
boulder-size clasts of red-brown andesite . . . . 4.00 m

Section 10. West flank of Cerro Prieto about 35 m above the surrounding valley floor (26°37′04″N; 111°38′54″W).

Upper Plioicene Infierno Formation

Cayuquitos Member
Cherty breccia, beige, poorly consolidated . . . . 2.00 m
Mudstone, red with pebble-size andesite clasts . . . . 2.00 m

Section 11. Eroded north bank of stream bed belonging to a tributary of the San Sebastián northeast of Cerro Prieto (26°36′55″N; 111°37′28″W).

Upper Pliocene Infierno Formation

Bahía Concepción Member
Lower limestone, gray, massive, fossiliferous with
common *Ostrea* sp. and *Aquepectin* sp. (bivalves) . . . . 4.25 m
Limestone coquina, white-gray, unarticulated
remains of *Ostrea* sp. . . . . 1.50 m
Limestone, beige-gray, very thick bedded, fossiliferous
with common *Ostrea* sp. and *Aquepectin* sp. . . . . 2.75 m
Calabaza Member
Mudstone, red . . . . 1.75 m

Section 12. Eroded north bank of stream bed belonging to a tributary of the Cayuquitos approximately 4 km west of the San Sebastián fish camp (26°37′26″N; 111°36′13″W).

Upper Pliocene Infierno Formation

Bahía Concepción Member
Lower limestone, beige weathering gray, very thick bedded,
very fossilferous with *Ostrea* sp. and *Aquepectin* sp. . . . . 2.80 m

REFERENCES CITED

Bray, N. A., and Robles, J. M., 1991, Physical oceanography of the Gulf of California, *in* Dauphin, J. P., and Simoneit, B. R. T., eds., The Gulf and Peninsular Province of the Californias: American Association of Petroleum Geologists Memoir 47, p. 511–553.

Dorsey, R. J., Umhoefer, P. J., and Renne, P. R., 1995, Rapid subsidence and stacked Gilbert-type fan deltas, Pliocene Loreto basin, Baja California Sur, Mexico: Sedimentary Geology, v. 98, p. 181–204.

Durham, J. W., 1950, 1940 E. W. Scripps cruise to the Gulf of California. Part II. Megascopic paleontology and marine stratigraphy: Geological Society of America Memoir 43, 216 p.

Haq, B. U., Hardenbol, J., and Vail, P. R., 1988, Mesozoic and Cenozoic chronostratigraphy and eustatic cycles, *in* Wilgus, C. K., Hastings, B. K., Posamentier, H., Wagoner, J. V., Ross, C. A., and Kendall, C.G.St.C., eds., Sea-level changes: An integrated approach: Society of Economic Paleontologists and Mineralogists Special Publication 42, p. 71–108.

Hoffmeister, J. E., and Multer, H. G., 1965, Fossil mangrove reef of Key Biscayne, Florida: Geological Society of America Bulletin, v. 76, p. 845–852.

Johnson, M. E., 1988a, Why are ancient rocky shores so uncommon?: Journal of Geology, v. 96, p. 469–480.

Johnson, M. E., 1988b, Hunting for ancient rocky shores: Journal of Geological Education, v. 36, p. 147–154.

Johnson, M. E., 1992, Studies on ancient rocky shores: A brief history and annotated bibliography: Journal of Coastal Research, v. 8, p. 797–812.

Johnson, M. E., and Simian, M. E., 1996, Discrimination between coastal ramps and marine terraces at Punta Chivato on the Pliocene-Pleistocene Gulf of California: Journal of Geoscience Education, v. 44, p. 569–575.

Ledesma-Vázquez, J., and Johnson, M. E., 1993, Neotectonica del area Loreto-Mulegé, *in* Delgado-Argote, L. A., and Martín-Barajas, A., eds., Contribuciones a la tectonica del Occidente de Mexico: Union Geofisica Mexicana Monografia 1, p. 115–122.

McFall, C. C., 1968, Reconnaissance geology of the Concepción Bay area, Baja California, Mexico: Stanford University Publications in Geological Sciences, v. 10, no. 5, 25 p.

Plaziat, J.-C., 1995, Modern and fossil mangroves and mangals: Their climatic and biogeographic variability, *in* Bosence, D. W. J., and Allison, P. A., eds., Marine palaeoenvironmental analysis from fossils: Geological Society of London Special Publication 83, p. 73–96.

Umhoefer, P. J., Dorsey, R. J., and Renne, P., 1994, Tectonics of the Pliocene Loreto basin, Baja California Sur, Mexico, and evolution of the Gulf of California: Geology, v. 22, p. 649–652.

Whybrow, P. J., and McClure, H. A., 1981, Fossil mangrove roots and palaeoenvironments of the Miocene of the eastern Arabian Peninsula: Palaeogeography, Palaeoclimatology, Palaeoecology, v. 32, p. 213–225.

Wilson, I. F., 1948, Buried topography, initial structures, and sedimentation in Santa Rosalia area, Baja California Mexico: American Association of Petroleum Geologists Bulletin, v. 32, p. 1762–1807.

Wilson, I. F., and Rocha, V. S., 1955, Geology and mineral deposits of the Boleo Copper District, Baja California, Mexico: U.S. Geological Survey Professional Paper 273, 134 p.

Manuscript Accepted by the Society December 2, 1996

Printed in U.S.A.

Geological Society of America
Special Paper 318
1997

El Mono chert: A shallow-water chert from the Pliocene Infierno Formation, Baja California Sur, Mexico

Jorge Ledesma-Vázquez
Facultad de Ciencias Marinas, Universidad Autónoma de Baja California, Ap. Postal 453, Ensenada, Baja California, Mexico 22800
Richard W. Berry
Department of Geological Sciences, San Diego State University, San Diego, California 92182-1020
Markes E. Johnson
Department of Geosciences, Williams College, Williamstown, Massachusetts 01267
Sonia Gutiérrez-Sanchez
Departamento de Geología Marina, Universidad Autónoma de Baja California Sur, La Paz, Baja California Sur, Mexico 23080

ABSTRACT

A well-bedded, 14-m-thick chert unit assigned to the Infierno Formation is located near the southeast corner of Bahía Concepción at the base of the Concepción Peninsula in Baja California Sur, Mexico. Previously unknown, the age of this chert is restricted by fossils that date conformably overlying limestones as late Pliocene in age. The chert beds, as well as the rest of the associated sequence, were deposited in an interconnected set of small basins flooded from the direction of present-day Bahía Concepción. A shallow-water environment for the chert is indicated by the occurrence of fossil mangrove roots preserved in life positon along the margin of one basin and the abundant presence of the ichnofossil Ophiomorpha in a laterally extensive horizon within the chert body. Stratigraphic bracketing of the chert unit by an underlying subaerial fan-deposit conglomerate and an overlying fossiliferous limestone confirms a likely water depth between intertidal and 10 m. X-ray diffraction and microscopic analyses of the chert suggest that a portion of the silica was deposited originally as a particularly silica-rich tuff. Basin-cutting faults acted as conduits, transporting silica-enriched hot water that eventually transformed most of the volcanic glass and all the original carbonates to opal-A and low cristobalite.

INTRODUCTION

The area of study is on the Concepción Peninsula at the southeast corner of Bahía Concepción, Baja California Sur (Fig. 1). It includes approximately 56 km^2 in the vicinity of the Santa Rosaliita ranch. Here, volcanic rocks of the Miocene Comondú Group formed the geologic framework, upon which a series of small sedimentary basins was developed during late Pliocene time. Some basin units was described by McFall (1968) as an interstratified set of sandstone, mudstone, and coquina and assigned to the Pliocene Infierno Formation. During our field survey, a general discrepancy was found with these previously reported lithologies. The authors of this chapter found that stratigraphic units in the study area consist mainly of three repetitive lithologies: fossiliferous limestones, reddish fanglomerates, and chert. None of these units has been described before from this region. The Infierno Formation was named and described by Wilson and Rocha (1955) as a fossiliferous marine sandstone sequence. At the Infierno Creek stratigraphic type section, an unconformity with the basal Gloria Formation is well exposed, and fossils present in the type section (*Ostrea hermanni*, *Turritela imperialis*) indicate the age of the Infierno Formation as late Pliocene.

In a more recent study, Ledesma and Johnson (1993) proposed that neotectonic deformation has occurred in three different areas along 130 km of the west margin of the California

Ledesma-Vázquez, J., Berry, R. W., Johnson, M. E., and Gutiérrez-Sanchez, S., 1997, El Mono chert: A shallow-water chert from the Pliocene Infierno Formation, Baja California Sur, Mexico, *in* Johnson, M. E., and Ledesma-Vázquez, J., eds., Pliocene Carbonates and Related Facies Flanking the Gulf of California, Baja California, Mexico: Boulder, Colorado, Geological Society of America Special Paper 318.

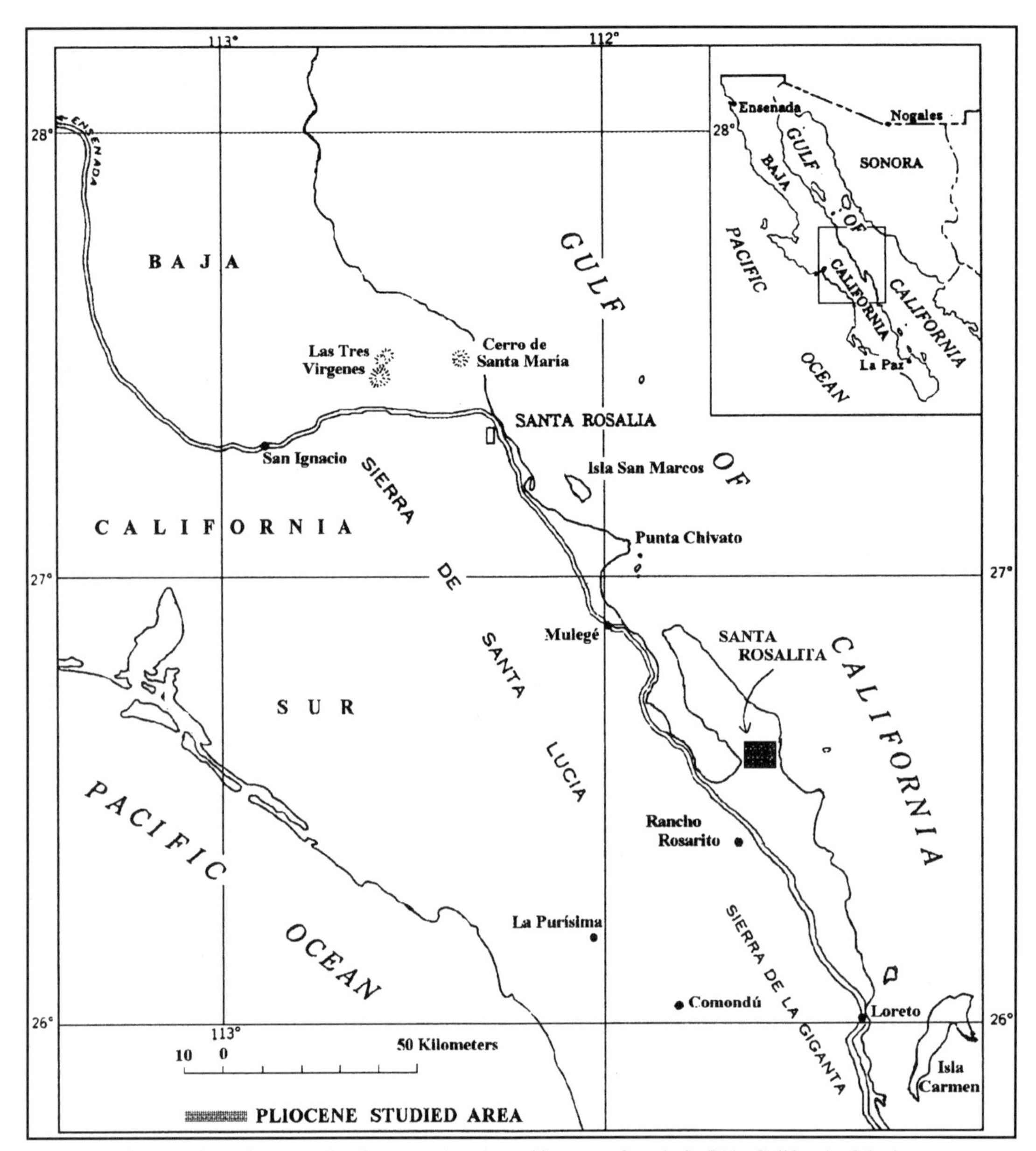

Figure 1. Location map for the general and specific area of study in Baja California, Mexico.

Gulf. Their work was based on an absolute baseline for sea level demarcated by rocky-shoreline facies. These facies are exposed in the study area of this report and are attributed to the Infierno Formation.

The Bahía Concepción area is faulted. The largest structure is the Bahía Concepción Fault Zone (McFall, 1968), with a northwest orientation parallel to Bahía Concepción. Sense of movement along this zone remains unclear. Neotectonic events in the study area are associated with extensional tectonics related to rifting in the Gulf of California.

Geological and structural analyses done by Zanchi (1993) established that tectonic structures related to the Basin and Range Province extensional event have been covered unconformably (at least in part) by Pliocene to Quaternary sedimentary units of the Loreto Basin. The tectonic analysis of an area 75 km to the south of the study area (Umhoefer et al., 1994) yielded conclusions in agreement with those reached by Ledesma and Johnson (1993) that the third stage of tectonic development in the southern Gulf of California, including the modern configuration of transform faults and spreading ridges, did not commence immediately at 3.5 Ma. It is clear that the east coast of the Baja California peninsula in the vicinity of Bahía Concepción may be classified as emergent, as characterized by uplifted marine terraces. In particular, the area of study is a Pacific Coast Type (Melendez-Fuster, 1981), because the structural and tectonic axes are parallel to the shoreline and are associated with lines of near-shore islands and elongated coastal coves, also parallel to the shoreline.

SILICA AND THE FORMATION OF CHERT

Silica and the development of chert deposits in the stratigraphic record are commonly linked to the availability of significant amounts of planktonic detritus (opal-A) in areas of high bioproductivity. Silica may also enter a depositional environment as weathering solutions in semiarid climates and as silicon supplied in solution by hydrothermal-volcanic systems (Hesse, 1988). Silica from all of these sources can accumulate in broad submarine volcanic areas (Calvert, 1974). The most common primary sedimentary form of silica, opal-A, is relatively unstable and may react with pore waters in a marine environment to form authigenic minerals under oxidizing or reducing conditions (Burley et al., 1985).

Much of our understanding of silica diagenesis comes from samples recorded during the Deep Sea Drilling Project (DSDP). For example, chert collected by Lancelot (1973) during DSDP Leg 17 consisted of porcellanite and layered nodules. The porcellanites are composed of disordered cristobalite, or opal-CT. Porcellanite was found in clay-rich sediments, whereas chert nodules were found only in carbonate-rich localities. Based on analyses of DSDP chert samples, Davies and Supko (1973) found that silica in nodular chert is biogenically derived, but they were unable to determine the origin of the silica in layered cherts. Also based on DSDP sample analyses, Greenwood (1973) presented evidence to support a diagenetic sequence: solution of biogenic silica → precipitation of poorly ordered cristobalite (opal-CT) → gradual inversion to microcrystalline quartz.

Williams et al. (1985) established that amorphous silica phases precipitate in nature because of the formation of dense colloids in supersaturated alkaline aqueous solutions. These solutions contain relatively low concentrations of other ions. Open structure polymers of silica develop in interstitial water that contains a high concentration of cations and silica. The polymers later flocculate to produce opal-CT. Opal-T (tridymite) increases its proportions relative to critobalite as the opal crystalites increase in size. Later, opal-CT dissolves to produce pore waters with a high concentration of silica. The presence of carbonate seems to favor the formation of opal-CT. Hesse (1988) reported low-temperature tridymite found as pore-filling cement, precipitated from groundwater or hydrothermal solutions in silicic volcanic rocks. Wise et al. (1973) analyzed a Miocene bentonite, derived from a glassy, tuffaceous rhyolite, and found that the original volcanic material was altered in part to opal by means of a dissolution-precipitation reaction. In normal marine conditions, opal-CT takes a minimum of about 10 m.y. to first appear. This process requires an elevated heat flow; otherwise its formation takes even longer (Hesse, 1988).

Chert nodules in carbonates can be explained as forming during diagenesis of biogenic silica contained in the host sediment. The origin of well-bedded chert, however, has repeatedly been disputed, even though siliceous organisms are sufficiently abundant in shallow waters with predominant carbonate deposition (Hesse, 1989).

Hesse (1989) indicated that the hypersaline and generally high-pH environment of evaporite deposition permits silicon concentration in solution to build up to high levels. Therefore, it is not surprising to find chert commonly replacing evaporites. He also pointed out that silicification of carbonate sediments requires pore solutions that are supersaturated with respect to the silica phase precipitated as well as undersaturated with respect to the carbonate minerals dissolved.

Silicification may also be associated with hydrothermal activity that alters sediment in isolated ponds and depressions in the median rift valley of mid-ocean ridges, where the fluids discharge. Silicified sediments may accumulate where the recharge of the silica-rich brine occurs in broad sediment-covered areas or where the discharge appears to be focused in narrow zones along fault scarps (Hesse, 1989).

According to Hesse (1989), there are siliceous sinter deposits in high-temperature geothermal areas on land, which also have low-temperature surface or near-surface alteration products. The sinter deposit may form terraces around hydrothermal vents and consist of light-colored, banded masses of friable or dense, well-cemented amorphous silica. The hydrothermal solutions at depth must contain sufficient dissolved silicon to become supersaturated with respect to Opal-A during ascent when they cool below 100° C. A good example of this type of deposit is located at the Beowawe geothermal system in Nevada, where the associated sinter deposit is 65 m thick and displays well-bedded layering on a meter-to-millimeter scale over a distance of 1.6 km. This deposit consists predominantly of opal-A with traces of quartz.

The goal of this chapter is to identify the processes involved in the formation of the chert units at Rancho Santa Rosaliita, Baja California Sur. A reevaluation of the complete stratigraphy exhibited in the Peninsula Concepción basins is dealt with by Johnson et al. (this volume).

METHODS OF INVESTIGATION

Field investigations were carried out during January 1994 and January 1995. Of the various chert beds exposed in the area, the thickest and most laterally persistent unit is named El Mono chert (Johnson et al., this volume) after its best exposure at a locality called the Notch on the south side of Arroyo El Mono (Fig. 2). A much thinner and laterally more restricted bed of finely fragmented, poorly consolidated, and reworked chert also was discovered on the north side of Cerro Prieto (see Fig. 2 in Johnson et al., this volume). Samples were taken from the aforementioned beds as well as from other, more limited exposures of chert beds found throughout the study area.

Hand samples were examined in the field for texture and composition, and the outcrops from which the samples were taken were described. In the laboratory, thin sections of the samples were studied by petrographic microscope in an attempt to find microfossils and any other evidence that would reveal the petrogenesis of the rock and its depositional environment. Samples were analyzed by X-ray diffraction in order to determine the mineral composition of the cherts. The instrument utilized at San Diego State University

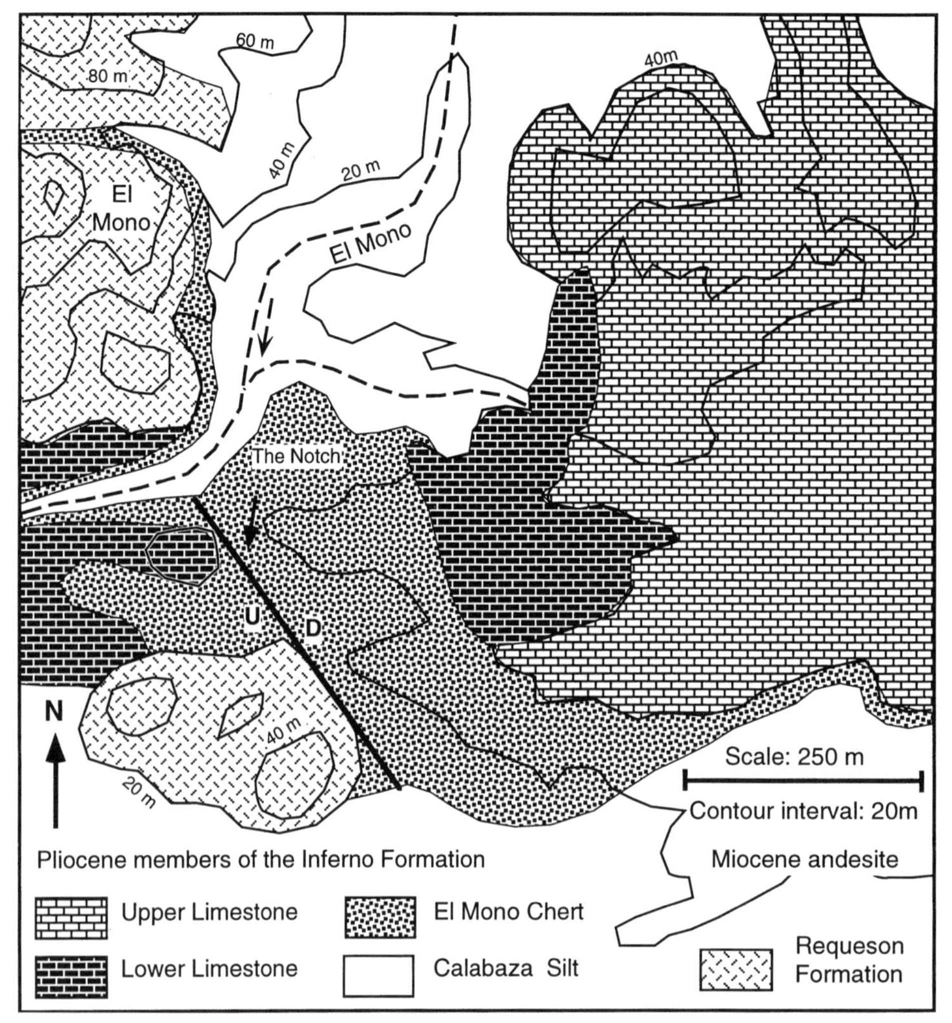

Figure 2. Simplified geological map of the Pliocene Santa Rosaliita area at the southern end of the Concepción Peninsula. The continuous thick line represents a fault. The most complete section through the El Mono chert is exposed at the Notch (arrow).

is a Diano microprocessor-controlled, computer-driven model. It was of particular interest to learn whether the chert consisted of opal-A, opal-C, opal-T, and/or opal-CT. Disaggregated material taken from exposure of reworked chert was separated into <200 mesh and >200 mesh-size fractions. The coarse fraction was analyzed microscopically using index of refraction oils. The fine fraction was subjected to X-ray diffraction analysis.

RESULTS

The approximately 14-m-thick chert-bearing sequence at the Arroyo El Mono locality contains two distinctly different chert lithologies that alternate with each other (Fig. 3). One type of chert is very well indurated and well bedded. It contains pebble-size brecciated clasts that are grain supported and surrounded by a sand- to silt-size chert matrix. This matrix is mottled with many micropores and fractures. The chert lithology varies from dark reddish-brown to milky white. The other type is detrital chert that consists of well-bedded, poorly indurated, coarse to fine sand-size chert grains that are invariably white (Fig. 4). Microscopic and X-ray diffraction analyses of the poorly indurated chert from the El Mono locality reveal that it consists primarily of a mixture of volcanic glass shards, opal-A, and opal-CT (Fig. 5). Halite is abundant and is accompanied by minor amounts of quartz and feldspar. X-ray diffraction analyses for the well-indurated chert from the upper part of the section at El Mono revealed that it consists primarily of opal-CT (Fig. 6), with no halite or glass. No clay minerals were detected. Index of refraction of the least devitrified glass shards from the poorly indurated chert indicates that the original lava was silica rich, in the range from rhyolite to dacite. Most grains appear to be partially or totally devitrified volcanic glass shards. Mangrove roots preserved in living position within the poorly consolidated chert are identified as belonging to the genus of black mangrove, *Avicennia* (Thomas Leslie, 1994, personal commun.).

The coarse, well-indurated breccia layers consist primarily of opal-CT with one-third of the samples showing measurable

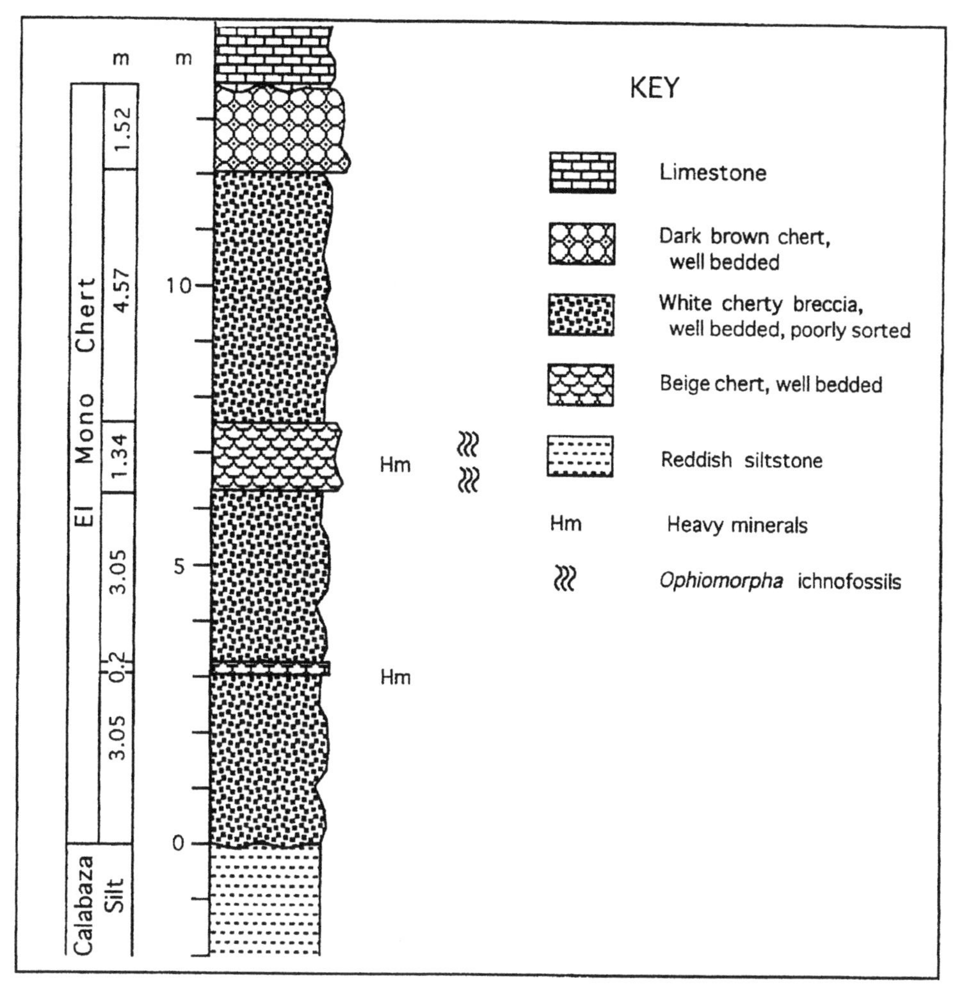

Figure 3. Stratigraphic section through the El Mono chert at the Notch. The chert lies unconformably above a reddish siltstone deposited in a subaerial environment, and it is conformably overlain by a shallow-marine limestone.

amounts of opal-A. Also present in the brecciated layers in minor amounts are silicified biogenic fragments (broken and corroded corals, sponges, gastropods, and bivalves). The matrix is almost exclusively opal-CT. Pores, vugs, and fractures are lined or filled with opal CT. Occasionally, vugs are filled with concentric layers of mircrocrystalline quartz, which resembles chalcedony and presumably grew during several episodes of precipitation. Detrital quartz, feldspar, and clay minerals, as well as some unidentified opaques, occur rarely as light-brown, fine silt-size fillings in vugs and fractures. The linings and fillings of some vugs and fractures are a combination of opal-CT with minor amounts of opal-A. Opal-CT may be replacing the opal-A. Some samples show a fine network of fractures.

Abundant ichnofossils identified as *Ophiomorpha* were found in a laterally extensive resistant marker bed of beige chert occurring in the middle of the El Mono Member at the Notch (Fig. 3). The burrows are 5 cm or more in diameter, oriented vertically to the bedding plane (Fig. 4c), and typically spaced approximately 10 cm from one another. A few diatoms were found in these samples during a careful examination of thin sections; all of them are corroded, and they are not identifiable. SEM (scanning electron microscope) imagery also revealed some biogenic evidence (10%) in the form of highly corroded diatom tests. SEM additionally shows grains of chert that are usually angular but occasionally subrounded (Fig. 7).

The best exposure of the El Mono Member (Fig. 2) is cut by a normal fault that imparts a slight dip to the beds. The topographic feature called the Notch is, in fact, a fault-controlled saddle on the ridge forming the otherwise continuous south border of Arroyo El Mono. At least three other faults were traced throughout the surrounding area (Johnson et al., this volume, their Fig. 2), all with the same general northwest-southeast strike. Dissolution cavities, thermal aureoles, and aggregates of secondary quartz crystals are commonly found along these faults, indicating hydrothermal hot-spring activity. No other evidence of hot springs was found, aside from that associated with faults.

SHALLOW-WATER ORIGIN

Individual chert beds within the El Mono Member range in thickness from a few centimeters to a maximum of 4.5 m (Fig. 3).

Figure 4. Aspects of the El Mono chert. a, basal part of the section featuring a well-bedded, white, cherty breccia. b, view of the area at the Notch looking north toward Cerro Prieto; the resistant capstone at the top of the section is a well-bedded, dark brown chert 1.52 m thick (person at lower right for scale). c, *Ophiomorpha* burrow casts in float at the notch originated from a layer of well-bedded beige chert in the middle of the section (pocket knife is 9 cm long). d, opalized sections from the root of a black mangrove, *Avicennia* sp. (scale in centimeters).

Bioturbation is noticeable in some chert layers. The extensive presence of the *Ophiomorpha* ichnofossils indicates deposition in the shallow intertidal to upper intertidal zones (Ekdale et al., 1984). Occurrence of in situ root casts belonging to a species of the black mangrove, *Avicennia*, corroborates the essentially coastal nature of the deposit. A continuous transgression is indicated by the fact that this chert member overlies a basal conglomerate and siltstone associated with a subaerial fan system developed by sediment discharge from the surrounding Miocene hills but underlies a conformable limestone rich in marine invertebrates indicative of shallow water (Johnson et al., this volume). A late Pliocene age for this transgressive event is suggested by the occurrence of the index fossil *Clypeaster marquerensis*, a sand dollar commonly found in the overlying limestone unit. The repetitive sequence of brecciated and massive chert beds in the El Mono Member of the Infierno Formation may also be related to high rates of evaporation, as suggested by the presence of halite in X-ray diffraction analysis.

PARAGENETIC SEQUENCE

The presence of in situ root casts belonging to a species of the black mangrove, *Avicennia*, the abundant *Ophiomorpha* ichnofossils, and the silicified invertebrate fragments all point to an original sediment layer deposited within a shallow basin at the Notch. This layer, which may have been carbonate sediment, tuffaceous sediment, or a mixture of these two, was eventually altered to chert. According to Hesse (1989), chertification of carbonates involves the replacement of carbonate by silica, before and/or after but not during carbonate cementation. Therefore, we

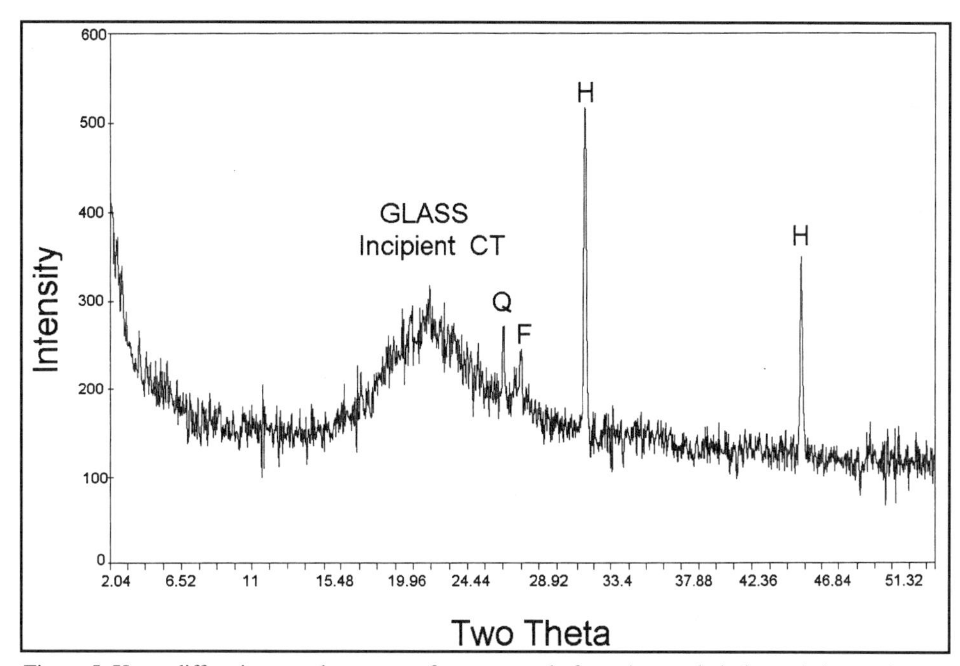

Figure 5. X-ray diffraction powder patterns for one sample from the poorly indurated chert at Arroyo El Mono. Halite (H), quartz (Q), and feldspar (F) are the crystalline constituents. Volcanic glass containing incipient opal-CT is also present. The sample was collected from the upper part of the section.

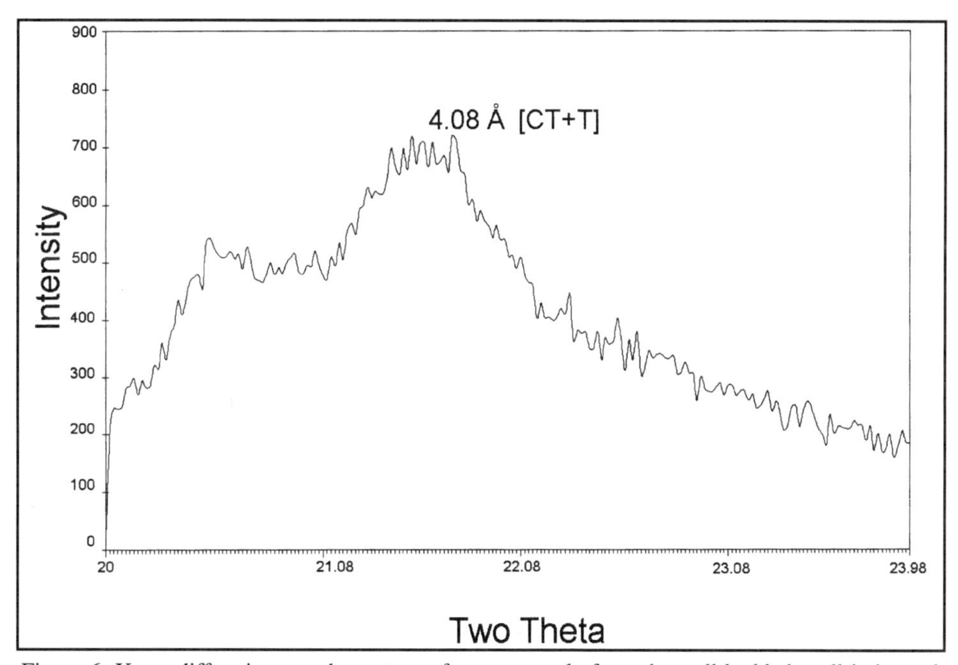

Figure 6. X-ray diffraction powder patterns for one sample from the well-bedded, well-indurated chert collected from the middle of the section, at Arroyo El Mono. Opal-CT is the main material present.

may surmise that the silica diagenesis of the original nonsiliceous sediments took place in a very early stage of the basin's history.

In a shallow-water setting of this kind, evaporation may have proceeded sufficiently to raise water salinity, which in turn promoted silica precipitation. Salinity may have increased to the point that halite precipitated along with silica in some of the beds. The basin did not completely evaporate, however, as indicated by the absence of saline crusts or evaporite deposits.

The foregoing interpretation is supported by modern deposition in Coorong Lake in South Australia and by results of research on the Tehuitzingo Formation of Mexico. Chert deposits were formed by contemporary inorganic precipitation in

Figure 7. SEM photograph of massive opal-A, with scattered small (partial and complete) lephisphers of opal-CT. Irregular plates and cracks may be related to desiccation during early stages of solidification of SiO_2.

Coorong Lake, according to Peterson and Von der Borch (1965). They found that water with a pH ranging from 9.5 to 10.2 dissolves detrital silicates and that chert precipitates out of the evaporating lake when pH falls within the range of 7.0 to 6.5. Bedded chert in the Tehuitzingo Formation from Oaxaca, Mexico, is interpreted to be inorganic in origin by Segura and Rodriguez-Torres (1970). The Tehuitzingo Formation chert is interbedded with halite and is overlain by gypsum.

Geothermal activity might also have been involved in the formation of the Bahía Concepción cherts. Evidence of geothermal activity along the basin-cutting faults of the Concepción Peninsula indicates that geothermal fluids rose through the underlying silica-rich volcanic rocks of the Miocene Comondú Group. Silica dissolved from this unit thus could have been deposited in the overlying Pliocene Infierno Formation. This is similar to the process suggested by Siever (1983) and Hesse (1989). Coastal hot springs at Santispac, within nearby Bahía Concepción, may serve as a contemporary model for such geothermal activity.

We may surmise that once the already silica-rich geothermal fluids entered the basin sediments of the Infierno Formation, they attacked and dissolved some of the previously deposited carbonate layers, thereby raising the pH of these fluids. The high pH fluids then could have dissolved some of the fine-grained tuffaceous volcanic material as well as the carbonates. As pH lowered with dissolution of silica, the saturated geothermal fluids began to redeposit secondary silica in the openings left by earlier solution activity. The newly deposited silica took the form of opal-A, which was transformed to opal-CT as the region continued to experience heating from geothermal sources. Iijima (1988) suggested a similar mechanism for production of chert where siliceous fossil remains are totally absent but where silica-rich geothermal fluids may dissolve volcanic material and reprecipitate it as chert or a variety of opal. Work by Iijima (1988) on the diagenetic transformation of volcanic ash to opal-A and then to opal-C presents a sequence of minerals similar to that observed to have formed in the study area. The same reaction pathways invoked by Iijima (1988) are suggested by the authors of this chapter. An important difference, however, is that Iijima (1988) relies on depth of burial to obtain sufficient heat to drive the reaction, whereas we rely on geothermal sources for the necessary heat. Williams et al. (1985) proposed that the opal-A to opal-CT transformation is not a single-step process but involves a series of dissolution-reprecipitation depth profiles of dissolved silica. Data derived from microscope studies suggest replacement of a carbonate unit and volcanic ash, rather than alteration of biosiliceous components, as the origin for the El Mono chert.

The texture of the brecciated bedded chert developed in several stages. The first stage was probably precipitation of secondary silica (opal-A), replacing earlier deposited carbonate beds or as the alteration product of volcanic ash. In the geothermally heated environment, opal-A transformed to opal-CT or opal-C. These beds may have been present as a gel that eventually collapsed, producing the brecciation. The finely broken material resulting from the brecciation process thus could have provided a matrix to fill in the spaces between the large angular clasts. In time, some of the openings were then filled or lined by secondary microcrystalline quartz resulting from ongoing geothermal activity. An alternative interpretation is that subaerial exposure to meteoric waters resulted in the dissolution of all the carbonate minerals in a chert-bearing limestone, leaving the chert nodules as a lure chert bubble (Knauth, 1994). The origin of these breccias, however, remains unresolved and should be investigated further.

CONCLUSIONS

The area of study on the Concepción Peninsula adjacent to Bahía Concepción involves a series of shallow basins that filled with terrestrial and marine sediments during late Pliocene time. The detrital basin fill was eroded from volcanic rocks belonging to the Miocene Comondú Group. Lithologies locally assigned to the upper Pliocene Infierno Formation include shallow-water marine carbonates and chert beds as well as conglomerates, siltstones, and mudstones that were part of a coalescing, subaerial fan system. The overall depositional environment for the Infierno Formation in the Peninsula Concepción basins was that of a transitional coastal zone.

No evidence was found to support a biogenic origin for the silica in the chert beds. Two sources of silica are suggested. One is from in situ alteration of carbonate layers and volcanic ash within the Infierno Formation. The other is from geothermal solutions that became enriched in silica as they passed through siliceous lavas of the Comondú Group. The geothermal solutions reacted with Pliocene sediments, particularly with carbonate-rich mudstones, to deposit secondary silica in the form of opal-A. The basin sediments were heated sufficiently by the associated geothermal activity to drive the transformation of opal-A to opal-CT. Some of the chert beds exhibit a brecciated texture. Brecciation is attributed either to the collapse of a silica gel or to dissolution of calcium carbonate limestone containing nodular cherts. Evaporation apparently proceeded sufficiently to raise water salinity, which in turn promoted silica precipitation and helped complete the diagenetic silicification of the original carbonate layers.

A shallow-water origin of bedded cherts is generally considered atypical or represents a very unusual environment (Hesse, 1989). In particular, this chapter suggests a unique combination of chertification processes in which sinter deposits and shallow submarine exhalations acted on the previously deposited carbonate muds. The degree of selectivity of the silicification process apparently was a function of a number of variables, including pH, dissolved silicon activity, and the extent of previous lithification of the rock reflected as loss of porosity and/or permeability (Hesse, 1989). This may account for the different degrees of silicification in the two distinctive cherts examined in this study.

ACKNOWLEDGMENTS

Fieldwork during 1994 and 1995 was supported by joint grants to Jorge Ledesma-Vázquez from Consejo Nacional de Ciencia y Tecnologia (CONACYT) (3171-T9307) and to M. E. Johnson from the National Science Foundation (INT-9313828). Jorge Ledesma-Vázquez and Sonia Gutierrez-Sanchez acknowledge Universidad Autónoma de Baja California and CONACYT as well as the Departamento de Geologia Marina at Universidad Autónoma de Baja California Sur for supporting the project Origen de capas de pedernal del Plioceno en Baja California Sur. Ledesma-Vázquez thanks Bryce Hoppie (University of California at Santa Cruz) for his help in reviewing some of the samples by petrographic observation. Sonia Gutierrez thanks Dr. John Minch for financial support during her share of the research. Richard W. Berry thanks Joan Kimbrough (San Diego State University) for assistance with X-ray diffraction of the chert samples. Francisca Staines-Urias is also acknowledged for her drafting and Ing. Israel Gradilla from the Physics Laboratory at Universidad Nacional Autonoma de México (UNAM) for the SEM analyses. We are most grateful to Dr. Robert E. Garrison (University of California at Santa Cruz), Dr. Donn S. Gorsline (University of Southern California), and Dr. Victor Camacho-Ibar for their insightful discussions and helpful comments regarding manuscript revision.

REFERENCES CITED

Burley, S. D., Kantorowics, J. D., and Waugh, B., 1985, Clastic diagensis, *in* Brenchley, D. A., and Williams, L. A., eds., Sedimentology, recent developments and applied aspects: London, Blackwell Science Publications, p. 112–123.

Calvert, S. E., 1974, Deposition and diagenesis of silica in marine sediments, *in* Pelagic sediments: On land and under the sea: International Association of Sedimentology Special Paper 1, p. 273–299.

Davies, T. A., and Supko, P. R., 1973, Oceanic sediments and their diagenesis: Some examples from deep sea drilling: Journal of Sedimentary Petrology, v. 43, p. 381–390.

Ekdale, A. A., Bromley, R. G., and Pemberton, S. G., 1984, Ichnology: Trace fossils in sedimentology and stratigraphy: Tulsa, Oklahoma, Society of Economic Paleontologists and Mineralogists, Short course 15, 317 p.

Greenwood, R., 1973, Cristobalite: Its relationships to chert formation in selected samples from the Deep Sea Drilling Project: Journal of Sedimentary Petrology, v. 43, p. 700–708.

Hesse, R., 1988, Origin of chert. I: Diagenesis of biogenic siliceous sediments: Geosciences Canada, v. 15, p. 171–192.

Hesse, R., 1989, Silica diagenesis: Origin of inorganic and replacement cherts: Earth Sciences Reviews, v. 26, p. 253–284.

Iijima, A., 1988, Silica diagensis, Part II, *in* Chilingarian, G. V., and Wolf, K. H., eds., Diagenesis II: Elsevier, Amsterdam, The Netherlands, Developments in Sedimentology, v. 43, p. 189–211.

Knauth, L. P., 1994, Petrogenesis of chert, *in* Heaney, P. J., Prewitt, C. T., and Gibbs, G. V., eds., Silica: Physical behavior, geochemistry and material applications: Reviews in Mineralogy (Mineralogical Society of America), v. 29, p. 233–258.

Lancelot, Y., 1973, Chert and silica diagensis in sediments from the Central Pacific, *in* Winterer, E. L. et al., eds., Initial reports of the Deep Sea Drilling Project, Volume 17: Washington, D.C.: U.S. Government Printing Office, p. 377–405.

Ledesma-Vázquez , J., and Johnson, M. E., 1993, Neotectonica del area Loreto-Mulegé, *in* Delgado-Argote, L. A., and Martín-Barajas, A., eds., Contribuciones a la tectonica del Occidente de Mexico: Union Geofisica Mexicana, Monografia 1, p. 115–122.

McFall, C. C., 1968, Reconnaissance geology of the Concepción Bay area, Baja California, Mexico: Stanford University Publications in Geological Sciences, v. 10, no. 5, 25 p.

Melendez-Fuster, J., 1981, Geologia: Madrid, Editorial Paraninfo, 912 p.

Peterson, M. N. A., and Von der Borch, C. C., 1965, Chert: Modern inorganic depositional in a carbonate precipitating locality: Science, v. 149, p. 1501–1503.

Segura, R. L., and Rodriguez-Torres, R., 1970, Libro guia de la excursion Mexico-Oaxaca: Mexico City, Sociedad Geologica Mexicana, p. 4–25.

Siever, R., 1983, Evolution of chert at active and passive continental margins, *in* Iijima, A., Hein, J. R., and Siever, R., eds., Siliceous deposits in the Pacific region: Developments in Sedimentology, v. 36, p. 7–25.

Umhoefer, P. J., Dorsey, R. J., and Rennee, P., 1994, Tectonics of the Pliocene Loreto basin, Baja California Sur, Mexico, and evolution of the Gulf of California: Geology, v. 22, p. 649–652.

Williams, L. A., Parks, G. A., and Crerar, D. A., 1985, Silica diagenesis. I: Solubility controls: Journal of Sedimentary Petrology, v. 55, p. 301–311.

Wilson, I., and Rocha, V., 1955, Geology and mineral deposits of the Boleo Copper District, Baja California, Mexico: U.S. Geological Survey Professional Paper 273, 134 p.

Wise, S. W., Jr., Weaver, F. M., and Guven, N., 1973, Early silica diagenesis in volcanic and sedimentary rocks: Devitrification and replacement phenomena: Proceedings, 31st Annual Meeting, Electronic Microscopy Society of America, New Orleans, August, p. 206–207.

Zanchi, A., 1993, Structural and geological analysis of the Loreto region (Baja California Sur, Mexico) during the opening of the Gulf of California: Proceedings, International Meeting on Geology of the Baja California Peninsula, 2nd, Ensenada, Baja California, Mexico, April, p. 83–84.

Manuscript Accepted by the Society December 2, 1996

Printed in U.S.A.

Geological Society of America
Special Paper 318
1997

Stratigraphy, sedimentology, and tectonic development of the southeastern Pliocene Loreto Basin, Baja California Sur, Mexico

Rebecca J. Dorsey, K. A. Stone,* and Paul J. Umhoefer
Department of Geology, Box 4099, Northern Arizona University, Flagstaff, Arizona 86011

ABSTRACT

The Pliocene Loreto basin formed as a westward-tilting asymmetric half graben that subsided rapidly in response to oblique dextral-normal slip on the Loreto fault during late Pliocene time. We recognize four stratigraphic sequences throughout the Loreto basin that record different phases and rates of fault-controlled subsidence, uplift, and input of siliciclastic sediment. The southeastern Loreto basin formed at the southern end of the eastern structural high, which is the uplifted portion of the hanging-wall tilt block of the Loreto fault. Sedimentary rocks of sequence 2 in the southeastern Loreto basin are divided into eight lithofacies associations that represent different depositional environments (in parentheses): (1) shelly sandstone and pebbly sandstone (siliciclastic shallow marine shelf), (2) sandy and pebbly bioclastic limestone (mixed bioclastic-siliciclastic shelf), (3) laterally continuous shell beds (sediment-starved shelf), (4) megaforesets and associated fan-delta facies (Gilbert-type fan deltas), (5) heterolithic facies (gravelly fan-delta plain), (6) cross-bedded conglomerate and sandstone (gravelly braid stream), (7) planar-stratified sandstone (siliciclastic shoreface), and (8) conglomerate and bioclastic limestone (rocky shoreline).

Detailed measured sections combined with 1:10,000-scale structural and facies mapping permit correlation of strata across numerous normal and strike-slip faults. A reconstructed north-south facies panel reveals complex stratigraphic architecture in sequence 2, which has a total thickness of 340 m and is organized into six parasequences bounded by hiatal marine shell beds. A prominent buttress unconformity on the northern margin of the basin records progressive subsidence of the basin relative to the southern flank of the eastern structural high during sequence 2 time. Paleocurrent data and lateral distribution of lithofacies show that gravelly and sandy siliciclastic sediments were shed into the southeastern Loreto basin from the footwall of the Loreto fault located south and west of the study area. The uplifted portion of the hanging wall tilt block, although acting as a structural high, remained mostly submerged below sea level during sequence 2 time and therefore was not a source of siliciclastic sediment.

The base of sequence 2 is a prominent, widespread marine flooding surface that coincides stratigraphically with a pronounced shift in paleocurrent directions. This sequence boundary records a rapid subsidence event and major structural reorganization within the basin that probably was controlled by a change in kinematic behavior of the Loreto fault. The lower half of sequence 2 was dominated by shelf-type fan deltas that prograded northward into the sandy siliciclastic shelf setting. The upper half of sequence 2 consists of four gravel-rich Gilbert-type fan deltas, 3 of which were derived

---

*Present address: Mobil Exploration & Producing U.S. Inc., P.O. Box 9989, Bakersfield, California 93389-9989.

Dorsey, R. J., Stone, K. A., and Umhoefer, P. J., 1997, Stratigraphy, sedimentology, and tectonic development of the southeastern Pliocene Loreto Basin, Baja California Sur, Mexico, *in* Johnson, M. E., and Ledesma-Vázquez, J., eds., Pliocene Carbonates and Related Facies Flanking the Gulf of California, Baja California, Mexico: Boulder, Colorado, Geological Society of America Special Paper 318.

from the footwall uplift of the Loreto fault to the southwest, and one of which consists of shelly debris derived from the ESH to the north. The upper boundary of sequence 2 is a low-angle unconformity marked by an abrupt shift to bioclastic carbonate deposition. This records a second major change in the structural behavior of the basin, in which the hanging wall tilt block experienced renewed uplift and erosion of carbonate shoals and reefs. The stratigraphic evolution of the southeastern Loreto basin thus provides insights into important details of the tectonic and structural evolution of this active basin and its margins.

INTRODUCTION

The Pliocene-Quaternary Loreto basin is located north of the town of Loreto, Baja California Sur (Fig. 1). The basin evolved during Pliocene time as a westward-tilting, rapidly subsiding asymmetric half graben that was controlled by oblique normal-dextral slip on the Loreto fault (Umhoefer et al., 1994). The sedimentary basin fill is over 1,200 m thick and comprises a diverse suite of gravelly nonmarine, deltaic, and marine strata that record very rapid subsidence controlled by faulting along the southwestern margin of the basin (Dorsey et al., 1995). Pliocene faulting and basin development occurred in response to regional oblique extension along the western margin of the Gulf of California transform-rift plate boundary (Moore and Buffington, 1968; Angelier et al., 1981; Stock and Hodges, 1989; Umhoefer et al., 1994; Zanchi, 1994). Since late Pliocene time, the southern and southeastern parts of the basin have been uplifted and dissected, creating excellent exposures of Pliocene sedimentary rocks, while the northern part of the basin has continued to subside, forming a broad flat alluvial valley in the north (Fig. 1; McLean, 1988, 1989).

This chapter presents a detailed analysis of the stratigraphy, sedimentology, and tectonic controls on sedimentary strata exposed in part of the southeastern Loreto basin (Fig. 1). This part of the basin formed at the southern termination of an uplifted hanging-wall fault block, and it therefore occupied a key position during the structural and tectonic development of the basin. Stratigraphic relationships in this area provide insights into the dynamic structural controls on sediment dispersal, shifting sedimentary environments, and interplay between siliciclastic and carbonate deposition during rapid subsidence along the Loreto fault. Correlation of sedimentary rocks in the southeastern basin with recently dated strata in the south-central part of the basin (Umhoefer et al., 1994) yields ages of about 2.6 to 2.4 Ma for strata that are described in this chapter. This study thus provides new information about paleogeography and sedimentation patterns during a phase of active faulting along the western margin of the Gulf of California during late Pliocene time.

GENERAL STRATIGRAPHY AND STRUCTURE

The Loreto basin forms a wedgelike geometry in plan view that is defined by diverging trends of the northwest-striking Loreto fault and the north-trending eastern structural high (Fig. 1). The eastern structural high represents the uplifted portion of a hanging-wall fault block that experienced westward tilting along the Loreto fault during Pliocene subsidence and basin filling (Dorsey et al., 1995). The western margin of this structural high therefore represents the approximate location of the fulcrum, or line of zero displacement, about which tilting of the hanging wall took place (e.g., Leeder and Gawthorpe, 1987). The southern margin of the eastern structural high, along with flanking sedimentary rocks to the south, marks the along-strike termination of this prominent structural feature.

Pliocene strata of the Loreto basin can be divided into four sequences that record four distinct stages of tectonic development within and around the margins of the basin (Figs. 1, 2; Dorsey and Umhoefer, 1996). Sequence 1 consists of nonmarine conglomerate and sandstone older than about 2.6 Ma (Umhoefer et al., 1994) that were deposited in alluvial fans derived from both the footwall and the hanging wall (eastern structural high) of the Loreto fault (Fig. 1). Sequence 2, the focus of this study, is a thick, diverse assemblage of nonmarine, Gilbert-delta, and marine strata that accumulated during a brief period of extremely rapid subsidence (~8 mm/yr) between about 2.46 and 2.36 Ma (Umhoefer et al., 1994; Dorsey et al., 1995). Sequence 3 was deposited during the latter part of the rapid-subsidence phase, and it consists of gravelly bioclastic and mixed volcaniclastic-bioclastic sediments produced by uplift and erosion of carbonate rocks in the hanging-wall tilt block (Fig. 1). These deposits interfinger to the west with thick, volcaniclastic, gravelly Gilbert-type fan deltas that were shed from the footwall of the Loreto fault. Sequence 4 (Fig. 2) is dominated by shallow-marine carbonates that accumulated during a final phase of reduced rates of faulting, subsidence, and input of siliciclastic material from the margins of the basin.

Pliocene sedimentary rocks in the southeastern Loreto basin are cut by a dense, complex array of antithetic (relative to the Loreto fault) normal, oblique-slip, and strike-slip faults that record approximately east-west extension in the hanging wall of the Loreto fault (Figs. 3, 4; Stone, 1994; Zanchi, 1994; Umhoefer and Stone, 1996). Most of the faulting in this area occurred shortly after deposition of sequence 2 strata and prior to deposition of sequence 4. Based on precisely dated equivalent strata in the south-central Loreto basin, sequence 2 was deposited between about 2.6 and 2.4 Ma, and most faulting took place between about 2.4 and 2.0 Ma (Umhoefer et al., 1994; Stone, 1994). The geologic map and cross section (Figs. 3, 4) reveal the highly faulted and dismembered nature of east-dipping Pliocene sedimentary rocks and underlying Miocene volcanic rocks. Despite this structural complexity, numerous laterally continuous shell beds and distinctive stratigraphic trends associated with them allowed us to make reliable stratigraphic correlations between fault blocks. In the westernmost part of the study area, sequence 1 strata rest unconformably on top of older Miocene vol-

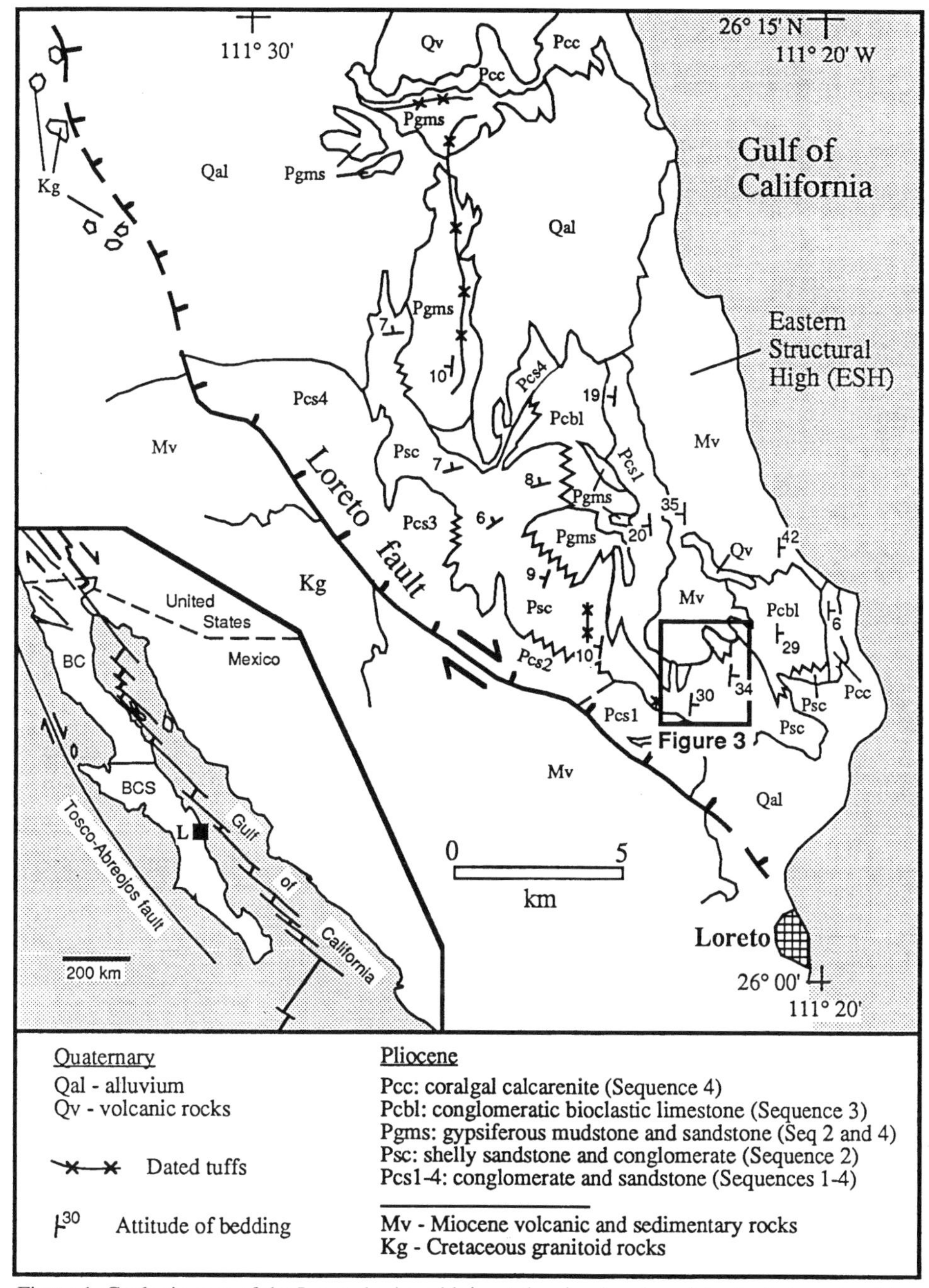

Figure 1. Geologic map of the Loreto basin, with inset showing location of the basin in regional tectonic framework. BC = Baja California; BCS = Baja California Sur; L = Loreto. Box shows location of study area (Fig. 3). Adapted from Umhoefer et al. (1994).

canic rocks; to the east we find sequence 2 strata and, farther east, sequence 3 rocks resting directly on top of Miocene basement rocks (Fig. 3). This trend results from a major buttress unconformity that defines the southern termination of the eastern structural high.

Figure 5 is a stratigraphic panel that was constructed from numerous detailed measured sections (Appendix 1) and detailed 1:10,000-scale facies mapping (Fig. 3). This panel reveals onlapping of sequences 1, 2, and 3 strata onto the prominent buttress unconformity at the southern margin of the eastern structural high hanging-wall tilt block. Sequence 2 overlies sequence 1 along a conformable but sharp contact (Fig. 5) that records rapid flooding of the basin by a marine seaway, probably in response to a sudden increase in the rate of fault-controlled subsidence (Dorsey et al., 1995). Sequence 2 is about 350 m thick in the study area, and it is subdivided into six parasequences (2A–2F) that are defined by laterally continuous shell beds (Fig. 5). In contrast, sequence 2 in the south-central Loreto basin is about 520 m thick and contains about 12 parasequences (Dorsey et al., 1995). A

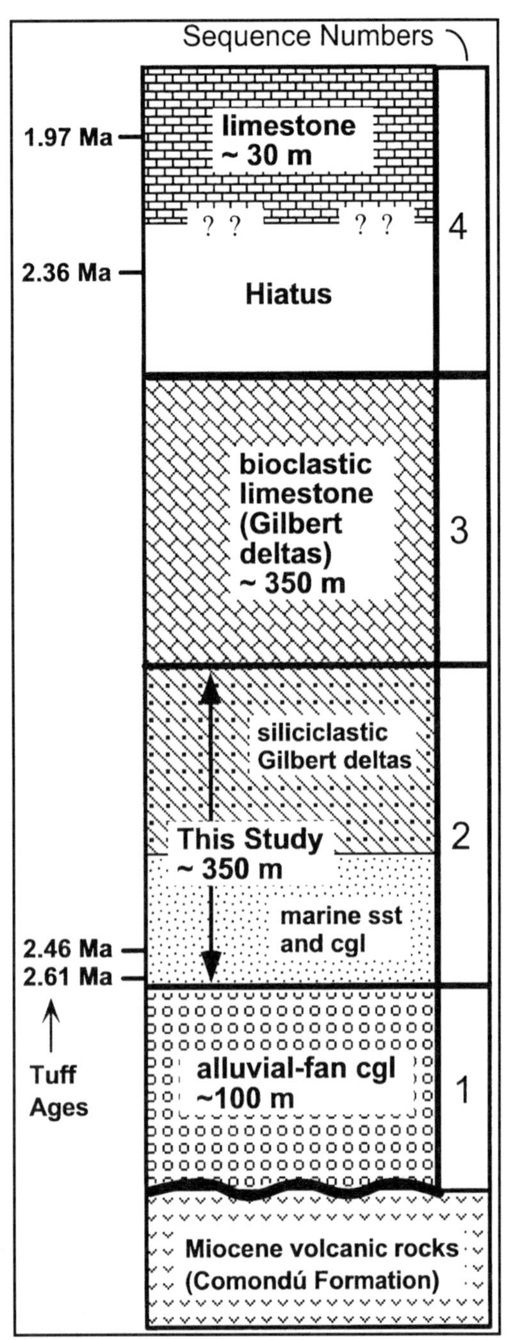

Figure 2. Simplified diagram showing sequence stratigraphy of the southeastern Loreto basin; sequence 2 is the focus of this study. Ages are based on interbedded tuffs extrapolated from the south-central Loreto basin (Umhoefer et al., 1994).

detailed discussion of sequence stratigraphic features shown in Figure 5 is presented in a later section of this chapter.

SEDIMENTARY LITHOFACIES

Sedimentary rocks in the southeastern Loreto basin contain a wide variety of lithofacies ranging from alluvial-fan conglomerate and Gilbert-delta foreset strata to marine shelly sandstone and bioclastic limestone. Lithofacies have been assigned to eight facies associations that are listed below, with inferred environment of deposition indicated in parentheses (Table 1): (1) shelly sandstone and pebbly sandstone (siliciclastic shallow marine shelf), (2) sandy and pebbly bioclastic limestone (mixed bioclastic-siliciclastic shelf), (3) laterally continuous shell beds (sediment-starved shelf), (4) megaforesets and associated fan-delta facies (Gilbert-type fan deltas), (5) heterolithic facies (gravelly fan-delta plain), (6) cross-bedded conglomerate and sandstone (braided stream system), (7) planar-stratified sandstone (siliciclastic shoreface), and (8) conglomerate and bioclastic limestone (rocky shoreline). These facies associations are represented in measured sections (Appendix 1) and the stratigraphic panel (Fig. 5). Below, each of these facies associations is described and interpreted, with the map units in Figure 3 indicated for each.

1. Shelly sandstone and pebbly sandstone (map unit ss)

This facies association is found in the central part of the study area (Fig. 3), and it dominates the upper part of parasequence 2B and the lower part of parasequence 2C (Fig. 5). These deposits consist of yellow to gray, massive to crudely bedded, poorly sorted shelly sandstone and shelly pebbly sandstone (Fig. 6). Shells are dominated by pectens, bivalves, oysters, and gastropods, typically making up a diverse assemblage of open-marine molluscan faunas (Durham, 1950; Smith, 1991). Shells are either scattered and isolated, or they form thin shell concentrations and small pockets (typical of oysters). Grain size of siliciclastic sediment ranges from fine sand to granules and small pebbles; granules and small pebbles commonly are found floating in massive fine- to medium-grained sandstone. Siliciclastic detritus consists of volcanic rock fragments and lesser amounts of feldspar, all derived from the Miocene Comondú basement rocks. The abundance of shells, granules, and pebbles varies considerably in this facies association; thus facies may include massive bioturbated sandstone with rare shells, shelly sandstone with no granules or pebbles, pebbly sandstone with rare shells, and shell-rich pebbly sandstone.

Relict stratification and bedding features in shelly pebbly sandstone facies indicate that the massive nature and mixing of coarse and fine detritus is due to pervasive bioturbation. Wide grain-size variations record fluctuations in the energy of marine currents as well as variations in faunal activity. Proximity to active sources of gravelly detritus (shelf-type and Gilbert-type fan deltas) is indicated by the common occurrence of granules and small pebbles.

2. Sandy and pebbly bioclastic limestone (map unit sbl)

This facies association is restricted to areas north of Arroyo de Arce in parasequence 2B, and it makes up the majority of rocks in sequence 3 (Figs. 3, 5). It comprises a mix of bioclastic carbonate and siliciclastic sediment (>50% carbonate material), reflecting input from different sources within and at the margins of the basin. These lithofacies are dominated by massive sandy and pebbly shell hash and sandy calcarenite; they can be classified as impure bioclastic limestone having variable grain size and variable percent of siliciclastic detritus. The siliciclastic component (volcanic detritus) varies from about 20 to 50% of the rock and ranges in size from fine sand to small pebbles. Pectens and oysters are abundant, ranging from whole and articulated shells to broken shell fragments to shell hash and calcarenite. Molds of bivalves and gastropods are rarely recognized. Abundant barnacles and bryozoans are observed encrusting onto many surfaces of pebbles and shells, and shells reveal abundant borings and other signs of taphonomic alteration. The massive nature of these deposits results from extensive and complete biogenic reworking.

This facies association is interpreted to record deposition in a shallow-marine shelf setting in which biologic productivity was high and siliciclastic input was slow. Bioclastic shell material was produced in situ and mechanically broken down by vigorous marine currents, probably in a high-energy lower-shoreface setting. Abundant encrustations and borings provide evidence for exposure and rolling of grains on the sea floor prior to burial.

3. Laterally continuous shell beds (map units dsb, sb1–5)

Shell beds in the study area provide valuable, laterally continuous marker horizons that enabled correlation across numerous faults and reliable reconstruction of the stratigraphic architecture. They are distributed throughout the map area and make useful stratigraphic horizons for subdividing sequence 2 (Figs. 3, 5). Shell beds are massive, dense shell concentrations, typically 0.5 to 2 m thick, that reveal a variety of paleontologic and taphonomic features that were not analyzed in detail for this study. Densely packed pectens, oysters, bivalves, and gastropods are the most abundant shell types recognized (see also Durham, 1950; Smith, 1991). Various borings and encrusting organisms (barnacles, bryozoans, and algae) are also present. Shells are found in a variety of conditions ranging from whole and articulated to smaller shell fragments and shell hash. Matrix consists of sandstone, pebbly sandstone, and less commonly sandy calcarenite. Shell beds in this area are found in a variety of stratigraphic positions: interbedded with nonmarine and delta-plain facies (dsb, sb1, sb2), interbedded with siliciclastic shelly marine sandstone (sb1–sb3), capping topset deposits of Gilbert-type fan deltas (sb4, sb5), and in one locality interbedded with and draped over Gilbert-delta foreset strata (sb5) (Fig. 5).

Shell beds are interpreted as hiatal shell concentrations that accumulated during times of severely reduced input of siliciclastic detritus to a shallow-marine shelf (e.g. Kidwell, 1988, 1991). We thus view shell beds as the sediment-starved equivalent of shelly sandstone and shelly pebbly sandstone facies (facies association 1: siliciclastic shallow-marine shelf). The reasons for episodic shutdown of siliciclastic input are not well understood, but they may include autocyclic processes such as lateral switching of delta lobes or allocyclic processes such as eustatic sea-level fluctuations and/or episodic movements on the Loreto fault (Dorsey et al., 1995). The origin of shell beds and their relation to other facies associations in the study area are discussed in a later part of this chapter.

4. Megaforesets and associated fan-delta facies (map units 1, 2, 3, 4, bs)

In the southeastern Loreto basin, we recognize deposits of Gilbert-type fan deltas on the basis of distinctive sedimentary lithofacies, vertical and lateral facies associations, and distinctive contacts between megaforesets and topset strata, which reveal angular discordance produced by the primary dips on foreset deposits. Lithofacies of Gilbert-type fan deltas dominate the central and eastern parts of the study area (Fig. 3); they comprise the upper part of parasequence 2C and most of parasequences 2D through 2F (Fig. 5). Gilbert-delta deposits are divided into topsets, foresets, and bottomsets (proximal and distal). In Figure 3, map units 1 through 4 include all topset, foreset, and proximal bottomset strata, and map unit bs represents distal bottomset strata. These components are subdivided and depicted in the stratigraphic panel (Fig. 5). For clarity of terminology, facies in this section are named and described according to their position within the known Gilbert-type fan-delta architecture, as determined from detailed facies mapping and stratigraphic analysis. The descriptions are followed by an explanation and interpretation of Gilbert-delta processes.

Topset facies (map units 1–4). Topset facies in the study area are highly variable in grain size, bed thickness, lithology, and sedimentary structures, ranging from siliciclastic shelly sandstone to channelized pebble-cobble conglomerate. These facies are recognized on the basis of angular truncations seen at the top of foreset packages and the presence of sedimentary lithofacies consistent with a gravelly delta-plain environment. Variations in topset facies are represented by descriptions of heterolithic facies (facies association 5, below) and shelly sandstone and pebbly sandstone (facies association 1, above).

Foreset facies (map units 1–4). Foresets of Gilbert-type fan deltas consist of tabular-bedded, sharp-based pebble-cobble conglomerate, shelly conglomerate, pebbly sandstone (± shells). Beds range from about 5 to 40 cm thick and show no obvious internal grading. In places, foreset strata comprise alternating conglomerate-sandstone couplets, and in other examples they consist dominantly of amalgamated conglomerate and shelly conglomerate. The abundance of pecten and oyster shells varies widely from 0 to ~20% (foreset packages gd1, gd2, and gd4) to over 50% (notably in gd3; Fig. 7A). Low-angle truncations and wedging geometries are sometimes seen within foreset packages. Foreset strata are best recognized by their steep (typically 15 to 25°) primary dips. In this study area, where steep tectonic dips of up to 45° are present, primary foreset dips were identified by comparison with (and/or truncation by) nearby topset or bottomset strata. Bioturbation is absent or rare, except in foresets of gd3, where shell bed sb5 is interbedded with shell-rich (transported) tabular foreset deposits.

Proximal bottomsets (map units 1–4). This facies consists of tabular- to lenticular-bedded, moderately to well stratified, ungraded sandy conglomerate and conglomeratic sandstone (Fig. 7B). This facies appears in measured section LA3,

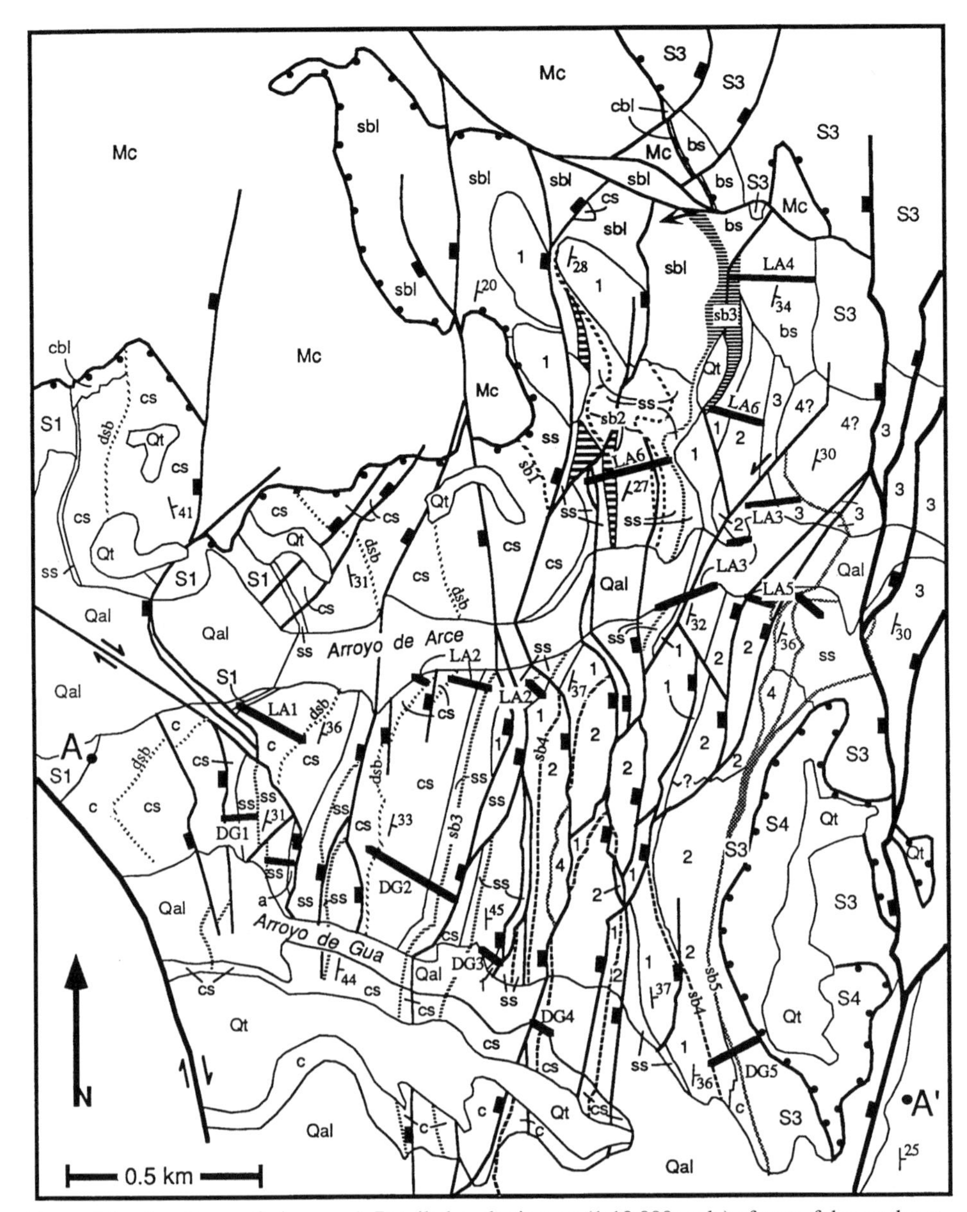

Figure 3 (explanation on facing page). Detailed geologic map (1:10,000 scale) of part of the southeastern Loreto basin. Thick lines with labels show the locations of measured sections (Appendix 1). Patterned lines represent marker shell beds. Points A and A′ indicate the line of section shown in Figure 4.

between 26 and 90 m in the section (Appendix 1). Transported whole and broken shells of pectens and oysters are common, often displaying a strong convex-up alignment produced by current transport. Conglomerate beds sometimes display hummocklike geometries (resembling hummocky cross-stratification) with sharp bases and tops, and they range in thickness from single-clast horizons to about 40 or 50 cm (Fig. 7B). Other conglomerate beds are internally structureless, average 1 to 2 m thick, and display downcutting channelized bases. Sandstone and pebbly sandstone beds are 5 to 30 cm thick, are moderately to well sorted, and commonly overlie discontinuous, planar, thin pebble conglomerate beds and single-clast stringers. Conglomerate and sandstone are typically interbedded as alternating sharp-based couplets, in which the conglomerate:sandstone ratio varies between about 1:1 and 1:4.

Distal bottomsets (map unit bs). This facies consists of tabular-bedded, thin- to thick-bedded well sorted sandy turbidites (Fig. 7c). Grain size is dominantly fine to coarse sand, but granules and small pebbles are seen at the base of some thicker beds. These turbidites have sharp bases and show subdued to well developed normal grading, in which medium- or coarse-grained sand fines up into fine- or very-fine-grained sand. Beds are pervasively planar stratified, and they almost completely lack bioturbation structures and ripple cross-lamination. Bed thicknesses mostly fall between about 5 and 30 cm, with fewer beds ranging up to 60 cm thick. Several weakly developed coarsening- and

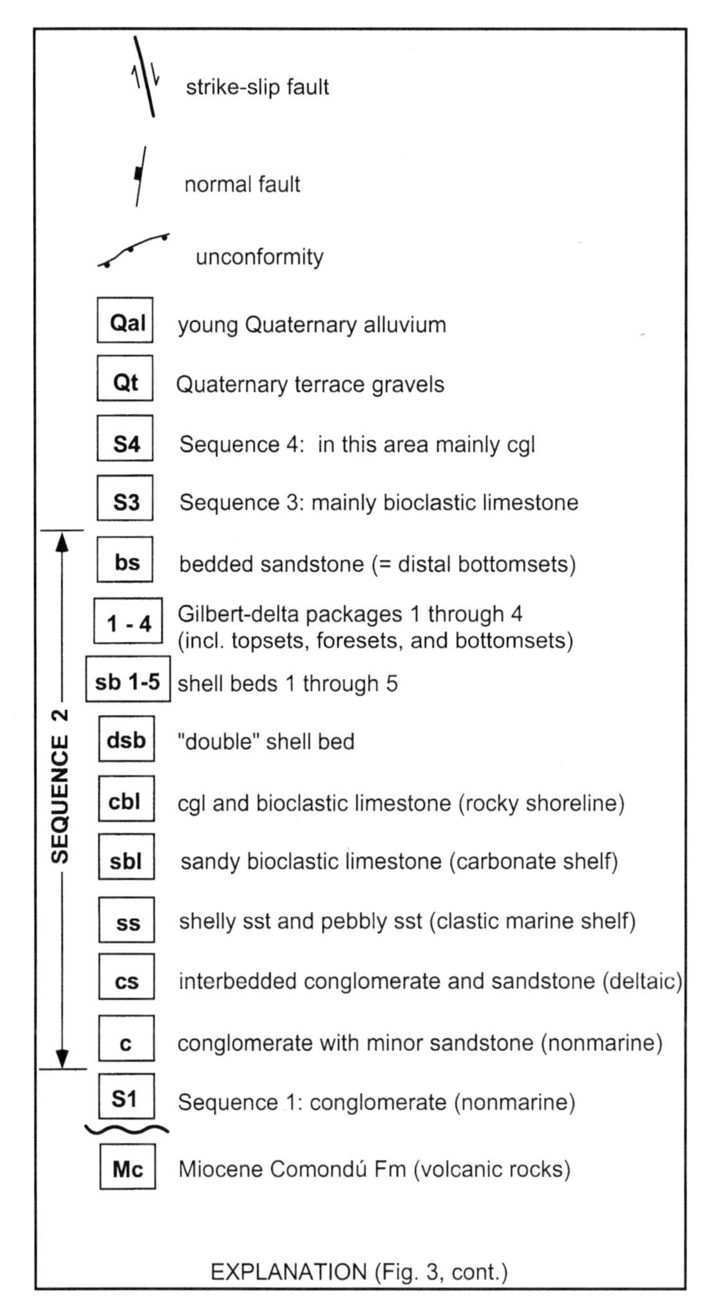

EXPLANATION (Fig. 3, cont.)

thickening-up intervals, 10 to 25 meters thick, were measured in section LA4 (Appendix 1).

Interpretation of Gilbert-delta facies. Marine Gilbert-type fan deltas have been recognized in a number of tectonically active basins, where they typically are controlled by rapid movements along active basin-bounding faults (Colella, 1988; Nemec and Steel, 1988; Colella and Prior, 1990). Conglomerate and sandstone facies of foreset strata in the southeastern Loreto basin were most likely deposited by high-density, gravelly sediment gravity flows, including cohesionless debris flows (grain flows) and gravelly turbidity currents (e.g., Lowe, 1982; Nemec and Steel, 1984; Nemec, 1990). Segregation of lithic and shell clasts provides evidence for hydraulic sorting during transport and deposition of these flows. Lenticular-bedded conglomerate and sandstone were deposited in a proximal bottomset position, within or just downslope of the transition from foresets to bottomsets. Well developed lenticular bedding and stratification suggest deposition by strongly turbulent and erosive currents, perhaps resulting from "hydraulic jump" within gravelly turbidity currents at the base of Gilbert-delta slopes (e.g. Nemec, 1990). Proximal bottomset facies pass laterally northward into distal bottomsets (Figs. 3, 5). Gravelly turbidity currents deposited their gravel load in foreset and proximal bottomset positions, and low-density turbulent overflows were transported northward to produce distal bottomsets (sandy turbidites). Coarsening-up intervals (or cycles) within distal bottomsets may have been produced by repeated progradation of Gilbert-delta foresets from the south, but this possibility was not tested in field mapping.

5. Heterolithic facies (map units cs and 1–4)

This is the most variable of all facies associations analyzed for this study. Facies consist of channel-fill conglomerate and shelly conglomerate, interbedded with interchannel sandstone, mudstone, and claystone. Conglomerate is massive to trough cross-stratified, is poorly to moderately well sorted, and contains pebble- to cobble-sized clasts up to about 20-cm diameter. Conglomeratic channel-fill units are sharp-based, ranging from about 2 to 10 m thick, and they commonly display normal grading through 0.5 to 1 m into overlying interchannel sandstone and mudstone facies. Interchannel (overbank) deposits consist of massive sandstone, mudstone, and claystone that are typically 2 to 10 m thick. Interchannel units commonly contain one or more of the following accessories: oyster and pecten shells, burrows, oxidized root casts, red to brown organic-rich paleosols, dispersed reddish organic matter, and gypsum. Gypsum-bearing deposits typically are mudstone and claystone, and they lack shells and burrows.

Interbedding of conglomerate, sandstone, and mudstone units at a scale of 2 to 10 m records lateral migration of channels on the subsiding gravelly deltaic plain of both shelf-type and Gilbert-type fan deltas. Shelf-type fan deltas (cf., Ethridge and Wescott, 1984) are recognized in parasequences 2A and 2B, and Gilbert-type fan deltas are found in parasequences 2C through 2F (gd1–gd4; Fig. 5). Root casts and paleosols record periods of soil development in overbank subenvironments, and shells and burrows provide evidence for interdistributary marine and brackish-water bays, probably analogous to those commonly seen in fine-grained deltas (Elliott, 1986). The presence of oyster and pecten shells in channel conglomerate provides additional evidence of lateral channel switching, which involved erosion into fossiliferous interdistributary-bay sands and muds, and reworking of marine shells into active channel gravels.

6. Cross-bedded conglomerate and sandstone (map unit c)

This facies association consists of poorly sorted, trough cross-stratified pebble and pebble-cobble conglomerate that is interbedded with poorly sorted, massive to trough cross-bedded pebbly sandstone and sandstone. Shells, burrows, and gypsum

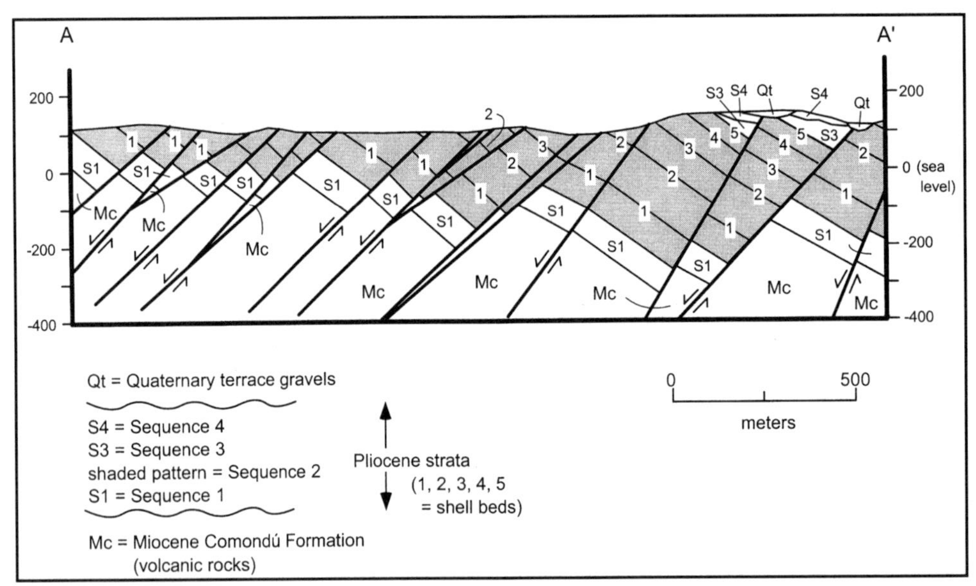

Figure 4. Geologic cross section across southern part of the study area, modified from Umhoefer and Stone (1996). See Figure 3 for line of section.

are notably absent. Thin caliche paleosols are rare and much less common than the oxidized organic-rich paleosols observed in overbank facies of the delta-plain facies association. Conglomerate beds typically range from 20 to 100 cm thick, and they commonly display shallow erosional relief on channelized bases. Sandstone and pebbly sandstone are closely interbedded with conglomerate, sometimes occurring in weakly developed fining-up intervals at the top of conglomerate beds.

We interpret these facies to record deposition in a gravelly braided stream system, probably in the distal parts of an alluvial fan. Vertical and lateral relationships with facies association 5 (fan-delta plain) indicate that these braided streams passed gradually downstream into gravelly delta-plain environments associated with both Gilbert-type and shelf-type fan deltas.

7. Planar-stratified sandstone (no map unit)

These lithofacies are areally restricted and thus do not appear in the detailed geologic map (Fig. 3); however, recognition of these deposits is important for reconstructing depositional systems for Sequence 2 time. They are divided into (1) thin- to medium-bedded bioturbated sandstone and siltstone and (2) well-sorted, planar-stratified fine- to medium-grained sandstone. Sandstone beds of facies 1 have sharp bases, bioturbated tops, and relict planar and ripple stratification. Facies 2 contains well-developed to obscured planar stratification, in which common overprinting of stratification is due to weathering or possibly bioturbation. Well-preserved sand dollars are common to abundant and are unique to this facies. In section LA1 (Appendix 1), between 7 and 21 m in the section, facies 1 and 2 (above) occur in sequence as transitional facies between marine shelly sandstone below and fan-delta plain facies above.

The systematic vertical associations described above, combined with diagnostic stratification and sorting, suggest a lower shoreface depositional environment (possibly sandy-shelf storm deposits) passing upward into an upper shoreface to beach setting. These shoreface deposits are found only in parasequences 2A and 2B, suggesting that they developed in nearshore settings adjacent to shelf-type (not Gilbert-type) fan deltas.

8. Conglomerate and bioclastic limestone (map unit cbl)

This facies association makes up a thin map unit that is found only along the buttress unconformity in the northern part of the study area (Figs. 3, 5). It consists primarily of calcarenite-matrix cobble-boulder conglomerate, sandy calcarenite and shell hash, and (locally) oyster reef deposits (Fig. 8). Conglomerate has a massive, clast-supported, open-framework texture in which cobbles and small boulders of Miocene volcanic rocks are set in a distinctive white matrix of pebbly calcarenite and molluscan shell hash. Conglomerate matrix is similar to bioclastic material in sandy calcarenite that commonly overlies basal conglomerate deposits (Fig. 8); contact relations suggest that the matrix may represent a sieve deposit derived from overlying bioclastic deposits. Calcarenite and shell hash consist of massive, thoroughly bioturbated, thick-bedded bioclastic limestone in which shells and shell fragments are mainly oysters, pectens, and other bivalves. Cobble-boulder conglomerate typically lies directly on top of variably weathered volcanic basement rocks of the Comondú Formation. Oyster reef deposits are small, restricted, moundlike oyster accumulations (Fig. 8B) that lie directly on top of basement rocks along a relatively flat portion of the buttress unconformity, in the lower part of parasequence 2B (Fig. 5).

This facies association was formed in a steep rocky shoreline setting (e.g., Johnson, 1988, 1992), in which there was very little input of siliciclastic detritus from the rocky headlands upon which these facies were deposited. Similar carbonate-rich, conglomeratic rocky shoreline deposits have been recognized both

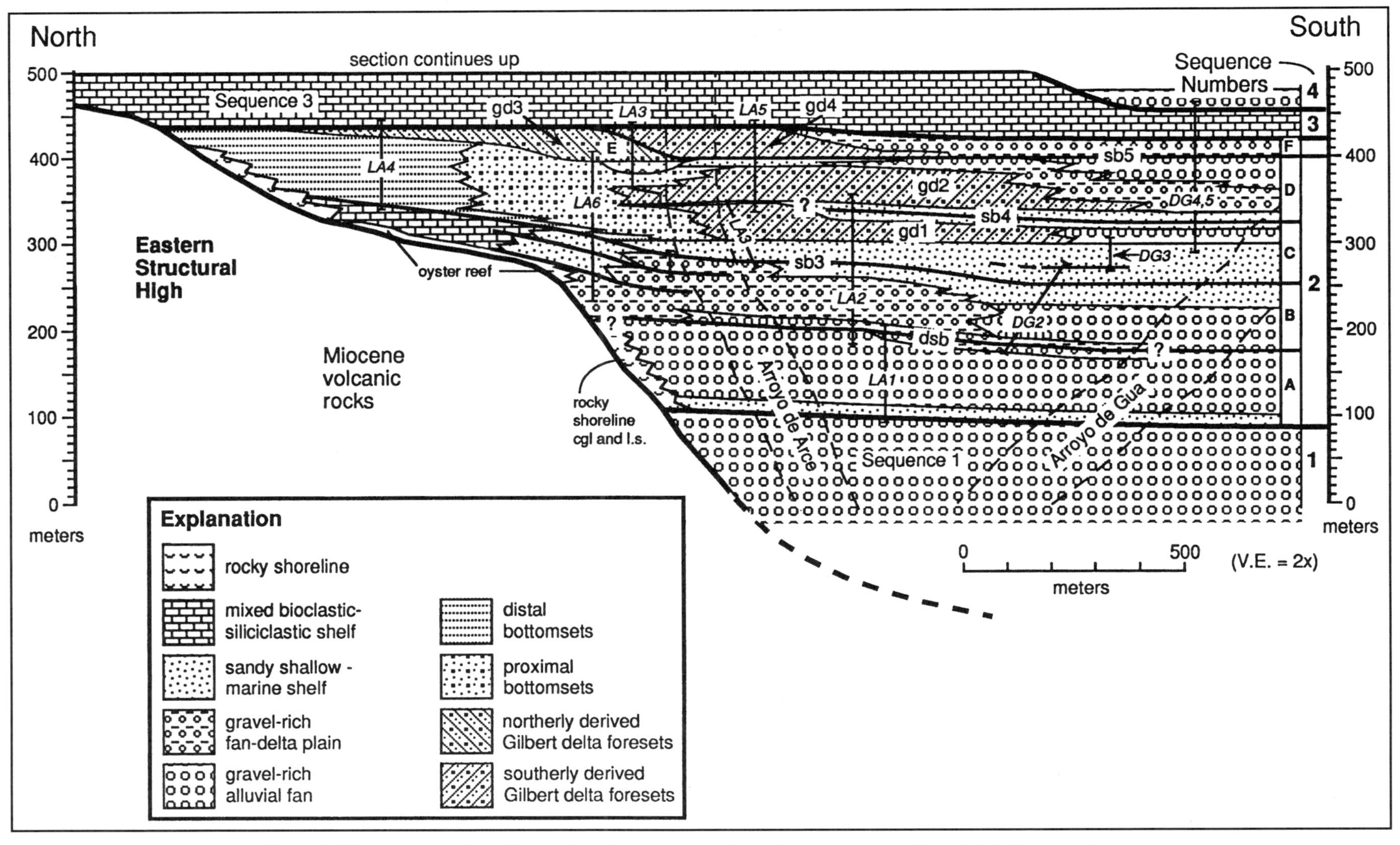

Figure 5. North-south stratigraphic panel for the southeastern Loreto basin, constructed from detailed measured sections and 1:10,000-scale structural and facies mapping. LA1, LA2, etc., represent measured sections in and near lower Arroyo de Arce; DG2, DG3, etc., represent measured section near Arroyo de Gua; thick lines represent sequence boundaries; medium lines (dsb, sb3–5) represent marker shell beds that define parasequence boundaries; letters A through F represent parasequences 2A–2F; gd1–gd4 are Gilbert-type fan deltas that correspond to parasequences 2C–2F. See Figure 3 for location of measured sections and arroyos. See text for further explanation and descriptions.

TABLE 1. MAJOR LITHOFACIES ASSOCIATIONS AND INTERPRETATIONS, SE LORETO BASIN

Lithofacies Association*	Summary Facies Description	Interpretation
1. Shelly sandstone and pebbly sandstone (ss).	Massive to crudely bedded, poorly sorted shelly sandstone and shelly pebbly sst.	Siliciclastic shallow-marine shelf.
2. Sandy and pebbly bioclastic limestone (sbl).	Massive sandy and pebbly shell hash and sandy calcarnite; abundant molluscan debris; encrusting barnacles and bryozoans.	Mixed bioclastic-siliciclastic marine shelf.
3. Laterally continuous shell beds (sb 1-5).	Thin, laterally extensive shell concentrations; massive; dominated by molluscan faunas; sand to pebble matrix.	Sediment-starved marine shelf.
4. Megaforesets and associated fan-delta facies (1, 2, 3, 4, bs).	Topsets, foresets, proximal bottomsets, and distal bottomsets; massive to stratified conglomerate and sandstone with transported shells.	Siliciclastic Gilbert-type fan deltas.
5. Heterolithic facies (cs; 1-4).	Conglomerate (channel-fill), sandstone, mudstone, and claystone, with accessory shells, burrows, root casts, paleosols, and gypsum.	Gravelly fan-delta plain.
6. Cross-bedded conglomerate and sandstone (c).	Poorly to moderately sorted, trough x-stratified conglomerate and sandstone; thin caliche horizons; no shells or bioturbation.	Braid stream system.
7. Planar stratified sandstone (no map unit).	Thin- to medium-bedded, planar stratified fine- to medium-grained sandstone; associated with bioturbated sandstone.	Wave-influenced marine shoreface.
8. Conglomerate and bioclastic limestone (cbl).	Shelly and calcarenite-matrix cobble-boulder conglomerate; interbedded with shelly calcarenite.	Rocky shoreline.

*Map unit in parentheses.

in ancient sequences of active continental margins (Harris and Frost, 1984; Burchette, 1988; Johnson and Hayes, 1993) and in a modern analog of the Loreto basin, which is located about 75 km to the north at Bahía Concepción (Hayes et al., 1993; Meldahl et al., this volume). The commonly observed upward transition from cobble-boulder conglomerate to sandy calcarenite and shell hash (Fig. 8A) is interpreted to record transgression of the sediment-starved shelf sequence over basal gravel-lag deposits.

STRATIGRAPHIC ANALYSIS

Methods

The complex facies architecture and stratigraphic evolution of the southeastern Loreto basin are revealed in a north-south facies panel (Fig. 5). The panel was constructed by compiling and correlating a series of detailed measured sections, which are presented in Appendix 1 and plotted on the geologic map (Fig. 3) and the stratigraphic panel (Fig. 5). Detailed structural and facies mapping between sections was a key element in constructing the panel. The presence of laterally continuous shell beds, combined with unique vertical facies trends, made it possible to obtain reliable correlations across numerous normal faults and thus restore the stratigraphy to its pre-faulting geometry. The panel was con-

Figure 6. Outcrop photograph of massive siliciclastic shelly pebbly sandstone. Note bioturbation features and small clusters of pecten shells. Hammer is 32.5 cm long.

Figure 7. Outcrop photographs of Gilbert-type fan delta facies. A, example of shell-rich, tabular-bedded foresets in Gilbert delta 3 (parasequence 2E, Fig. 5). Note high concentration of oriented pecten shell debris. Other foreset packages in the study area (Gilbert deltas 1, 2, and 4) have similar tabular bedding style, but they are composed primarily of siliciclastic gravel and sand with minor, variable amounts of transported shell material. Measuring staff is 1.8 m long. B, proximal bottomset facies, consisting of lenticular-bedded, well-stratified conglomerate and sandstone with sharp lower and upper contacts. Hammer is 32.5 cm long. C, distal bottomset facies, consisting of tabular-bedded, planar stratified coarse- to fine-grained sandstone. Hammer is 32.5 cm long.

structed by hanging all sections from shell bed 5 (sb5), which is used as the stratigraphic datum. This shell-bed datum is inclined in the area of section LA3 (Fig. 5), because it lies within (drapes over) inclined foreset strata belonging to Gilbert delta 3 (gd3, Fig. 5). The slope and stratigraphic thickness of the inclined portion of shell bed 5 were constrained by field observations. The spatial distribution of architectural elements and lateral facies contacts was determined from an orthogonal map grid, which allowed measured sections and map contacts to be projected into a single north-south plane.

Overall geometry

The stratigraphic panel is dominated by a major buttress unconformity along which strata of sequences 1, 2, and 3 onlap northward onto older Miocene volcanic rocks (Fig. 5). About 500 m of paleorelief is documented along this unconformity within the study area; the total thickness of strata (sequences 1-4) that onlap onto this buttress unconformity throughout the southeastern Loreto basin is estimated at 800 to 1000 m (Umhoefer and Dorsey, unpublished data). The gross stratigraphy of sequence 2 can be divided into a lower half (parasequences 2A and 2B) and an upper half (parasequences 2C–2F). The base of the sequence is a prominent, widespread flooding surface, covering approximately 20 to 25 km^2 (based on this study and unpublished map data elsewhere in Loreto basin), that records the first incursion of marine environments into the Loreto basin shortly before 2.6 Ma (Figs. 2, 5). This was followed by progradation of alluvial fans and shelf-type fan deltas into the southeastern basin (Appendix 1, Section LA1). Whereas Gilbert-type fan deltas are completely lacking in the lower half of sequence 2, they dominate the stratigraphy of the upper half.

The basal unconformity has an irregular geometry that

Figure 8. Outcrop photographs of rocky shoreline facies. A, basal cobble-boulder lag conglomerate overlain by pebbly calcarenite and shell hash. Base of area shown in photograph is weathered Miocene volcanic basement rock. Hammer is 32.5 cm long. B, thin basal oyster reef deposit situated directly on basal unconformity, north of section LA6 (Fig. 5). Hammer is 32.5 cm long.

includes a subhorizontal ramp, or platform, in the upper part of parasequence 2B (Fig. 5). This platform is overlain by a prominent oyster reef, which is in turn overlain by an areally restricted unit of sandy bioclastic limestone that covers about 0.5 to 0.6 km^2 on the geologic map (Fig. 4). This limestone unit represents a narrow carbonate shelf, about 0.5 km wide, that existed on the north flank of the basin for a short time prior to the onset of Gilbert-delta sedimentation (Fig. 5). It appears that the platform was cut on Miocene volcanic bedrock during the transgression that produced the flooding surface within parasequence 2B, below shell bed 3 (sb3, Fig. 5). Mixed carbonate and siliciclastic sedimentation was terminated by rapid northward progradation of distal (sandy) siliciclastic bottomsets of Gilbert delta 1 (gd1) in the northern part of parasequence 2C (Fig. 5).

The upper boundary of sequence 2 is marked by an abrupt change to widespread sandy and pebbly bioclastic limestone of sequence 3, which is similar to bioclastic limestone (mixed carbonate and siliciclastic shelf facies) of sequence 2 (Fig. 5). The contact between sequences 2 and 3 appears conformable in outcrop, but one large cliff exposure reveals very low angle truncations at the base of sequence 3 that may represent a slight angular unconformity.

Parasequences

Sequence 2 in the study area is divided into six packages, or parasequences, which are defined by laterally continuous shell beds. A parasequence is a "relatively conformable succession of genetically related beds or bedsets bounded by marine-flooding surfaces and their correlative surfaces" (Van Wagoner et al., 1988, p. 39). Parasequences typically consist of shallowing-upward successions (high-stand systems tracts), and the flooding surfaces that cap them are produced by a rapid rise in relative sea level. An important distinction can be made between flooding surfaces, which directly cap progradational packages, and surfaces or zones of maximum transgression, which may be marked by marine condensed intervals and commonly lie some thickness above flooding surfaces at the top of transgressive intervals (Arnott, 1995). Arnott (1995) argued that flooding surfaces are more easily recognized than surfaces of maximum transgression, and therefore parasequences should be defined by flooding surfaces. Shell beds in the southeastern Loreto basin are hiatal concentrations that record conditions of maximum transgression during periods of severely reduced siliciclastic input; they commonly lie 5 to 10 m above the flooding surfaces that cap nonmarine or marginal marine strata (Fig. 5; Appendix 1). Because shell beds are the most easily recognized, lithologically prominent, and laterally continuous markers within sequence 2, we have used them (rather than flooding surfaces) to define parasequence boundaries. This represents an interesting exception to the general case described by Arnott (1995).

The upper half of sequence 2 contains 4 Gilbert-type fan deltas (gd1–gd4; parasequences 2C–2F; Fig. 5). The facies architecture of Gilbert deltas gd1 and gd2 reveals a systematic distribution of subenvironments, passing laterally from nonmarine and gravelly delta-plain facies in the south, through foreset and proximal bottomset facies in the central part of the area, to distal sandy bottomset deposits in the north (Fig. 5). In contrast, Gilbert delta gd3 was transported to the south or southeast (Fig. 9); it contains abundant transported shell material that was derived from carbonate shoals and reefs located on the eastern structural high to the north. This may have been a precursor to the major input of bioclastic material from the north and northwest, which dominated deposition of sequence 3 strata. Gilbert delta gd4 contains a subequal mix of bioclastic and siliciclastic material.

PALEOCURRENTS

Paleocurrents in the southeastern Loreto basin were measured using clast imbrications and restored primary dips on Gilbert-delta foresets (Fig. 9). Clast imbrications in sequence 1 indicate consistent transport toward the west. A substantial shift is seen at the boundary between sequences 1 and 2, and conglomerates in parasequences 2A and 2B show transport toward the northwest, north, and northeast (Fig. 9). In parasequences 2C and 2D, the paleocurrent data reveal a distinct clustering: Clast imbrications record transport to the north and northeast, and primary dips on foreset strata indicate transport to the east-southeast. Primary dips on foreset strata in parasequences 2E and 2F show transport toward the east, southeast, and south (Fig. 9). The paleocurrent data are interpreted and discussed below.

PALEOGEOGRAPHIC EVOLUTION

Paleogeographic maps (Fig. 10) depict our interpretation of the tectonic development of the southeastern Loreto basin during sequence 2 time. These interpretive maps were constructed using inferred sedimentary environments (Appendix 1), map relationships within and adjacent to the study area (Fig. 3; unpublished maps), architectural information from the stratigraphic panel (Fig. 5), and grouped paleocurrent data (Fig. 9). The overall geometry of the southeastern basin was controlled by its position between the northwest-striking Loreto fault and the southern termination of the eastern structural high, which represents the uplifted portion of the hanging wall of the Loreto fault (Figs. 1, 10). This resulted in a wedge-shaped embayment that opened to the east (Fig. 10). Tectonically related subsidence and uplift, occurring in different parts of the basin at different times, exerted the primary controls on sedimentation patterns. Autocyclic (internal) sedimentary processes, such as lateral channel migration and in situ carbonate production, also contributed to the devolution of environments and facies. During sequence 2 time there was little or no input of sediment from the eastern structural high (Fig. 10).

Deposition of parasequence 2A and the lower part of parasequence 2B took place in gravelly alluvial fans, braided streams, and shelf-type fan deltas (e.g., Ethridge and Wescott, 1984; Fig. 10A). Paleocurrent data (Fig. 9) indicate that these gravelly systems prograded northward into a siliciclastic marine setting, which is represented by the thin marine sandstone interval at the base of sequence 2 (Fig. 5). The progradational, coarsening-up nature of gravel-rich facies in the lower part of sequence 2 indicates that the shoreline was displaced away from source areas to the south and west and probably was situated somewhere to the east and southeast of the study area during parasequence 2A time (Fig. 10A).

By the end of parasequence 2B time, the southeastern Loreto embayment was flooded by a major marine transgression, resulting in deposition of shelly and pebbly siliciclastic sandstone in most of the basin (Figs. 5, 10B). Mixed carbonate and siliciclastic sediments were deposited on a small platform that formed on the southern flank of the eastern structural high. Because this narrow platform was sufficiently removed from a southern source of clastic material, marine organisms (molluscan fauna) were able to thrive, producing abundant bioclastic sediment that was mixed with siliciclastic sand derived from the south. Northerly transport of sediment continued during sequence 2B time (Fig. 9). Shell bed sb3 (Fig. 5) was deposited during a marine hiatus that represents the culmination of this transgression. The boundary between parasequences 2B and 2C (Fig. 5) marks an important transition from shelf-type fan deltas to Gilbert-type fan deltas. This transition, also recognized in the south-central Loreto basin, resulted from increased rates of basin subsidence (Dorsey et al., 1995). We assign a similar mechanism to the upward transition from shelf-type to Gilbert-type fan deltas seen in the southeastern Loreto basin.

During the remainder of sequence 2 time there were four episodes of progradation followed by transgressive drowning of successive gravelly Gilbert-type fan deltas (parasequences 2C–2F; Figs. 5, 10C). Fan deltas prograded toward the north and east, with the notable exception of parasequence 2E, which contains abundant transported shell material derived from carbonate shoals and reefs to the north (Fig. 9). These Gilbert-type fan deltas record a period of voluminous and rapid input of gravelly detritus from the footwall of the Loreto fault to the south and southwest, with almost no input from the hanging-wall tilt block to the north (Fig. 10C).

DISCUSSION

Paleocurrents

Paleocurrent data (Fig. 9), combined with paleogeographic reconstructions (Fig. 10), provide important insights into late Pliocene sediment dispersal patterns in the southeastern Loreto basin. The shift in transport directions between sequences 1 and 2 (Fig. 9) is especially interesting. Westerly directed paleocurrents in sequence 1 are similar to paleocurrent data collected from sequence 1 strata farther north along the western flank of uplifted Miocene volcanic rocks in the eastern structural high (Figs. 1, 10; Dorsey, unpublished data). It thus appears that during sequence 1 time, the uplifted portion of the hanging-wall tilt block projected southward into the study area and possibly far-

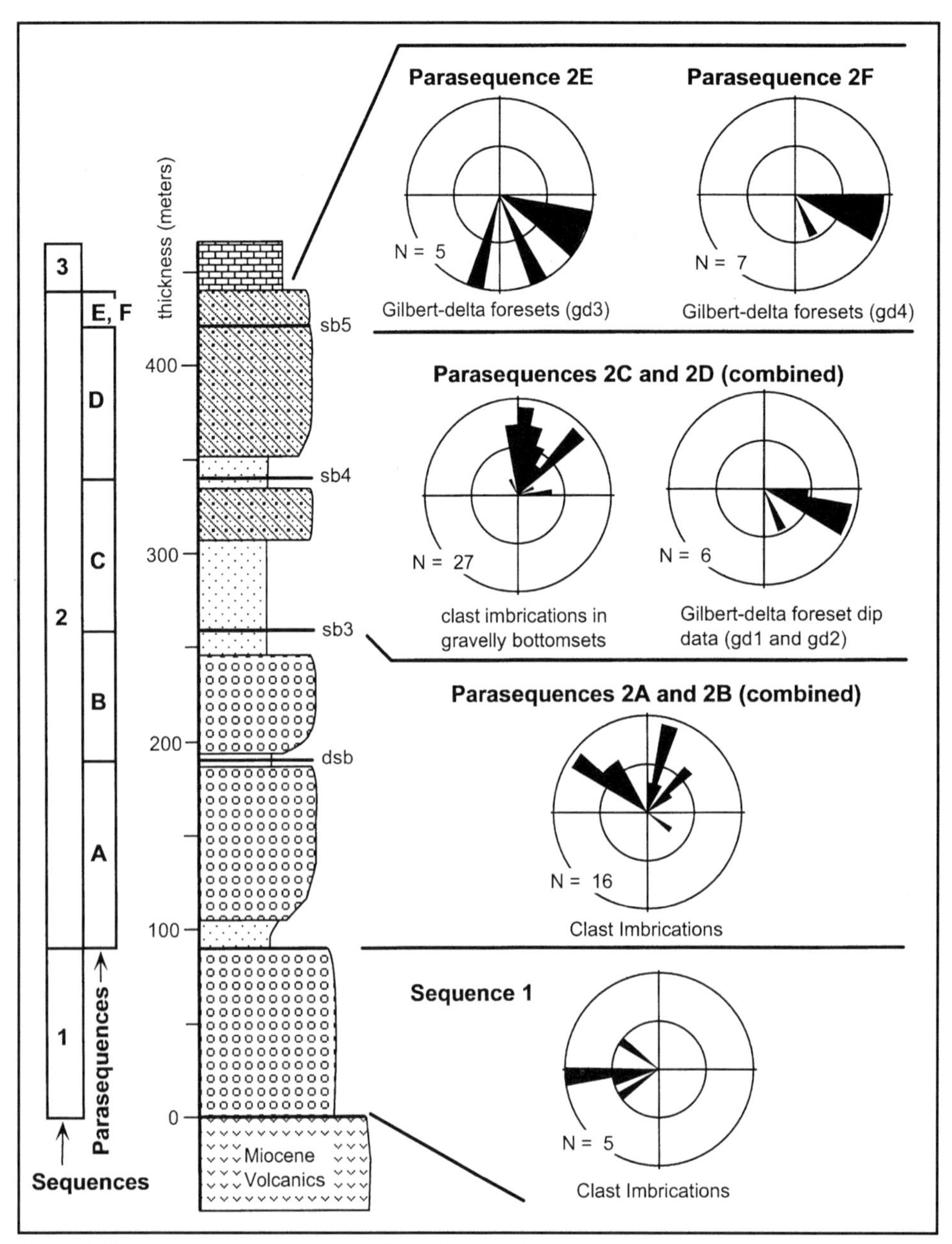

Figure 9. Paleocurrent data for sequence 2, southeastern Loreto basin. N represents the number of data points per rose diagram; the type of data is indicated for each plot. Note abrupt shift in paleocurrent directions between sequences 1 and 2 and the discrepancy between clast-imbrication data versus foreset-dip data in parasequences 2C and 2D. Patterns are the same as in Figure 5.

ther to the south, shedding gravelly detritus westward into the basin. The shift to sequence 2, and the associated abrupt shift in sediment dispersal patterns, records an important structural reorganization within the basin that caused the previously high-standing hanging-wall tilt block to become submerged. The initiation of this new basin configuration coincided with a widespread marine flooding event, which covered about 20 to 25 km^2 and produced the oldest known incursion of marine environments into the Loreto basin. We interpret these changes to record an abrupt subsidence event that was probably caused by an increase in rate of slip along the Loreto fault at the beginning of sequence 2 time.

Paleocurrents in parasequences 2C and 2D reveal an interesting discrepancy between clast imbrication data and primary dips on foreset strata (Fig. 9). North-directed transport recorded in bottomset clast imbrications is consistent with northward fining and transition to more distal facies that onlap onto Miocene volcanic basement rocks in the north (Figs. 5, 10). East- and southeast-directed paleocurrents, which are recorded in foreset dip data, apparently resulted from dispersal of sediment along the

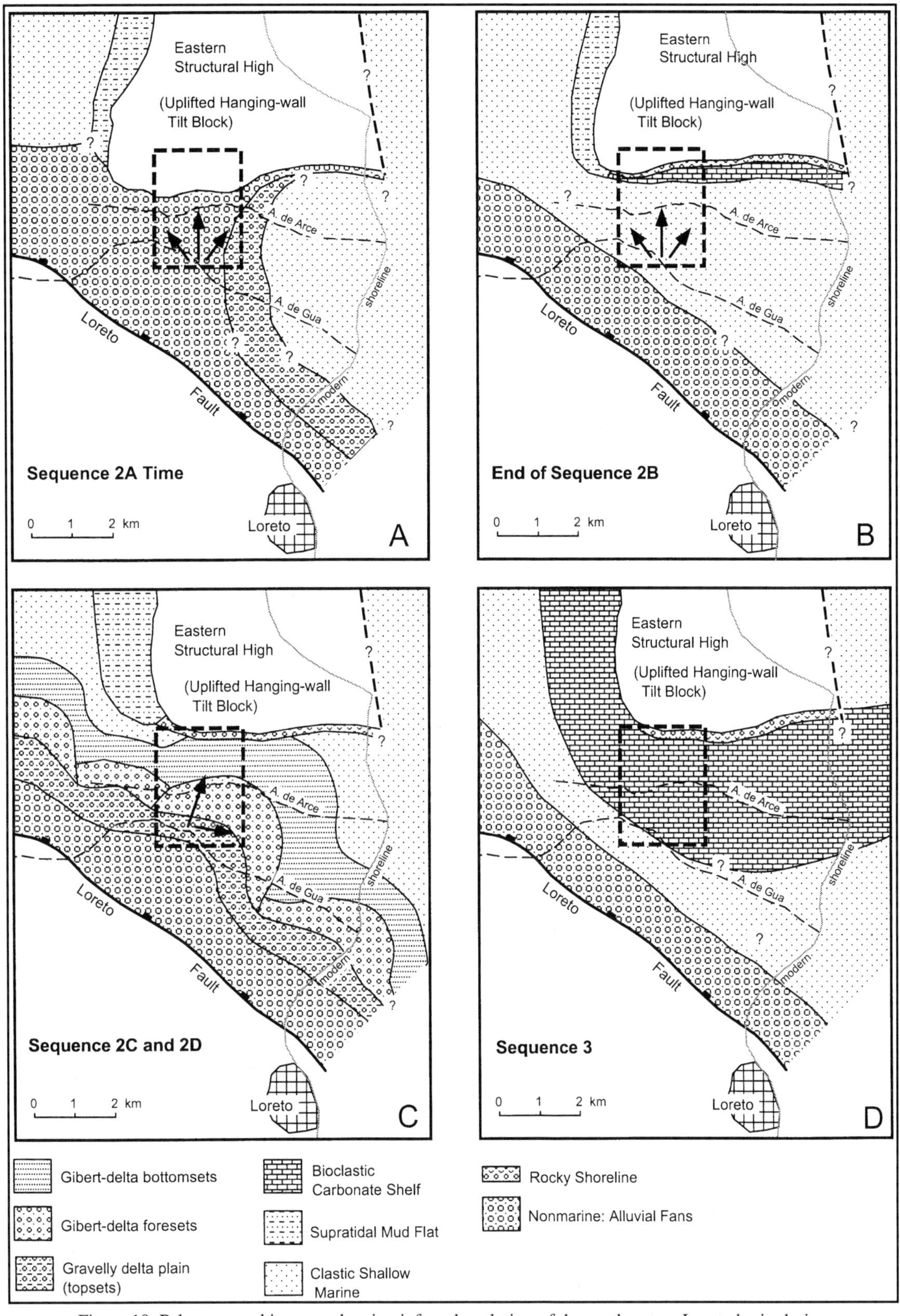

Figure 10. Paleogeographic maps showing inferred evolution of the southeastern Loreto basin during sequence 2 time. Dashed box shows location of study area. Arrows indicate overall paleocurrent directions, from Figure 9. See Figure 3 for detailed map distribution of parasequences. Dashed box indicates location of study area.

axis of the approximately east-west–trending marine embayment (Fig. 10). We infer that Gilbert-delta foresets prograded eastward along this embayment, whereas gravelly and sandy bottomsets were transported laterally to the north, east, and south, away from the core of advancing foreset units. In this interpretation, our data collection sampled only bottomsets that were transported northward toward the basin-bounding structural high.

Subsidence and sedimentation patterns

Following early basin reorganization, there was virtually no input of siliciclastic detritus from the uplifted hanging wall of the Loreto fault during most of sequence 2 time. The lack of detrital input, combined with onlapping of strata onto the buttress unconformity and consistently shallow water depths through time, requires that the southeastern basin experienced subsidence relative to the hanging-wall tilt block, thus producing accommodation space for sediment accumulation. Alternatively, if a high-standing mountain range had simply been buried by sediment during a progressive rise in relative sea level, there should have been substantial input of volcaniclastic detritus in the early stages when high topographic relief existed, but there is no evidence for this. This implies either progressive tilting toward the south or synbasinal offset along a fault between the southeastern basin and the eastern structural high. Excellent exposures of the buttress unconformity along the northern margin of the basin reveal that there is not a fault in this position, and southward tilting therefore seems required. However, available stratigraphic constraints reveal little or no thickening toward the south (Fig. 5), which would be expected for progressive southward tilting during deposition of sequence 2. We infer that some southward tilting did take place during sequence 2 time and that stratal thickening would be observed if the panel in Figure 5 could be extended farther south. This represents a subtle component of pronounced southwestward tilting that is known to have occurred during rapid asymmetric subsidence in sequence 2 time (Umhoefer et al., 1994; Dorsey et al., 1995).

Taking into account the thickness of sequence 2 (~350 m) and a reasonable range of values for the duration of deposition (0.1 to 0.2 m.y.; Fig. 2), we estimate the average rate of subsidence and sediment accumulation to be 1.8 to 3.5 mm/yr for sequence 2. This is only a rough estimate. Data from the south-central Loreto basin indicate that the subsidence rate increased dramatically early in sequence 2 time (Dorsey et al., 1995).

Origin of parasequence cyclicity

The origin of cyclicity seen in parasequences of the southeastern Loreto basin is not well understood. In other tectonically active basins, repeated vertical stacking of marine Gilbert-type fan deltas has been attributed to episodic movements along basin-bounding faults; this can abruptly terminate progradation of foresets, drown the deltaic plain, and provide accommodation space for progradation and infilling by the next Gilbert delta (e.g., Colella, 1988). In these models, the back edge of the foreset unit *is* the fault or fault zone that controls episodic basin-margin subsidence events. It is difficult to apply this model to the southeastern Loreto basin, because foreset packages pass laterally upstream into gravelly delta-plain, braid-stream, and nonmarine alluvial-fan deposits, with no evidence for fault displacements at their back edge (Fig. 5). Other possible controls on Gilbert-delta cyclicity include steady (constant rate) background subsidence overprinted by high-frequency eustatic sea-level fluctuations or steady subsidence combined with autocyclic channel avulsions and associated delta lobe switching (Dorsey et al., 1995). Our study of the southeastern Loreto basin has not provided any new insights into this problem, which therefore remains unresolved.

End of sequence 2

A second major basin reorganization is recorded in the abrupt transition from sequence 2 to sequence 3 (Figs. 5, 10). Large volumes of bioclastic material in sequence 3 were derived from the eastern structural high, the tilting history of which was directly controlled by oblique, dextral-normal displacement along the Loreto fault (Figs. 1, 2; Umhoefer et al., 1994). This stratigraphic shift is also recognized in the south-central Loreto basin on the western flank of the eastern structural high (Fig. 1), where transported bioclastic carbonates of sequence 3 abruptly overlie siliciclastic sandstone and conglomerate of sequence 2 (Dorsey and Umhoefer, unpublished field data). The onset of sequence 3 deposition thus provides a basinwide record of renewed uplift and erosion in the hanging-wall tilt block of the Loreto fault, which followed a period (sequence 2) when this structural high was mostly at or below sea level. Transported bioclastic material in sequence 3 was derived from carbonate reefs and shoals that covered much of the shallowly submerged eastern structural high. This abrupt change in structural behavior was probably controlled by changes in the rate and direction of movement along the Loreto fault, combined with initiation of a new normal fault on the east side of the eastern structural high (Dorsey and Umhoefer, 1996). Very low angle truncations observed at the base of sequence 3 suggest that slight tilting occurred within the southeastern basin during this change in structural behavior.

Modern analog for the Loreto basin

Our study of Pliocene strata in the southeastern Loreto basin can be compared with recent detailed studies of the modern Bahía Concepción, which is located about 75 km north of the town of Loreto (Hayes et al., 1993; Meldahl et al., this volume). Bahía Concepción is a shallow-marine bay that formed by drowning of an alluvial rift basin during the Holocene transgression (Meldahl et al., this volume). It therefore resembles the greater Loreto basin (Fig. 1) as it existed shortly after the earliest marine transgression, as represented by the thin marine sandstone unit at the base of sequence 2 (Fig. 5). The orientation of the two basins is different, as is their structural asymmetry: The Loreto basin has its bounding fault and thickest section on the southwest side (Fig. 1; Dorsey et al., 1995), whereas Bahía Concepción has the youngest fault and deepest water on the east side (Meldahl et al., this volume).

In spite of the above structural differences and the greater size of Bahía Concepción, some notable similarities in sedimentary lithofacies and depositional environments are found in Bahía Concepción and the southeastern Loreto basin (this study). First, volcaniclastic sand and gravel in both basins are (or were) transported into the marine realm via prograding alluvial fans and fan deltas. It appears, however, that wave and current reworking of sand and gravel in gravelly delta-front environments is more vigorous in the modern Bahía Concepción than it was in the Pliocene Loreto basin. This is probably because Bahía Concepción faces north and is open to northerly winds that sweep down the Gulf of California, whereas the Pliocene Loreto basin was mostly protected from northerly winds by structural barriers. Second, in Bahía Concepción carbonate sands and mixed carbonate-siliciclastic sands are found in shallow coastal areas along the steep western margin of the basin, which makes up steep topographic headlands along the structurally inactive (?) side of the basin (Meldahl et al., this volume). These facies resemble calcarenite and mixed carbonate-siliciclastic facies that accumulated on the steep northern margin of the southeastern Loreto basin, geographically removed from the source of active detrital input, during late parasequence 2B and parasequence 3 time (Figs. 5, 10B). Third, Pliocene rocky shoreline deposits in the southeast Loreto basin bear strong resemblance to deposits of modern rocky shore environments documented from Bahía Concepción (Hayes et al., 1993), indicating that similar processes of shoreline development are (and were) active in both basins. These and other similarities suggest that tectonically active basins along the eastern margin of the Baja California peninsula have exhibited similar aspects of sedimentation from late Pliocene time to the present. Future detailed comparisons between Pliocene and modern sedimentary basins in Baja California Sur will be important for better understanding the tectonic evolution of this active continental margin.

CONCLUSIONS

Pliocene strata exposed in the southeastern Loreto basin were deposited in a wedge-shaped, east-west trending structural embayment that was bounded by the active Loreto fault to the southwest and the southern termination of the eastern structural high (uplifted portion of the hanging-wall tilt block) to the north. During deposition of sequence 2 (~2.6 to 2.4 Ma), this subbasin experienced sedimentation in marine, nonmarine, gravelly fan-delta, and rocky shoreline environments, each of which is represented by unique sedimentary lithofacies and facies associations. This embayment and many of the observed lithofacies can be compared to a modern analog in the present-day Bahía Concepción to the north. In spite of numerous normal and oblique-slip faults that cut the sequence, the presence of marker shell beds and unique stratigraphic trends makes it possible to correlate strata across all faults.

A detailed north-south stratigraphic panel reveals complex lateral and vertical facies relationships in sequence 2. The stratigraphic architecture is the product of (1) voluminous input of sandy and gravelly detritus from the footwall of the Loreto fault to the south and west, (2) lack of detrital input from the eastern structural high to the north, and (3) evolving patterns of subsidence and uplift, which controlled the lateral shifting of depositional environments through time. Strata of sequence 2 are divided into six parasequences that are defined by laterally continuous shell beds. These shell beds record transgressive marine hiatuses during which input of siliciclastic detritus virtually ceased. Possible mechanisms that may have produced these hiatuses include tectonic, eustatic, and autocyclic processes.

Integrative stratigraphic analysis provides insights into the evolution of tectonically controlled paleogeography and sediment-dispersal patterns during sequence 2 time. The earliest marine transgression into the Loreto basin occurred at the beginning of sequence 2 time and was accompanied by a pronounced shift in paleotransport directions. This event, which was probably related to changing fault kinematic behavior of the Loreto fault, caused the eastern structural high to become mostly submerged, effectively cutting off all input of siliciclastic detritus from the hanging-wall tilt block. Following initial transgression, alluvial fans and shelf-type fan deltas prograded into the basin from the footwall of the Loreto fault to the south and west. Following a second marine transgression, Gilbert-type fan deltas prograded northward and eastward into the basin; this produced four Gilbert-delta parasequences with a vertical stacking pattern that is characteristic of rapidly subsiding, tectonically active basins.

At the end of sequence 2 time there was an abrupt shift to widespread deposition of bioclastic and mixed carbonate-siliciclastic sediments. This records a second structural reorganization, in which the eastern structural high experienced renewed uplift and erosion of molluscan carbonate reefs and shoals to the north. The physical stratigraphy and depositional history of the southeastern Loreto subbasin thus provide detailed information about subsidence and uplift patterns within and around the margins of the Loreto basin during a period of active faulting on the Loreto fault.

ACKNOWLEDGMENTS

Research presented in this paper was supported by grants from the National Science Foundation (EAR-9117269 and EAR-9296255) and Northern Arizona University. We thank Keith Meldahl, Jorge Ledesma-Vázquez, Markes Johnson, and Sue Kidwell for insightful conversations and the generous exchange of ideas. Thanks go to Peter Falk for his able assistance in the field. Cathy Busby and Joann Stock provided helpful reviews of the manuscript. Markes Johnson and Jorge Ledesma-Vázquez are thanked for inspiring and editing this volume.

APPENDIX 1.

Measured sections collected in the southeastern Loreto basin.

See Figures 3 and 4 for map and stratigraphic position of sections. See main text and Figures 5 - 7 for descriptions and interpretations of sedimentary lithofacies. All thicknesses are in meters. gd-1 to gd-4 refers to Gilbert deltas 1 through 4 (Fig. 4). All sections are within sequence 2, except where noted on left side of graphic log.

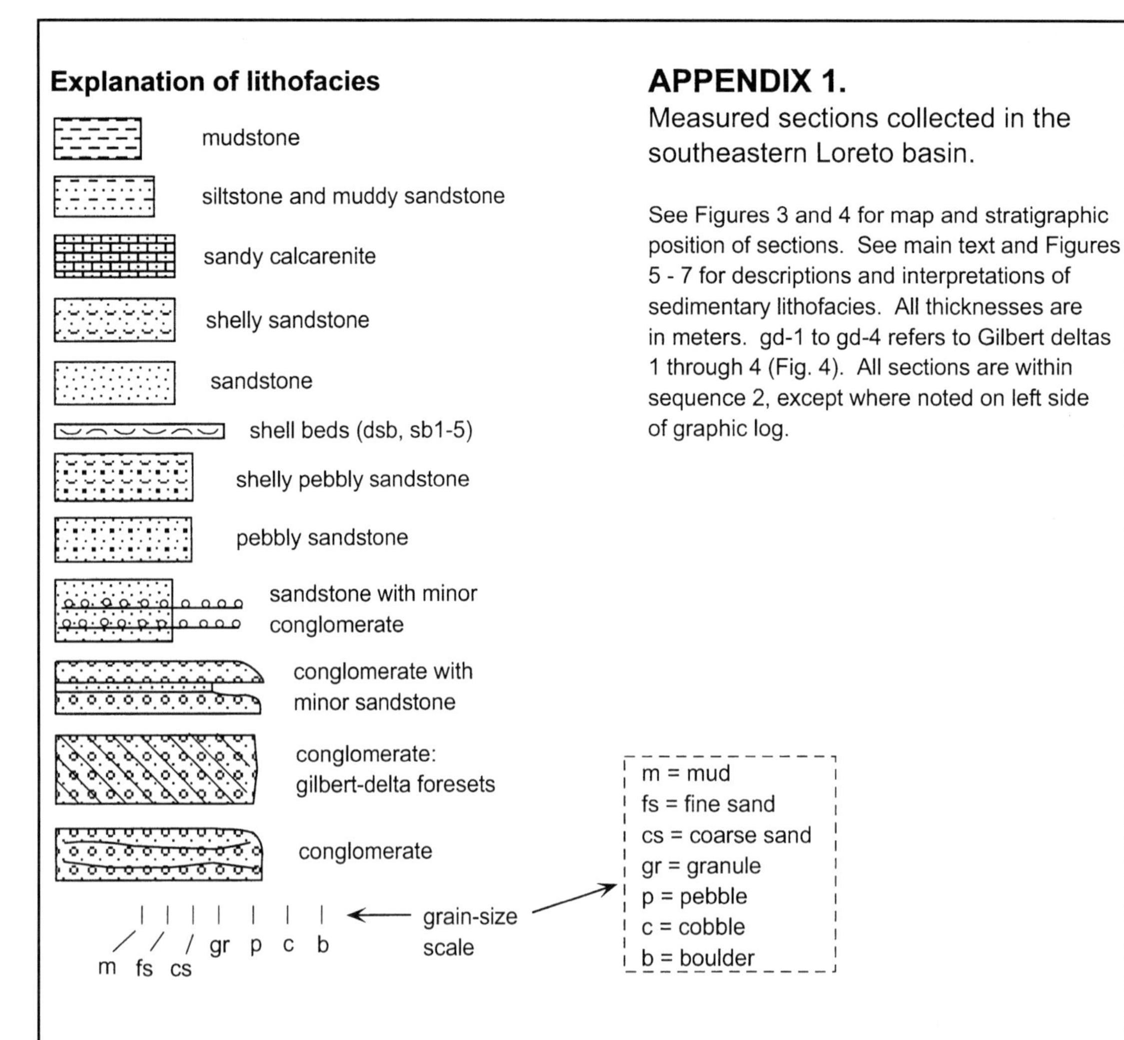

Explanation of sedimentary structures (strx) and accessory materials (access.)

gd-a gilbert delta - a

sb3 shell bed 3

dsb double shell bed

calcarenite component, minor

shell hash component, minor

disarticulated shells, usually broken and transported, dom. pectens and oysters

articulated shells, usually in situ, dom. pectens and oysters

sand dollars

bioturbation

p paleosol

concentrated plant matter

calcite nodules (caliche)

root casts

gypsum

gilbert-delta foresets

planar stratification

trough cross-bedding

low-angle, tabular cross-bedding

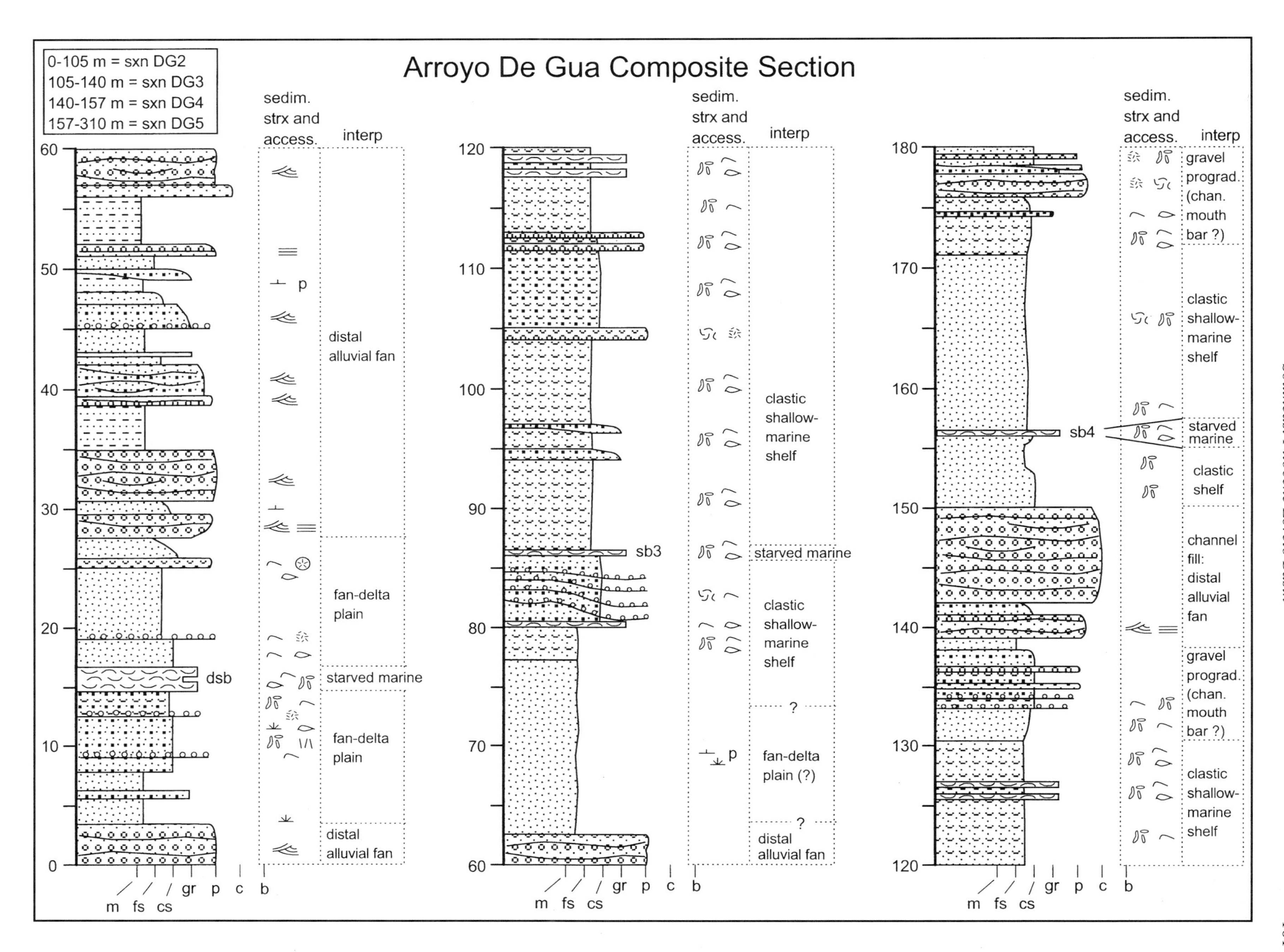

Arroyo De Gua Composite Section
0-105 m = sxn DG2
105-140 m = sxn DG3
140-157 m = sxn DG4
157-310 m = sxn DG5
sedim. strx and access.
interp
distal alluvial fan
fan-delta plain
starved marine
fan-delta plain
distal alluvial fan
dsb
m fs cs gr p c b
clastic shallow-marine shelf
starved marine
clastic shallow-marine shelf
?
fan-delta plain (?)
?
distal alluvial fan
sb3
gravel prograd. (chan. mouth bar ?)
clastic shallow-marine shelf
starved marine
clastic shelf
channel fill: distal alluvial fan
gravel prograd. (chan. mouth bar ?)
clastic shallow-marine shelf
sb4

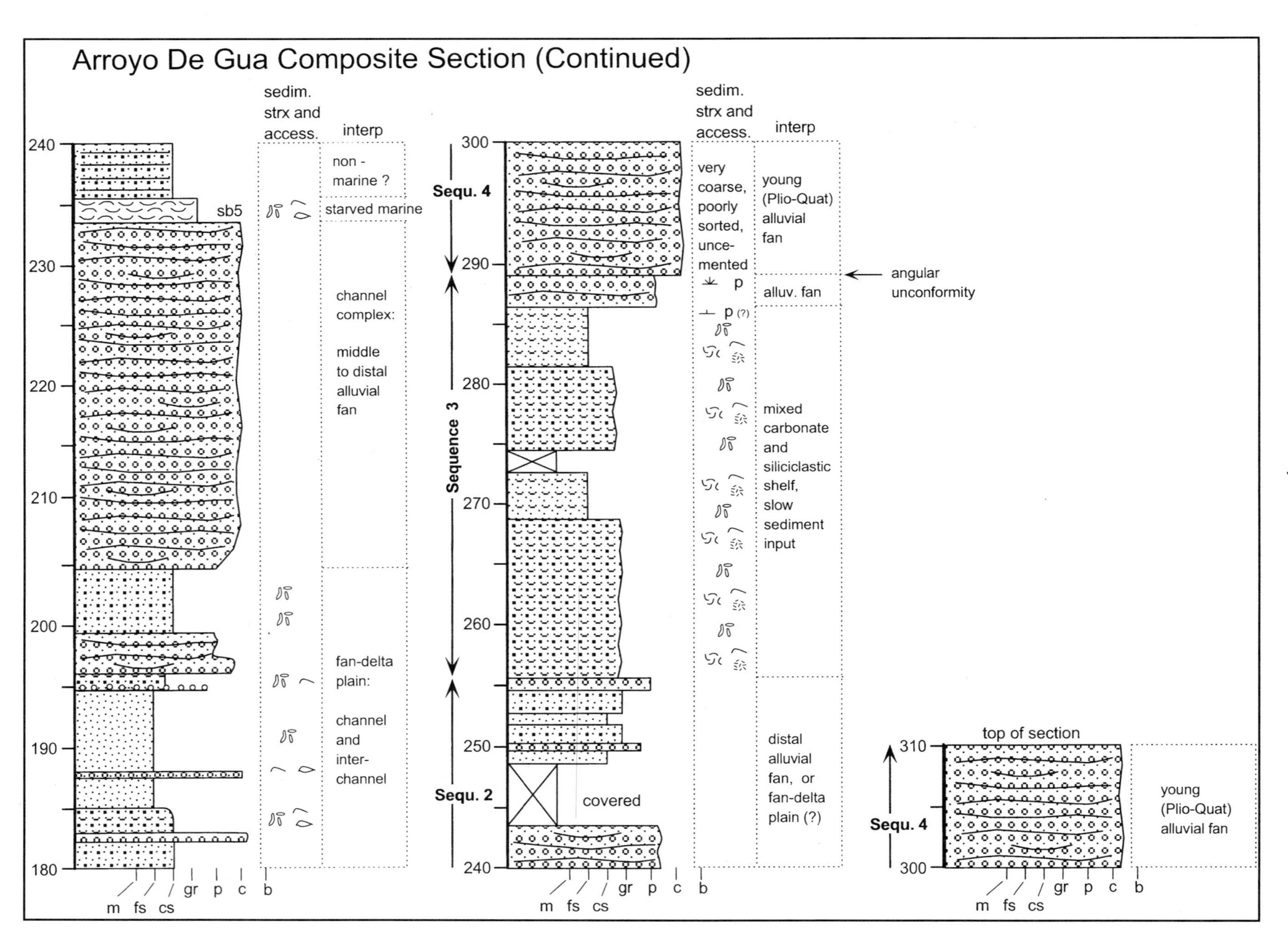

Arroyo De Gua Composite Section (Continued)
sedim. strx and access.
interp
240
230
220
210
200
190
180
sb5
non - marine ?
starved marine
channel complex:
middle to distal alluvial fan
fan-delta plain:
channel and inter-channel
m fs cs gr p c b
300
290
280
270
260
250
240
Sequ. 4
Sequence 3
Sequ. 2
covered
very coarse, poorly sorted, unce-mented
p
p (?)
young (Plio-Quat) alluvial fan
alluv. fan
angular unconformity
mixed carbonate and siliciclastic shelf, slow sediment input
distal alluvial fan, or fan-delta plain (?)
top of section
310
300
Sequ. 4
young (Plio-Quat) alluvial fan

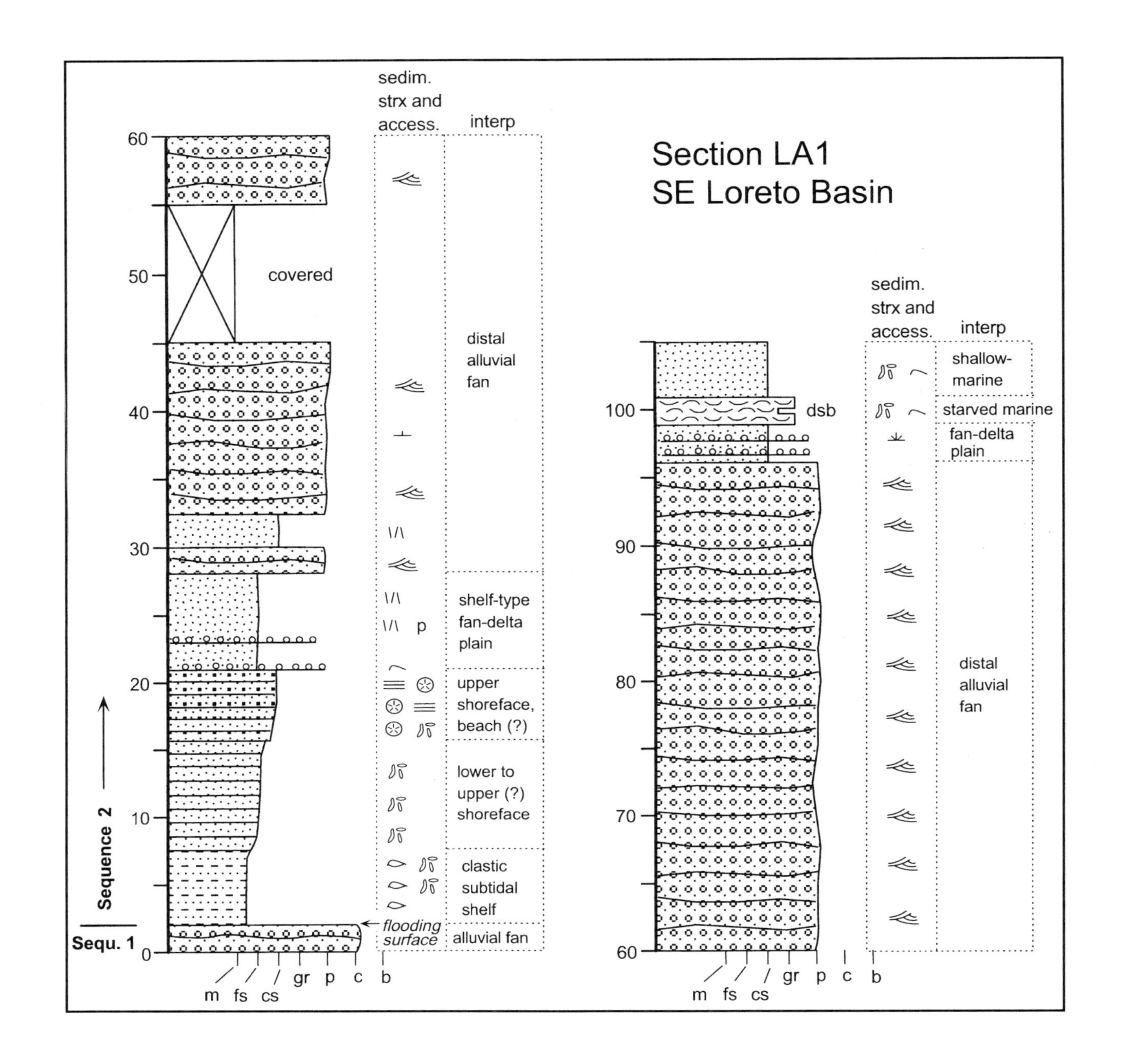

Section LA1
SE Loreto Basin
sedim.
strx and
access.
interp
60
50
40
30
20
10
0
covered
distal
alluvial
fan
p
shelf-type
fan-delta
plain
upper
shoreface,
beach (?)
lower to
upper (?)
shoreface
clastic
subtidal
shelf
flooding
surface
alluvial fan
Sequence 2
Sequ. 1
m fs cs gr p c b
100
90
80
70
60
dsb
shallow-
marine
starved marine
fan-delta
plain
distal
alluvial
fan
m fs cs gr p c b

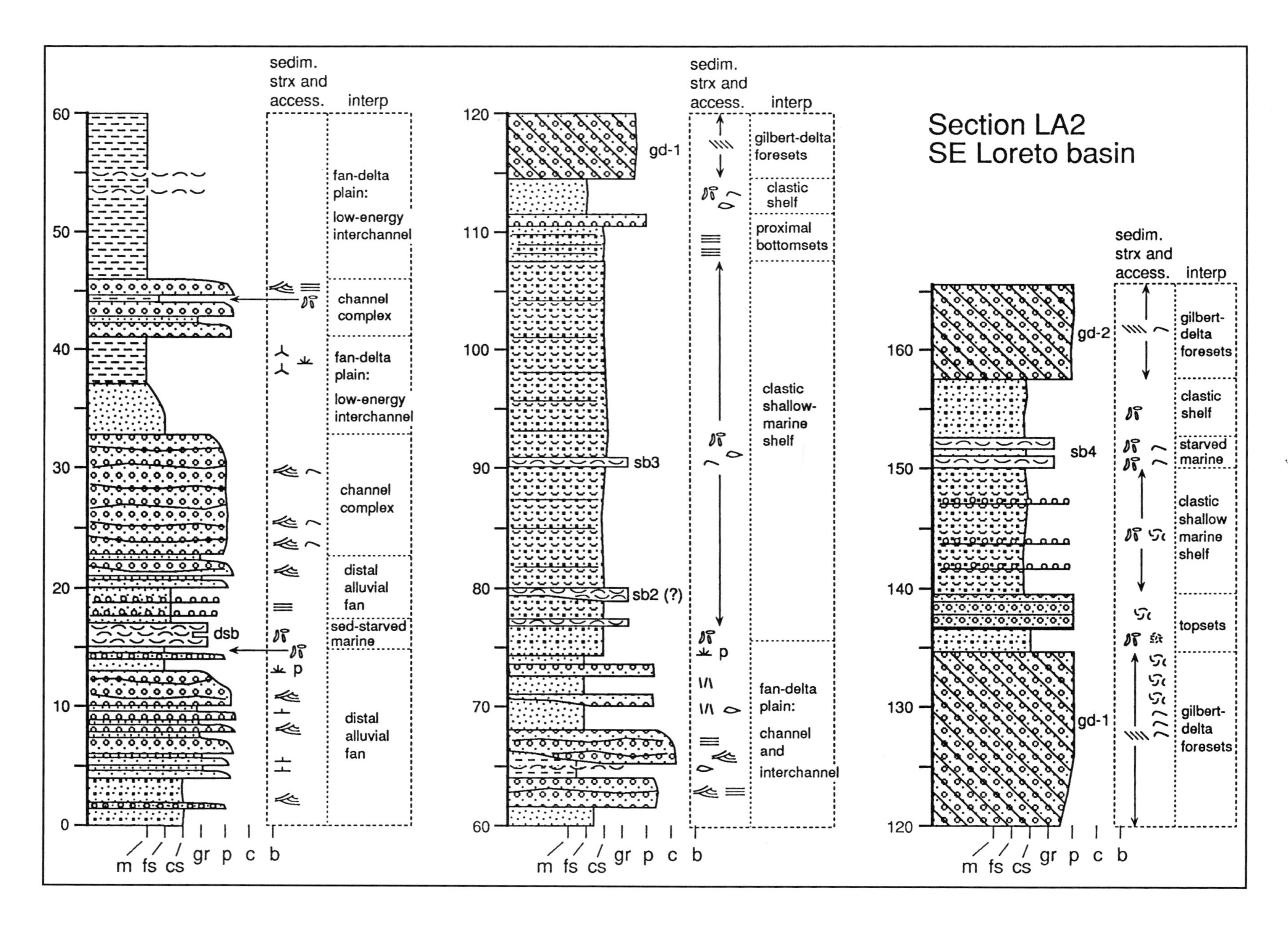

Section LA2
SE Loreto basin
sedim. strx and access.
interp
fan-delta plain: low-energy interchannel
channel complex
fan-delta plain: low-energy interchannel
channel complex
distal alluvial fan
sed-starved marine
distal alluvial fan
dsb
p
m fs cs gr p c b
gilbert-delta foresets
clastic shelf
proximal bottomsets
clastic shallow-marine shelf
fan-delta plain: channel and interchannel
gd-1
sb3
sb2 (?)
gilbert-delta foresets
clastic shelf
starved marine
clastic shallow marine shelf
topsets
gilbert-delta foresets
gd-2
sb4
0
10
20
30
40
50
60
70
80
90
100
110
120
130
140
150
160

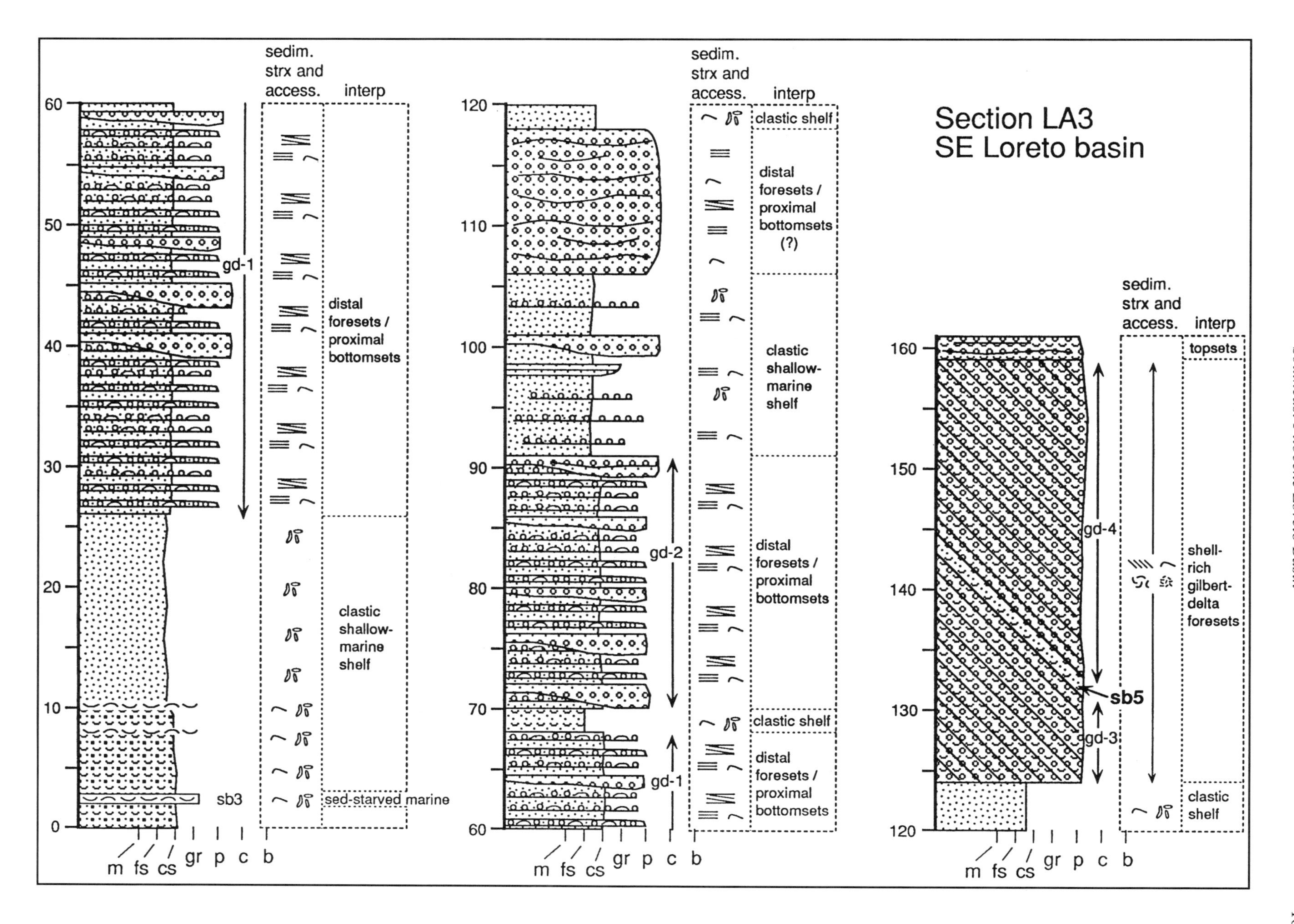
Section LA3
SE Loreto basin
sedim. strx and access.
interp
60
50
40
30
20
10
0
gd-1
sb3
distal foresets / proximal bottomsets
clastic shallow-marine shelf
sed-starved marine
m fs cs gr p c b
120
110
100
90
80
70
60
gd-2
gd-1
clastic shelf
distal foresets / proximal bottomsets (?)
clastic shallow-marine shelf
distal foresets / proximal bottomsets
clastic shelf
distal foresets / proximal bottomsets
m fs cs gr p c b
160
150
140
130
120
gd-4
sb5
gd-3
topsets
shell-rich gilbert-delta foresets
clastic shelf
m fs cs gr p c b

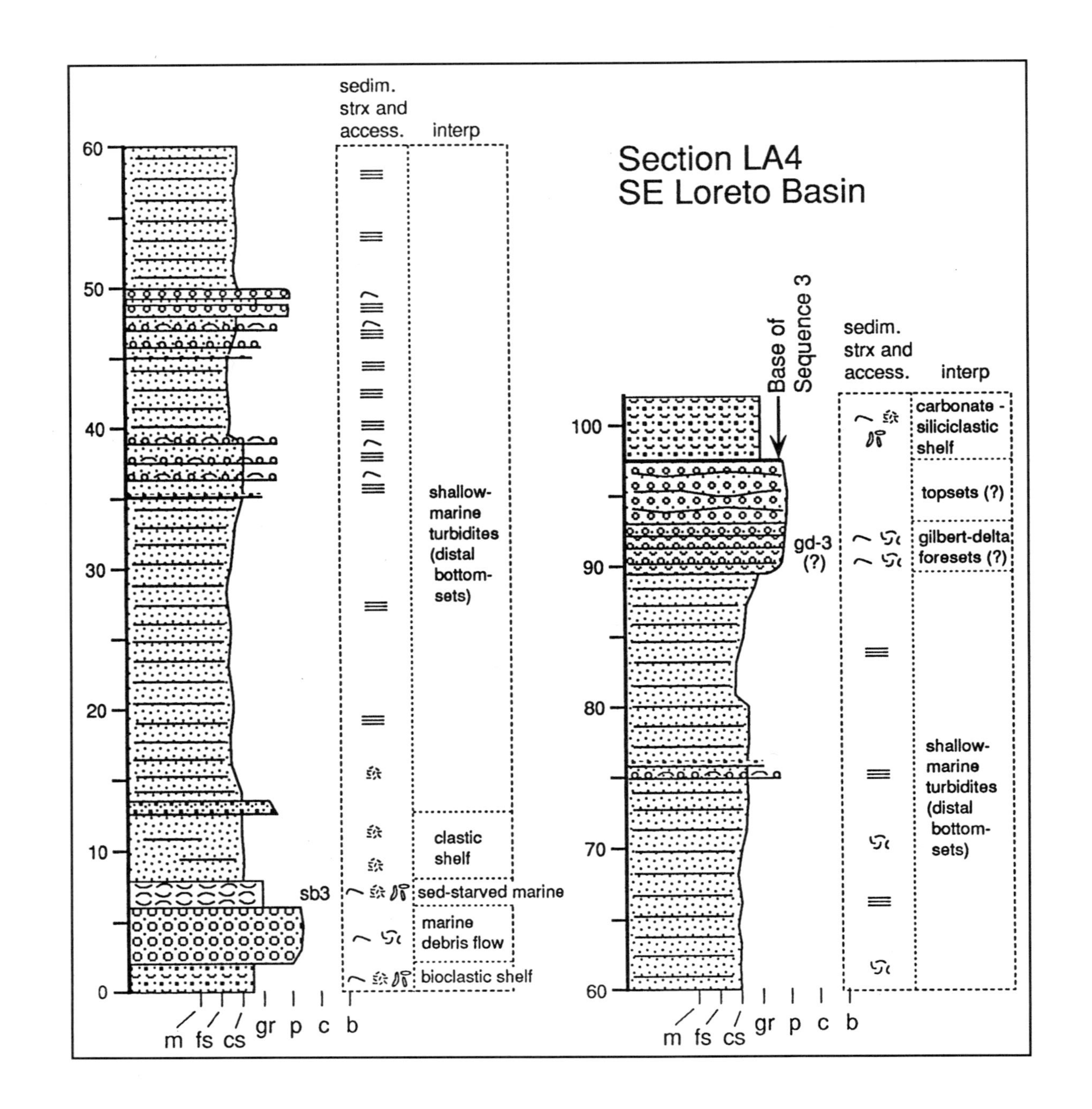
Section LA4
SE Loreto Basin
sedim.
strx and
access.
interp
60
50
40
30
20
10
0
shallow-marine turbidites (distal bottom-sets)
clastic shelf
sb3
sed-starved marine
marine debris flow
bioclastic shelf
m fs cs gr p c b
Base of Sequence 3
sedim.
strx and
access.
interp
100
90
80
70
60
carbonate - siliciclastic shelf
topsets (?)
gd-3 (?)
gilbert-delta foresets (?)
shallow-marine turbidites (distal bottom-sets)
m fs cs gr p c b

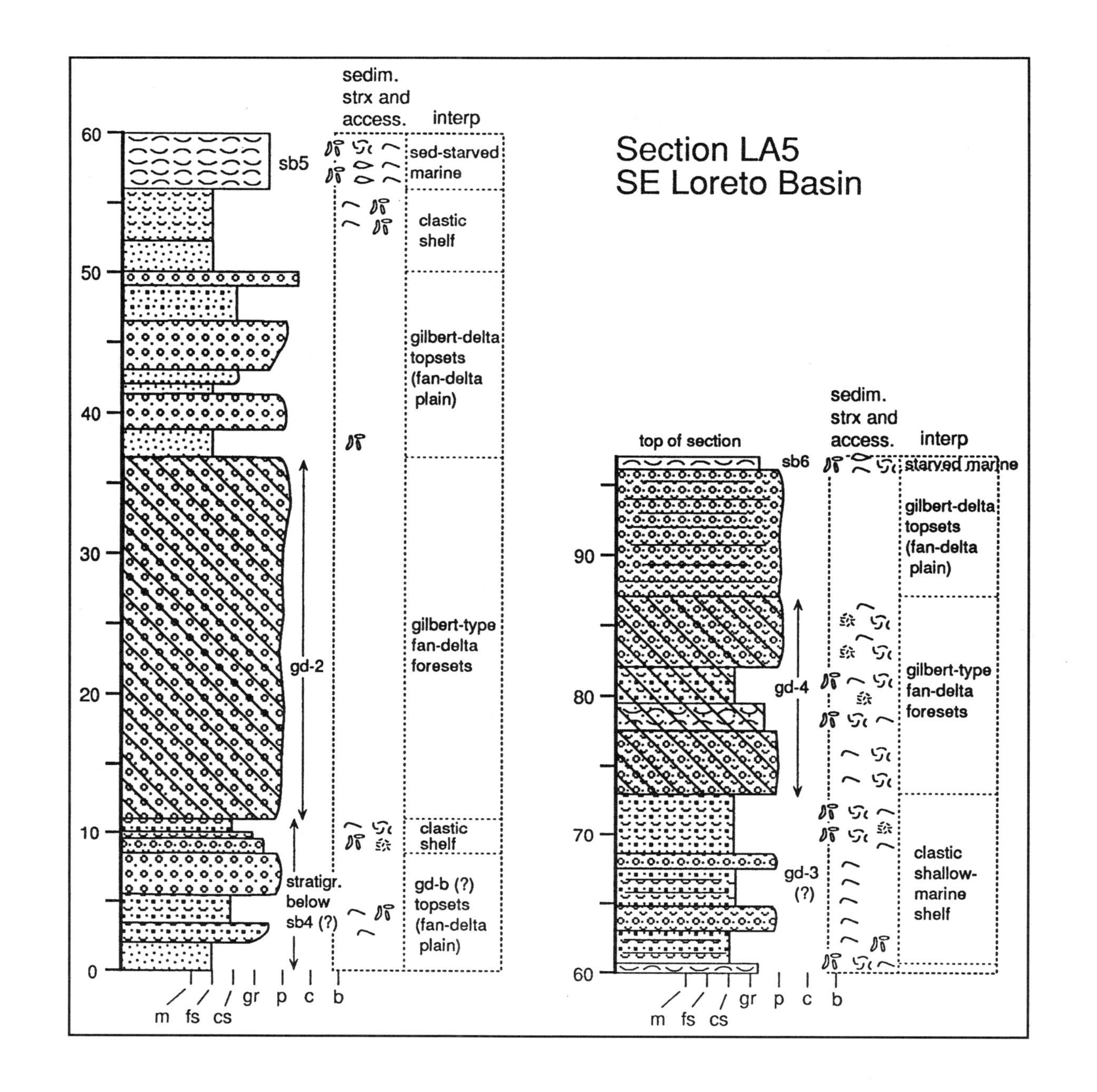
Section LA5
SE Loreto Basin
sedim. strx and access.
interp
sed-starved marine
clastic shelf
gilbert-delta topsets (fan-delta plain)
gilbert-type fan-delta foresets
clastic shelf
gd-b (?) topsets (fan-delta plain)
sb5
gd-2
stratigr. below sb4 (?)
m fs cs gr p c b
0
10
20
30
40
50
60
top of section
sb6
gd-4
gd-3 (?)
starved marine
gilbert-delta topsets (fan-delta plain)
gilbert-type fan-delta foresets
clastic shallow-marine shelf
m fs cs gr p c b
60
70
80
90

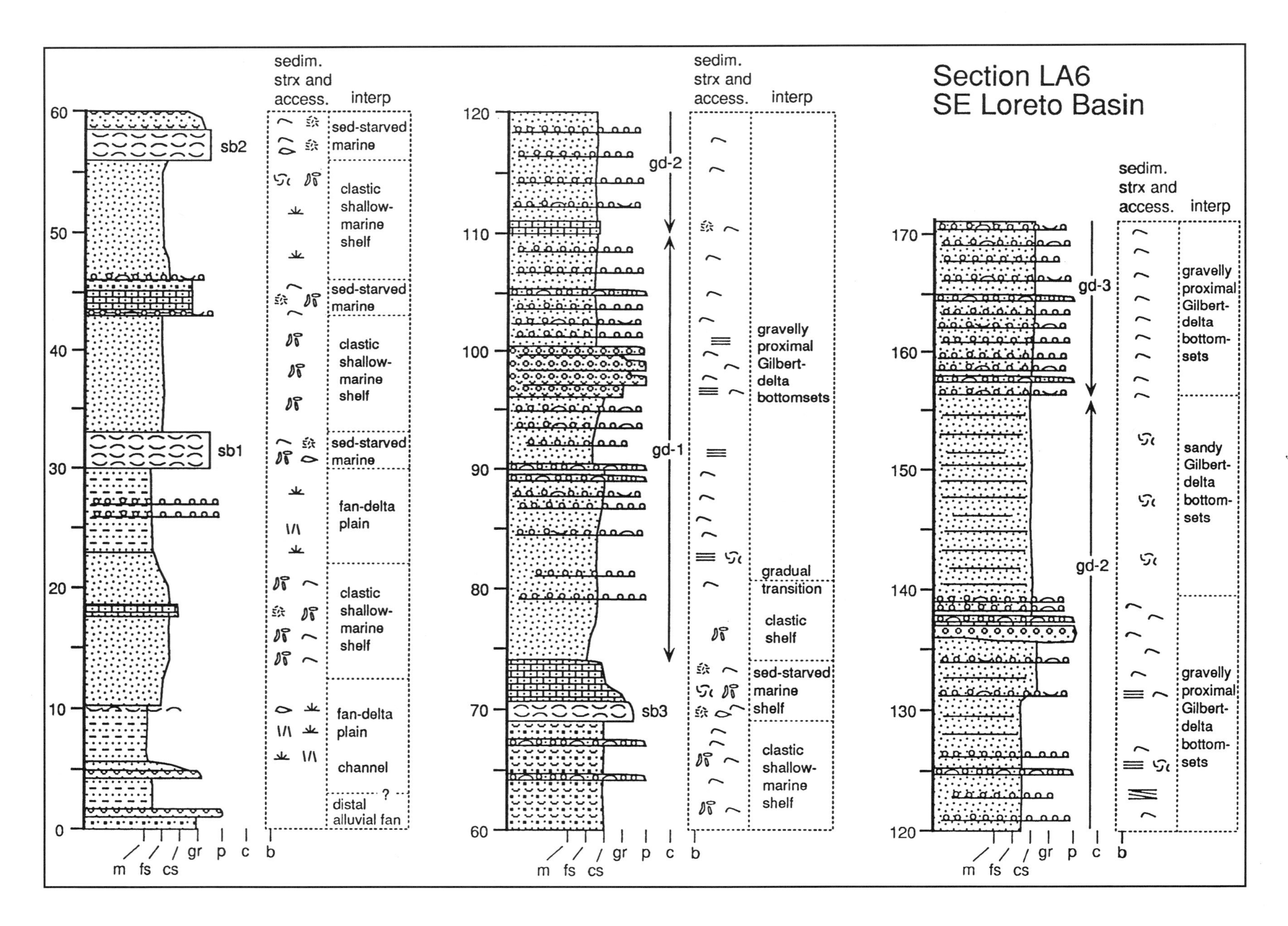
Section LA6
SE Loreto Basin
sedim. strx and access.
interp
sed-starved marine
clastic shallow-marine shelf
sed-starved marine
clastic shallow-marine shelf
sed-starved marine
fan-delta plain
clastic shallow-marine shelf
fan-delta plain
channel
?
distal alluvial fan
sb2
sb1
60
50
40
30
20
10
0
m fs cs gr p c b
gd-2
gd-1
sb3
gravelly proximal Gilbert-delta bottomsets
gradual transition
clastic shelf
sed-starved marine shelf
clastic shallow-marine shelf
120
110
100
90
80
70
60
gd-3
gd-2
gravelly proximal Gilbert-delta bottom-sets
sandy Gilbert-delta bottom-sets
gravelly proximal Gilbert-delta bottom-sets
170
160
150
140
130
120

REFERENCES CITED

Angelier, J., Colletta, B., Chorowicz, J., Ortlieb, L., and Rangin, C., 1981, Fault tectonics of the Baja California Peninsula and the opening of the Sea of Cortez, Mexico: Journal of Structural Geology, v. 3, p. 347–357.

Arnott, R. W. C., 1995, The parasequence definition—Are transgressive deposits inadequately addressed?: Journal of Sedimentary Research, v. B65, p. 1–6.

Burchette, T. P., 1988, Tectonic control on carbonate platform facies distribution and sequence development: Miocene Gulf of Suez: Sedimentary Geology, v. 59, p. 179–204.

Colella, A., 1988, Fault-controlled marine Gilbert-type fan deltas: Geology, v. 16, p. 1030–1034.

Colella, A., and Prior, D. B., eds., 1990, Coarse-grained deltas: International Association of Sedimentologists Special Publication 10, 357 p.

Dorsey, R. J., Umhoefer, P. J., and Renne, P. R., 1995, Rapid subsidence and stacked Gilbert-type fan deltas, Pliocene Loreto basin, Baja California Sur, Mexico: Sedimentary Geology, v. 98, p. 181–204.

Dorsey, R. J., and Umhoefer, P. J., 1996, Sequence stratigraphy and structural evolution of the Pliocene Loreto basin, Baja California Sur, Mexico: American Association of Petroleum Geologists 1996 Annual Convention Abstracts with Program, p. A38.

Durham, J. W., 1950, 1940 E.W. Scripps cruise to the Gulf of California. Part II: Megascopic paleontology and marine stratigraphy: Geological Society of America Memoir 43, 216 p.

Elliott, T., 1986, Deltas, *in* Reading, H. G., ed., Sedimentary environments and facies (second edition): Oxford, Blackwell Scientific, p. 113–154.

Ethridge, F. G., and Wescott, W. A., 1984, Tectonic setting, recognition and hydrocarbon potential of fan-delta deposits, *in* Koster, E. H., and Steel, R. J., eds., Sedimentology of gravels and conglomerates: Canadian Society of Petroleum Geologists Memoir 10, p. 217–235.

Harris, P. M., and Frost, S. H., 1984, Middle Cretaceous carbonate reservoirs, Fahud Field and northwestern Oman: American Association of Petroleum Geologists Bulletin, v. 68, p. 649–658.

Hayes, M. L., Johnson, M. E., and Fox, W. T., 1993, Rocky-shore biotic associations and their fossilization potential: Isla Requeson (Baja California Sur, Mexico): Journal of Coastal Research, v. 9, p. 944–957.

Johnson, M. E., 1988, Why are ancient rocky shorelines so uncommon?: Journal of Geology, v. 96, p. 469–480.

Johnson, M. E., 1992, Studies on ancient rocky shores: A brief history and annotated bibliography: Journal of Coastal Research, v. 8, p. 797–812.

Johnson, M. E., and Hayes, M. L., 1993, Dichotomous facies on a Late Cretaceous rocky island as related to wind and wave patterns (Baja California, Mexico): Palaios, v. 8, p. 385–395.

Kidwell, S. M., 1988, Taphonomic comparison of passive and active continental margins: Neogene shell beds of the Atlantic coastal plain and northern Gulf of California: Palaeogeogeography, Palaeoclimatology, Palaeoeceanography, v. 63, p. 201–223.

Kidwell, S. M., 1991, The stratigraphy of shell concentrations, *in* Allison, P. A., and Briggs, D. E. G., eds., Releasing the data locked in the fossil record: Topics in Geobiology, v. 9: New York, Plenum Press, p. 211–290.

Leeder, M. R., and Gawthorpe, R. L., 1987, Sedimentary models for extensional tilt-block/half-graben basins, *in* Coward, M. P, Dewey, J. F., and Hancock, P. L., eds., Continental extensional tectonics: Geological Society of London Special Publication 28, p. 139–152.

Lowe, D. R., 1982, Sediment gravity flows. II: Depositional models with special reference to the deposits of high-density turbidity currents: Journal of Sedimentary Petrology, v. 52, p. 279–297.

McLean, H., 1988, Reconnaissance geologic map of the Loreto and part of the San Javier quadrangles, Baja California Sur, Mexico: U.S. Geological Survey Map MF-2000, 1:50,000 scale.

McLean, H., 1989, Reconnaissance geology of a Pliocene marine embayment near Loreto, Baja California Sur, Mexico: Pacific Section, Society of Economic Paleontologists and Mineralogists, Field Trip Guidebook, v. 63, p. 17–25.

Moore, D. G., and and Buffington, E. C., 1968, Transform faulting and growth of the Gulf of California since late Pliocene: Science, v. 161, p. 1238–1241.

Nemec, W., 1990, Aspects of sediment movement on steep delta slopes, *in* Colella, A., and Prior, D. B., eds., Coarse-grained deltas: International Association of Sedimentologists Special Publication 10, p. 29–73.

Nemec, W., and Steel, R. J., 1984, Alluvial and coastal conglomerates: Their significant features and some comments on gravelly mass-flow deposits, *in* Koster, E. H., and Steel, R. J., eds., Sedimentology of gravels and conglomerates: Canadian Society of Petroleum Geologists Memoir 10, p. 1–31.

Nemec, W., and Steel, R. J., eds., 1988, Fan deltas: Sedimentology and tectonic settings: London, Blackie and Son, 444 p.

Smith, J. T., 1991, Cenozoic marine mollusks and paleogeography of the Gulf of California, *in* Dauphin, J. P., and Simoneit B. R. T, eds., The Gulf and Peninsular Province of the Californias: American Association of Petroleum Geologists Memoir 47, p. 637–666.

Stock, J. M., and Hodges, K. V., 1989, Pre-Pliocene extension around the Gulf of California and the transfer of Baja California to the Pacific plate: Tectonics, v. 8, p. 99–115.

Stone, K. A., 1994, Geology and structural evolution of part of the southeast Loreto basin, Baja California Sur, Mexico [M.S. Thesis]: Flagstaff, Northern Arizona University, 131 p.

Umhoefer, P. J., and Stone, K. A., 1996, Description and kinematics of the SE Loreto basin fault array, Baja California Sur, Mexico: A positive field test of oblique-rift models: Journal of Structural Geology, v. 18, p. 595–614.

Umhoefer, P. J., Dorsey, R. J., and Renne, P., 1994, Tectonics of the Pliocene Loreto basin, Baja California Sur, Mexico, and the evolution of the Gulf of California: Geology, v. 22, p. 649–652.

Van Wagoner, J. C., Posamentier, H. W., Mitchum, R. M., Vail, P. R., Sarg, J. F., Loutit, T. S., and Hardenbol, J., 1988, An overview of the fundamentals of sequence stratigraphy and key definitions, *in* Wilgus, C. K., Hastings, B. S., Posamentier, H. W., Van Wagoner, J. C., Ross, C. A., and Kendall, C. G. St.C., eds., Sea-level changes: An integrated approach: Society of Economic Paleontologists and Mineralogists Special Publication 42, p. 39–45.

Zanchi, A., 1994, The opening of the Gulf of California near Loreto, Baja California, Mexico: From basin and range extension to transtentional tectonics: Journal of Structural Geology, v. 16, p. 1619–1639.

Manuscript Accepted by the Society December 2, 1996

Geological Society of America
Special Paper 318
1997

Bryozoan nodules built around andesite clasts from the upper Pliocene of Baja California: Paleoecological implications and closure of the Panama Isthmus

Roger J. Cuffey
Department of Geosciences, 412 Deike Building, Pennsylvania State University, University Park, Pennsylvania 16802
Markes E. Johnson
Department of Geosciences, Williams College, Williamstown, Massachusetts 01267

ABSTRACT

Fossil bryozoan nodules ("bryoliths," "ectoproctaliths," or "bryozoan macroids") are found in upper Pliocene deposits estimated to have formed at 2.4 Ma in the Loreto Basin, located two-thirds of the way down the gulf coast of Baja California (at 26°05′ 10″ N, 111°21′05″W). The nodules occur abundantly, along with numerous pectinid bivalves, in a shelly sandy conglomerate interpreted as part of a marginal-marine fan delta now exposed in Arroyo Arce. The nodules, mostly between 4 and 8 cm in diameter, range up to 19 cm long by 12 cm wide by 9 cm high. Smooth surfaced and well rounded in form, each consists of many thin sheetlike laminae encrusting an andesite-cobble core. Except for the largest specimen recovered, growth was concentric as a result of frequent overturning of the nodules by storm waves, tidal currents, and perhaps bioturbators. Bryozoan crusts are composed of laminae all belonging to the anascan cheilostome *Conopeum commensale*.

Nodules of this species today are concentrated in tidal inlets along barrier-island coasts, such as the Delmarva Peninsula of Virginia (where it may possibly be intergrown with an encrusting variety of *Membranipora arborescens*). *Conopeum commensale* ranges farther south along the coasts of Brazil and west Africa. Its modern Atlantic distribution, compared to its occurrence in the upper Pliocene of Baja California, shows that this species migrated to the Pacific Ocean before closure of the Panama Isthmus no later than about 3.5 Ma. Invasion of the Pliocene Gulf of California by this Atlantic species substantiates the influence of a northerly directed coastal current in the east Pacific related to the flow-through of marine waters from the Atlantic to the Pacific.

INTRODUCTION

Water depth and other related aspects of the physical environment are factors that paleoecological analysis using fossil bryozoans has the potential to resolve in considering the depositional setting of rock units. With this application in mind, Schopf (1969) reviewed the distribution of Recent bryozoans off the coast of Massachusetts, particularly those with erect as opposed to encrusting growth forms. For nodule-forming crustose forms at the shallower end of the spectrum, the term "ectoproctaliths" (or more succinctly "bryoliths") was proposed by Rider and Enrico (1979). More recently, the textural nomenclature applied to reef rocks was expanded to include bioclasts formed by reef-dwelling bryozoans (Cuffey, 1985). Many of the growth forms recorded in the fossil record, especially those in reefs, involve bryozoans encrusting, binding, or nucleating on and around other organic surfaces. An alternate term for all these features might be "bryozoan macroids," as extended from the terminology of Hottinger (1983).

Cuffey, R. J., and Johnson, M. E., 1997, Bryozoan nodules built around andesite clasts from the upper Pliocene of Baja California: Paleoecological implications and closure of the Panama Isthmus, *in* Johnson, M. E., and Ledesma-Vázquez, J., eds., Pliocene Carbonates and Related Facies Flanking the Gulf of California, Baja California, Mexico: Boulder, Colorado, Geological Society of America Special Paper 318.

In their systematic review of hard-ground faunas, Wilson and Palmer (1992) discuss the diverse range of trepostome, cyclostome, and cryptostome bryozoans encrusting nonorganic surfaces during the Paleozoic. By comparison, the less diverse Mesozoic hard-ground dwellers among the bryozoans are mostly cyclostomes. Representatives of Cenozoic hard-ground dwellers are said to be "poorly known," usually "cryptic" in habitat, and "rare" (Wilson and Palmer, 1992, p. 26).

The purpose of this chapter is to describe an unusual example of bryoliths from the upper Pliocene of the Loreto area in Baja California Sur, Mexico. This occurrence is especially noteworthy for two reasons. First, it represents the encrustation of andesite pebbles and cobbles in an embayment of the Pliocene Gulf of California. Second, the bryozoan crusts enclosing the clasts belong to an extant species of anascan cheilostomes heretofore known from the barrier-island coasts of the Atlantic Ocean (Cook, 1968; Monod and D'Hondt, 1978; Dade and Cuffey, 1984). These fossils not only provide evidence for the depositional environment of the rocks in which they locally occur but also lend credence to the existence of an anomalous northerly flowing current hypothesized by Weaver (1990) to have existed along the Californias prior to closure of the Panama Isthmus at 3.5 Ma.

COLLECTION SITE AND PREVIOUS STUDIES

The bryoliths were collected in Baja California Sur about 8 km north of Loreto on the north side of Arroyo Arce, near locality 2 as previously mapped by McLean (1989, p. 19). His field site 86JS2 is located "500 m upstream from the prominent cliff-forming coquina that constricts the width of the arroyo" (McLean, 1989, p. 23). Our locality is at the conspicuous narrows in the arroyo (Fig. 1). All study materials are reposited with the Colección Paleontologica de Referencia de la Universidad Autonoma de Baja California (UABC) in Ensenada, Mexico.

Pliocene strata in Arroyo Arce and nearby Arroyo de Gua are first mentioned in the published literature that resulted from the 1940 Scripps expedition to the Gulf of California. Anderson (1950) subdivided strata from this area into the San Marcos, Carmen, and Marquer Formations based on correlations with type sections for lower, middle, and upper Pliocene units on some of the larger gulf islands. Durham (1950, table 5) sampled Pliocene molluscs and echinoids from several spots in Arroyo Arce. McLean (1989, p. 21) conducted a reconnaissance study of the Loreto marine embayment and pointed out the "onlapping relations between Pliocene strata and Miocene volcanic and volcaniclastic rocks" in the vicinity of Arroyo Arce. Steeply dipping (30°E) alluvial fan facies exposed upstream grade into yellow calcareous sandstones interbedded with buff shell coquinas exposed downstream. The shell beds found at McLean's field site 86JS2 contain a diverse fauna of bivalves, including *Argopecten abietis*, *A. revellei*, *Leopecten backeri*, *"Aequipecten" dallasi*, *Nodipecten arthriticus*, *Undulostrea megodon*, and *Placunanomia cumingii*. This assemblage is discussed and partly illustrated by Smith (1991).

Abrupt changes in the rates of Pliocene subsidence for this area are calculated on the basis of $^{40}Ar/^{39}Ar$ dates collected from a succession of four tuff beds by Umhoefer et al. (1994). Ages range between 2.61 ± 0.01 Ma almost 200 m above the base of the succession to 1.97 ± 0.02 Ma 1,000 m higher in the succession dominated by stacked Gilbert-type fan deltas (Dorsey et al., 1995). The best age estimate for the bryolith-bearing beds in Arroyo Arce is about 2.4 Ma, based on physical correlation of a sequence boundary from the south-central part of the basin (where the dated tuffs occur) to the lower Arroyo Arce and interpolation between the ages of the second (2.46 Ma) and third tuffs (2.36 Ma), respectively (Rebecca Dorsey, personal communication, 1995). In any case, the full range of dated tuffs suggests correlation to the upper Pliocene Piacenzian Stage and possibly the coastal onlap cycle TB3.7 of Haq et al. (1988). This means that the San Marcos and probably the Carmen Formations as originally attributed to this area by Anderson (1950) based on the fossil data of Durham (1950) are improperly correlated. Much of the succession in Arroyo Arce is essentially upper Pliocene in position. New formation names warranted for this area on the basis of extensive fan-delta systems have not yet been suggested.

OBSERVATIONS REGARDING BRYOZOANS

Clast-encrusting bryozoans were detected at only a single stratigraphic horizon in Arroyo Arce, at the narrows, but the associated fauna is comparable to that reported by McLean (1989) from slightly older strata upstream at site 86JS2. Visually, the continuity of the host strata is easy to trace, but the lateral extent of the bryoliths is difficult to determine because of the steepness of the canyon walls and the high dip angle of the beds. At arroyo level, however, the bryoliths are abundant and mixed together with even more prolific pectinid bivalves typically encrusted by barnacles. The presence of andesite clasts, derived from the Miocene Comondú Group, is not obvious unless the bryoliths are cracked open. Most of the bryoliths are between 4 and 8 cm in diameter. Except for the largest specimen recovered, they completely enclose very coarse pebbles sized according to the Wentworth grade scale. Thicknesses of the concentric bryozoan crusts range between 0.5 and 1.5 cm. The largest bryolith represents a colony with a length of 19 cm, a width of 12 cm, and a height of 9 cm (Fig. 2a). This particular colony encrusts an andesite cobble on all but one flat side (Fig. 2b), and its maximum thickness is 5 cm (Fig. 2c).

The surface of the largest Loreto bryolith shows several rounded mounds, some of which exhibit large central holes or depressions. A cross section through this specimen reveals that the holes are the result of occasional barnacles encrusting early growth surfaces on the colony (Fig. 2c). Those mounds lacking barnacles mostly result from piling up of extra zooecial layers or obliquely oriented zooecia around local frontal-budding centers during bryolith development. The smaller holes also visible on the outer surface are openings of serpulid-worm or vermetid-gastropod tubes intergrown with bryozoan laminae.

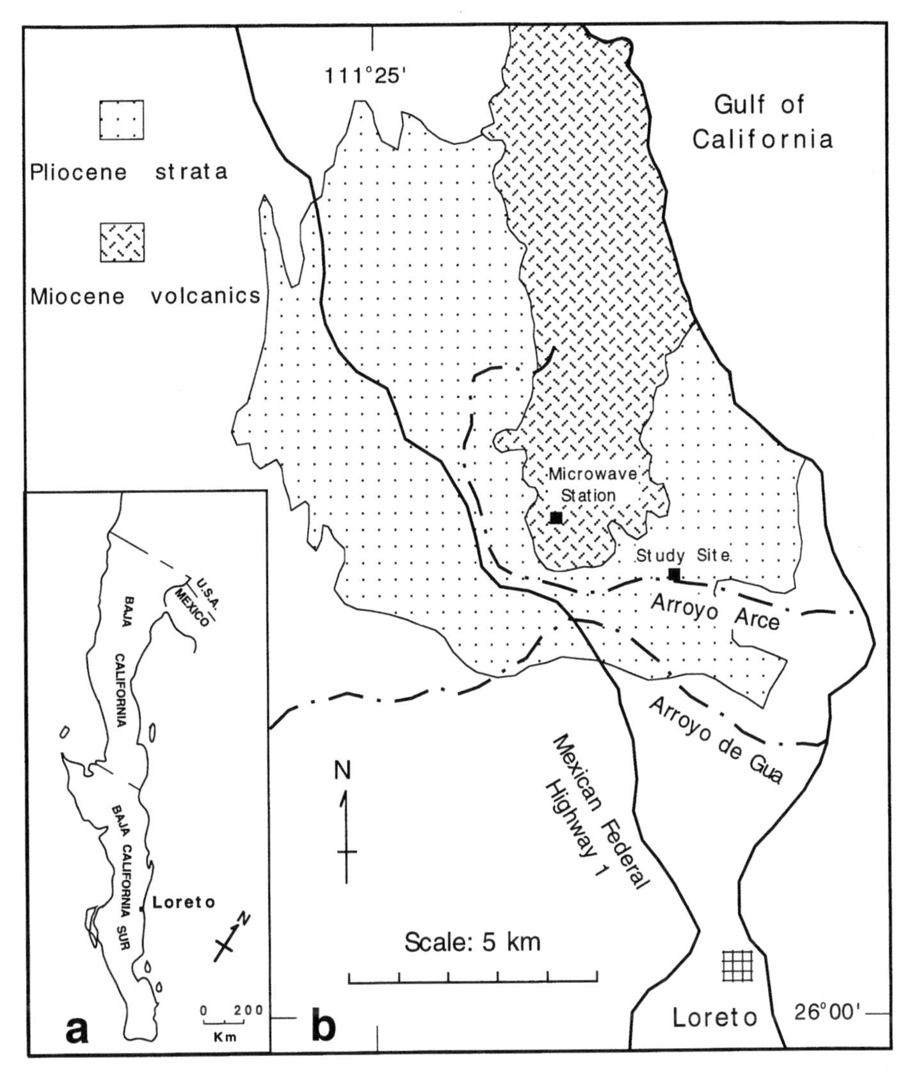

Figure 1. Location of study site. a, map of the Baja California peninsula, showing the town of Loreto in Baja California Sur. b, enlargement of the Loreto area, showing the location of the site on Arroyo Arce where Pliocene bryoliths were found nucleated around andesite clasts derived from a Miocene shoreline (modified from McLean, 1989, figs. 1 and 2).

More commonly, the bryozoan colonies are smooth surfaced and well rounded in shape, consisting of concentric sheetlike laminae (Figs. 2e and 2d). On the basis of all samples examined, the laminae are composed of a single species: the anascan cheilostome *Conopeum commensale*. Each thin lamina or crust may represent a fresh colony or may have budded upward from the lamina below. At one extreme, an entire nodule could represent multiple regenerations of the same colony. This can be determined only by dissecting the individual nodules, layer by layer. In short, each nodule may involve many different colonies, a few colonies, or only a single, long-lived colony.

Two andesite clasts (Figs. 2c and 2f) are shown in cross sectional views. Completely encrusted by bryozoan laminae, only the smaller clast (7 cm long and 1.5 cm thick) was clearly circumrotatory. Measuring 14 cm long and 4.8 cm thick, the larger clast retained an unencrusted base and remained stationary. As both clasts are flat sided, it must have been the size and weight of the larger clast that limited its movement on the seabed.

DISCUSSION

Issues concerning species composition, comparative ecology and paleoecology, geographic distribution, and paleoceanography are all relevant to the upper Pliocene bryoliths from Loreto.

Taxonomy

Conopeum commensale was first described by Kirkpatrick and Metzelaar (1922) as a fundamentally Atlantic species with a geographic range spanning both the northern and southern hemi-

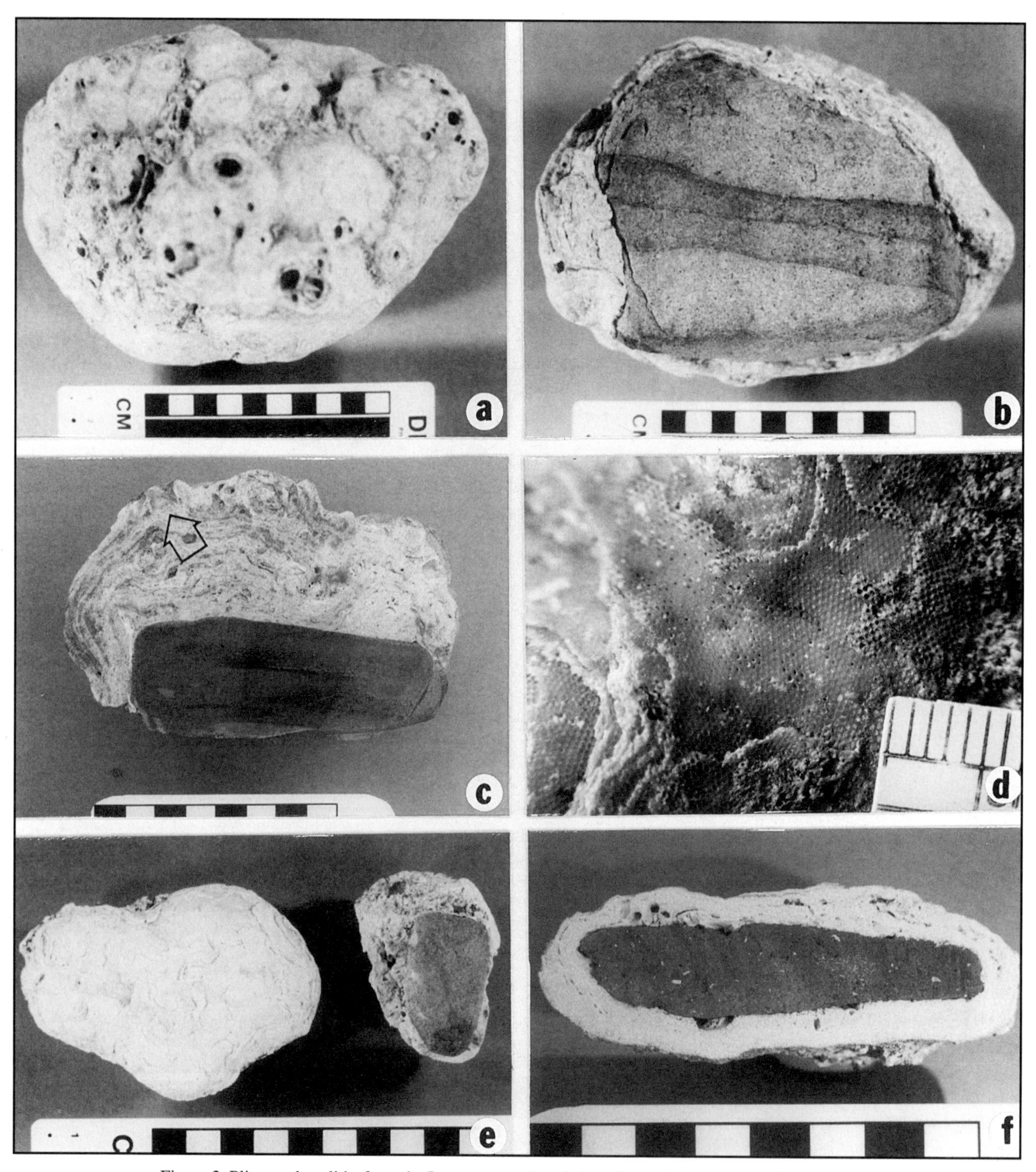

Figure 2. Pliocene bryoliths from the Loreto area; all scale bars in centimeters except for d, which is in millimeters. a, upper surface of large bryolith belonging to the species *Conopeum commensale* (UABC collection number 1935); small openings are for barnacles. b, bottom view of same bryolith, showing flat surface of andesite cobble. c, cross-sectional view of same bryolith, showing maximum thickness of laminae (5 cm) on the upper surface and the inclusion of a barnacle (arrow). d, close-up view of same specimen showing surface view of sheetlike laminae comprising the anascan zoarium. e, two smaller bryoliths fully enclosing andesite pebbles (UABC collection numbers 1936, left specimen; 1937, right specimen, which is broken). f, cross-sectional view of medium-sized bryolith (UABC collection number 1936), showing concentric laminations.

spheres as well as both the eastern and western shores of that ocean. Subsequently, Cook (1968) reassigned the species to another genus, as *Membranipora commensale* (Kirkpatrick and Metzelaar, 1922), based on specimens from west Africa. Because the genus *Membranipora* is widely applied to embrace many bryozoan species, the revision may be considered less precise and potentially more confusing than retention of the genus *Conopeum*. It is further argued by Cook (1968), however, that *Membranipora commensale* is entirely restricted to west Africa and that all western Atlantic species of the coastal United States and Brazil belong to another species: *Membranipora arborescens*.

Dade (1983) and Dade and Cuffey (1984) discuss the occurrence of bryoliths from coastal Virginia, where Recent nodules are interpreted as composed primarily of *Conopeum commensale* but with a few layers approaching the morphology of *Membranipora tenuis* and others the encrusting but not erect form of *Membranipora arborescens*. The differences between these species are found in the minute granulation and denticulation around the zooecial opesium. Variation is very subtle, and one possible taxonomic conclusion may be that *Conopeum commensale*, *Membranipora arborescens*, and *M. tenuis* are all synonyms for the same species. For the most part, such subtleties are not preserved in the Loreto specimens. Where observable, however, they fit best with the original description of *Conopeum commensale* (Kirkpatrick and Metzelaar, 1922). Previous recognition in the eastern Pacific of extant species belonging to *Conopeum* has been burdened by these taxonomic problems.

Ecology and paleoecology

Holocene bryoliths from the coastal zone of Virginia's Delmarva Peninsula and associated islands of Wallops and Chincoteague form as concentric layers around mollusc shells subjected to daily turbulence in tidal inlets between barrier islands but also on the adjacent near-shore shelf (Dade, 1983; Dade and Cuffey, 1984). The nodules are often found tossed up as flotsam on barrier beaches as the the result of storm waves. Depth of water over the inlets ranges between intertidal conditions to 8 m. Agitation of "free-rolling" nodules in this setting is frequent, as indicated by the nodular shape typically exhibiting growth on all sides. Daily tides, occasional storms, and the activity of bioturbators in shallow coastal waters contribute to reorientation on a regular basis. Barnacles sometimes encrust the nodules. The Delmarva bryoliths thrive in a euhaline or normal-marine setting, although they may be tolerant of and possibly even prefer slightly reduced salinities that result from runoff from the nearby land in a humid climate.

Similar bryozoan nodules constructed by *Conopeum commensale* are reported by Monod and d'Hondt (1978) from Mauritania, west Africa. In this case, however, the material was recovered from an archeological site occupied by a Neolithic culture 10,000 years ago. Bryozoan nodules referred to the genus *Biflustra* from the lower Pliocene Imperial Formation of southeastern California are also briefly reported by Gyllenhaal and Kidwell (1989). In both these examples, the bryozoans are fully nucleated around gastropod shells.

The fundamental difference shown by the Loreto specimens is their extensive encrustation of andesite pebbles and cobbles instead of shells. Thin patches of bryozoan crusts on mobile Paleozoic and Mesozoic pebbles, cobbles, and boulders are reviewed by Wilson (1987), but nothing comparable from the Cenozoic has been previously recognized. Eocene to Recent bryozoan encrustations on nonorganic hardgrounds previously discussed by Wilson and Palmer (1992) tend to represent cryptic forms living on the undersides of eroded beachrock slabs.

Although ostensibly in a more humid climate, the best contemporaneous analog to the paleoecological setting of the late Pliocene Loreto embayment is the Delmarva Peninsula of Virginia with its tidal channels through barrier islands. Domination of the Loreto Basin by stacked Gilbert-type fan deltas (Dorsey et al., 1995), however, does imply that local runoff was significant through the late Pliocene.

Geographic distribution

Conopeum commensale and its closely allied if not synonymous species, *Membranipora arborescens*, are known today to flourish almost exclusively in the Atlantic Ocean (Cook, 1968). Reports of Recent *Conopeum* from the Pacific coasts of Mexico and Ecuador exist (Osburn, 1950; Soule, 1959), but they are questionable in the opinion of Cook (1968). Fossil occurrences of *C. commensale* also are known from the late Cenozoic of Virginia (Dade and Cuffey, 1984) and Africa (Cook, 1968). Lower Pliocene bryoliths recovered from the Salton Sea extension of the Gulf of California in southeast California by Gyllenhall and Kidwell (1989) are attributed to the cheilostome genus *Biflustra*. Unfortunately, this is an old and possibly invalid name. Thus, it appears that the upper Pliocene bryoliths from the Loreto embayment represent the first well-documented occurrence of *Conopeum commensale* in the Gulf of California.

By necessary geographic extension, the Loreto material also must signify the first record of *Conopeum commensale* entering the eastern Pacific Ocean from its more thriving Atlantic habitats. Closure of the Isthmus of Panama is believed to have occurred about 3.5 Ma, based on careful biostratigraphic correlation of near-shore marine faunas between upper Pliocene deposits on the Caribbean and Pacific sides of the Isthmus (Coates et al., 1992). Because the marine connection between the Atlantic and Pacific oceans was permanently closed from about that time to the present, it appears that *C. commensale* survived at least locally in the Gulf of California until the bryolith-bearing deposits in the Loreto embayment accumulated at approximately 2.4 Ma. Should the material briefly described by Gyllenhall and Kidwell (1989) turn out to be *C. commensale*, it would represent a significantly older, early Pliocene record, but one certainly consistent with these paleogeographic patterns.

Paleoceanographic ramifications

Expansion of the geographic range of *Conopeum commensale* from the eastern Atlantic to the Caribbean region via an equatorial current and hence northward to coastal Virginia and southward to coastal Brazil is readily accounted for by circulation gyres in the North and South Atlantic Oceans. Once *C. commensale* entered the eastern Pacific Ocean, however, its movement north along Californian shores or in the opposite direction along South American shores presently would be hindered by the southerly flowing California current as well as the northerly flowing Peru current. These currents are well known for their vigorous upwelling in the eastern Pacific Ocean, not unlike other eastern boundary currents.

Contrary to expectations, the Leeuwin current off the west coast of Australia is an anomalous eastern boundary current that flows poleward from a subtropical position in the southern hemisphere, instead of northerly toward the equator. Weaver (1990) suggested that the Leeuwin current is driven by a deep alongshore density gradient in the Indian Ocean, influenced by the through-flow of warm equatorial waters from the western Pacific Ocean between Indonesia and New Guinea. The mechanics of this anomalous current are explained in more detail by Smith (1992). It is further proposed by Weaver (1990) and Smith (1992) that a mirror-image situation may have existed in the eastern Pacific Ocean prior to the closure of the Panamanian Isthmus, as a result of the through-flow of waters from the Atlantic. In this case, the California current would have flowed poleward and not toward the equator as it does today. The relict fossil distribution of *C. commensale* in the late Pliocene Gulf of California provides tangible evidence that such a reverse coastal current may have existed prior to about 3.5 Ma.

CONCLUSIONS

Intricate nuances of bryozoan taxonomy formulated by marine zoologists for the description of living species sometimes are awkward to apply to fossil species. In our opinion, the upper Pliocene bryoliths from the Loreto embayment on the Gulf of California best fit the description of the extant Atlantic species *Conopeum commensale* originally offered by Kirkpatrick and Metzelaar (1922). The Loreto bryoliths are significant on several accounts. In contrast to other fossil and Recent examples (Monod and d'Hondt, 1978; Rider and Enrico, 1979; Dade and Cuffey, 1984; Gyllenhaal and Kidwell, 1989; McKinney, 1995), the Loreto bryoliths are thickly encrusted on inorganic pebbles and cobbles. Heretofore, bryozoan rinds on inorganic hard grounds have been considered rare for the Cenozoic (Wilson and Palmer, 1992). Morphologically comparable to other bryoliths encrusted on shells, however, the Loreto specimens clearly depict a moderately high-energy environment in which the nodules were free to roll around and grow evenly concentric colonies. Such activity would be normal on the distal margin of Gilbert-type fan deltas, where eroded Miocene clasts were carried by streams to the Gulf of California, and tidal inlets probably developed in very shallow water.

This example of *Conopeum commensale,* beyond its significance as a local paleoecological indicator, represents the first well-documented case of a relict bryozoan population in the Pliocene Gulf of California with strong affinities to the Atlantic Ocean, although earlier Cenozoic bryozoan species from California have been identified as Caribbean forms (Cuffey et al., 1981). This species migrated from the Atlantic Ocean, where it still thrives today, to the eastern Pacific Ocean before closure of the Panama Isthmus sometime prior to 3.5 Ma. Establishment of this Atlantic species in the Gulf of California well north of the connecting strait between the Caribbean and Pacific oceans provides strong evidence for the predicted existence of a polar-flowing coastal current comparable to the present-day Leeuwin current of Western Australia (Weaver, 1990).

ACKNOWLEDGMENTS

Fieldwork was supported by a grant to Johnson from the National Science Foundation (INT-9313828). We are very grateful to Rebecca J. Dorsey and Paul J. Umhoefer (Northern Arizona University) for hosting Johnson and a contingent of Williams College students at Arroyo Arce in the Loreto area during January 1994. The largest example of a cobble-encrusting bryozoan was discovered by Maximino E. Simian (Williams College), alerting us to the potential for this project. Mark A. Wilson (College of Wooster) and Joseph F. Pachut Jr. (Indiana University–Purdue University) offered reviews that helped improve the manuscript.

REFERENCES CITED

Anderson, C. A., 1950, 1940 E.W. Scripps cruise to the Gulf of California. Part I: Geology of the islands and neighboring land areas: Geological Society of America Memoirs 43, 53 p.

Coates, A. G., Jackson, J. B. C., Collins, L. S., Cronin, T. M., Dowsett, H. H., Bybell, L. M., Jung, P., and Obando, J. A., 1992, Closure of the Isthmus of Panama: The near-shore marine record of Costa Rica and western Panama: Geological Society of America Bulletin, v. 104, p. 814–828.

Cook, P. L., 1968, Polyzoa from west Africa; the Malacostega, Part I: British Museum (Natural History), Bulletin (Zoology), v. 16, p. 113–160.

Cuffey, R. J., 1985, Expanded reef-rock textural classification and the geologic history of bryozoan reefs: Geology, v. 13, p. 307–310.

Cuffey, R. J., Stadum, C. J., and Cooper, J. D., 1981, Mid-Miocene bryozoan coquinas on the Aliso Viejo Ranch, Orange County, southern California, *in* Larwood, G. P., and Nielsen, C., eds., Recent and fossil bryozoa: Fredensborg, Denmark, Olsen and Olsen, p. 65–72.

Dade, W. B., 1983, Bryozoans of the modern Wallops-Chincoteague coast, Virginia [M.S. thesis]: University Park, Pennsylvania State University, 217 p.

Dade, W. B. and Cuffey, R. J., 1984, Holocene multilaminar bryozoan masses—The "rolling stones" or "ectoproctaliths" as potential fossils in barrier-related environments of coastal Virginia: Geological Society of America Abstracts with Programs, v. 16, p. 132.

Dorsey, R. J., Umhoefer, P. J., and Renne, P. R., 1995, Rapid subsidence and stacked Gilbert-type fan deltas, Pliocene Loreto basin, Baja California Sur, Mexico: Sedimentary Geology, v. 98, p. 181–204.

Durham, J. W., 1950, 1940 E. W. Scripps cruise to the Gulf of California.

Part II: Megascopic paleontology and marine stratigraphy: Geological Society of America Memoirs 43, 216 p.

Gyllenhaal, E. D., and Kidwell, S. M., 1989, Growth histories of subspherical bryozoan colonies: Ordovician and Pliocene evidence for paleocurrent regimes and commensalism: Geological Society of America Abstracts with Programs, v. 21, (6), p. A112–A113.

Haq, B. U., Hardenbol, J., and Vail, P. R., 1988, Mesozoic and Cenozoic chronostratigraphy and eustatic cycles, *in* Wilgus, C. K., Hastings, B. K., Posamentier, H., Wagoner, J. V., Ross, C. A., and Kendall, C.G.St.C., eds., Sea-level changes: An integrated approach: Society of Economic Paleontologists and Mineralogists Special Publication 42, p. 71–108.

Hottinger, L., 1983, Neritic macroid genesis, an ecological approach, *in* Peryt, T. M., ed., Coated grains: Berlin, Springer-Verlag, p. 38–55.

Kirkpatrick, R., and Metzelaar, J., 1922, On an instance of commensalism between a hermit crab and a polyzoon: Proceedings of the Zoological Society of London, [for] 1922, p. 983–990.

McKinney, F. K., 1995, Circumrotatory growth of bryozoans: A disturbance-driven growth habit influenced by complexly interacting physical conditions: Geological Society of America Abstracts with Programs, v. 27, no. 2, p. 73.

McLean, H., 1989, Reconnaissance geology of a Pliocene marine embayment near Loreto, Baja California Sur, Mexico, *in* Abbott, P. L., ed., Geologic studies in Baja California: Pacific Section, Society of Economic Paleontologists and Mineralogists, Los Angeles, Book 63, p. 17–25.

Monod, T., and d'Hondt, J.-L., 1978, A propos d'un echantillon de *Conopeum commensale* (Kirkpatrick & Metzelaar, 1922) (Bryozoa Cheilostomata) trouvé dans un site archéologique mauritanien: Bulletin de'Institut Fondamental d'Afrique Noire, series A, v. 40, p. 423–427.

Osburn, R. C., 1950, Bryozoa of the Pacific Coast of America. Part I: Cheilostomata Anasca: Reports of the Allan Hancock Pacific Expedition, v. 14, no. 1, p. 1–269.

Rider, J., and Enrico, R., 1979, Structural and functional adaptations of mobile anascan ectoproct colonies (ectoproctaliths), *in* Larwood, G. P., and Abbott, M. B., eds., Advances in bryozoology: London, Academic Press, p. 297–319.

Schopf, T. J. M., 1969, Paleoecology of ectoprocts (bryozoans): Journal of Paleontology, v. 43, p. 234–244.

Smith, J. T., 1991, Cenozoic marine mollusks and paleogeography of the Gulf of California, *in* Dauphin, J. P., and Simoneit, B. R. T., eds., The Gulf and Peninsular Province of the Californias: American Association of Petroleum Geologists Memoir 47, p. 637–666.

Smith, R. L., 1992, Coastal upwelling in the modern ocean, *in* Summerhayes, C. P., Prell, W. L., and Emelis, K. C., eds., Upwelling systems: Evolution since the early Miocene: Geological Society of London Special Publication 64, p. 9–28.

Soule, J. D., 1959, Anascan Cheilostomata (Bryozoa) of the Gulf of California; report of the Puritan American Museum Expedition to western Mexico: American Museum Novitates, no. 1969, p. 1–54.

Umhoefer, P. J., Dorsey, R. J., and Renne, P., 1994, Tectonics of the Pliocene Loreto basin, Baja California Sur, Mexico, and evolution of the Gulf of California: Geology, v. 22, p. 649–652.

Weaver, A. J., 1990, Ocean currents and climate: Nature, v. 347, p. 432.

Wilson, M. A., 1987, Ecological dynamics on pebbles, cobbles, and boulders: Palaios, v. 2, p. 594–599.

Wilson, M. A., and Palmer, T. J., 1992, Hardgrounds and hardground faunas: Aberystwyth, University of Wales, Institute of Earth Studies Publications, no. 9, 131 p.

Manuscript Accepted by the Society December 2, 1996

Printed in U.S.A.

Geological Society of America
Special Paper 318
1997

Origin and significance of rhodolith-rich strata in the Punta El Bajo section, southeastern Pliocene Loreto basin

Rebecca J. Dorsey
Department of Geology, Box 4099, Northern Arizona University, Flagstaff, Arizona 86011

ABSTRACT

Pliocene sedimentary rocks at Punta el Bajo accumulated in the southeastern Loreto basin between about 2.3 and 2.0 Ma, during a period of relatively slow subsidence along the Loreto fault. Sediments in this section unconformably overlie Miocene volcanic rocks and consist of interbedded cobble to cobble-boulder conglomerate and rhodolith-bearing massive calcarenite. Based on the presence of boulder-size clasts, shelly calcarenite matrix, sharp erosional bases, and one lithofagid-bored unconformity, conglomerate beds are interpreted to have been deposited along a late Pliocene rocky shoreline. Interbedding of conglomerate and calcarenite beds preserves a record of four episodes of deposition of coarse shoreline gravels followed by submergence of the shoreline (transgression) and deposition of shoreface and shallow offshore calcarenites. The section displays a distinctive upsection decrease in bedding dips accommodated by nonerosional stratal wedging and one internal erosional unconformity. Repeated episodes of erosion and submergence of this rocky shoreline (as interpreted from vertical facies trends) and the well-expressed fanning-dip geometry are both attributed to discrete episodes of fault slip and related tilting along active intrabasinal faults.

Calcarenite deposits in this section are rich in rhodoliths (coralline red algae) with accessory molluscan shell fragments and siliciclastic sand, thus comprising a rhodalgal lithofacies. Rhodoliths are present in three forms: (1) quasi-spherical, concentrically laminated calcareous nodules 1 to 3 mm in diameter; (2) cylindrical fragments 3 to 5 mm long, some displaying relict branches; and (3) broken to nearly complete ball-shaped plants 0.5 to 4 cm in diameter. Forms 1 and 2 are dominant, and form 3 is found only high in the section. The highly fragmented nature of the rhodoliths and their close association with rocky-shore conglomerate beds indicate that they were deposited in a moderate- to high-energy shoreface setting. Rare whole rhodoliths high in the section suggest a slightly deeper, more offshore setting, possibly in a current-influenced tidal channel. Globally, rhodalgal lithofacies are typical of climate belts transitional between tropical and temperate zones. The Loreto basin currently is situated between a coral reef–dominated carbonate shelf to the south and a siliciclastic shelf with echinoids, bivalves, bryozoans, and barnacles to the north. It is likely that, in spite of some modifications due to post–2.0 Ma rifting, late Pliocene regional paleogeography of the Gulf of California resembled that of today. Thus, the occurrence of rhodolith-rich facies in the Punta el Bajo section appears consistent with facies models for carbonate deposition on climatically zoned continental shelves.

Dorsey, R. J., 1997, Origin and significance of rhodolith-rich strata in the Punta El Bajo section, southeastern Pliocene Loreto basin, *in* Johnson, M. E., and Ledesma-Vásquez, J., eds., Pliocene Carbonates and Related Facies Flanking the Gulf of California, Baja California, Mexico: Boulder, Colorado, Geological Society of American Special Paper 318.

INTRODUCTION

Pliocene sedimentary basins exposed along the eastern margin of Baja California preserve a critical record of syn-rift tectonism and sedimentation along this young, obliquely rifted continental margin (e.g., Moore and Buffington, 1968; Angelier et al., 1981; Stock and Hodges, 1989; Zanchi, 1994; Umhoefer et al., 1994a). The Loreto basin (Fig. 1; McLean, 1988, 1989) is an excellent laboratory within which to study the structural and tectonic controls on sedimentation within this transtensional setting. Sedimentary rocks of the Loreto basin accumulated in a westward-tilting half graben that formed by dextral-normal slip on the Loreto fault. These sediments were deposited in nonmarine, marine, deltaic, and coastal environments during several episodes of moderate to very rapid subsidence (Umhoefer et al., 1994a; Dorsey et al., 1995; Dorsey et al., this volume).

This chapter describes a sedimentologic study of rhodolith-rich bioclastic and gravelly siliciclastic strata exposed at Punta el Bajo de Tierre Firme, ~10 km north of the town of Loreto, in the south-eastern Loreto basin (Fig. 1). Strata exposed in the Punta el Bajo

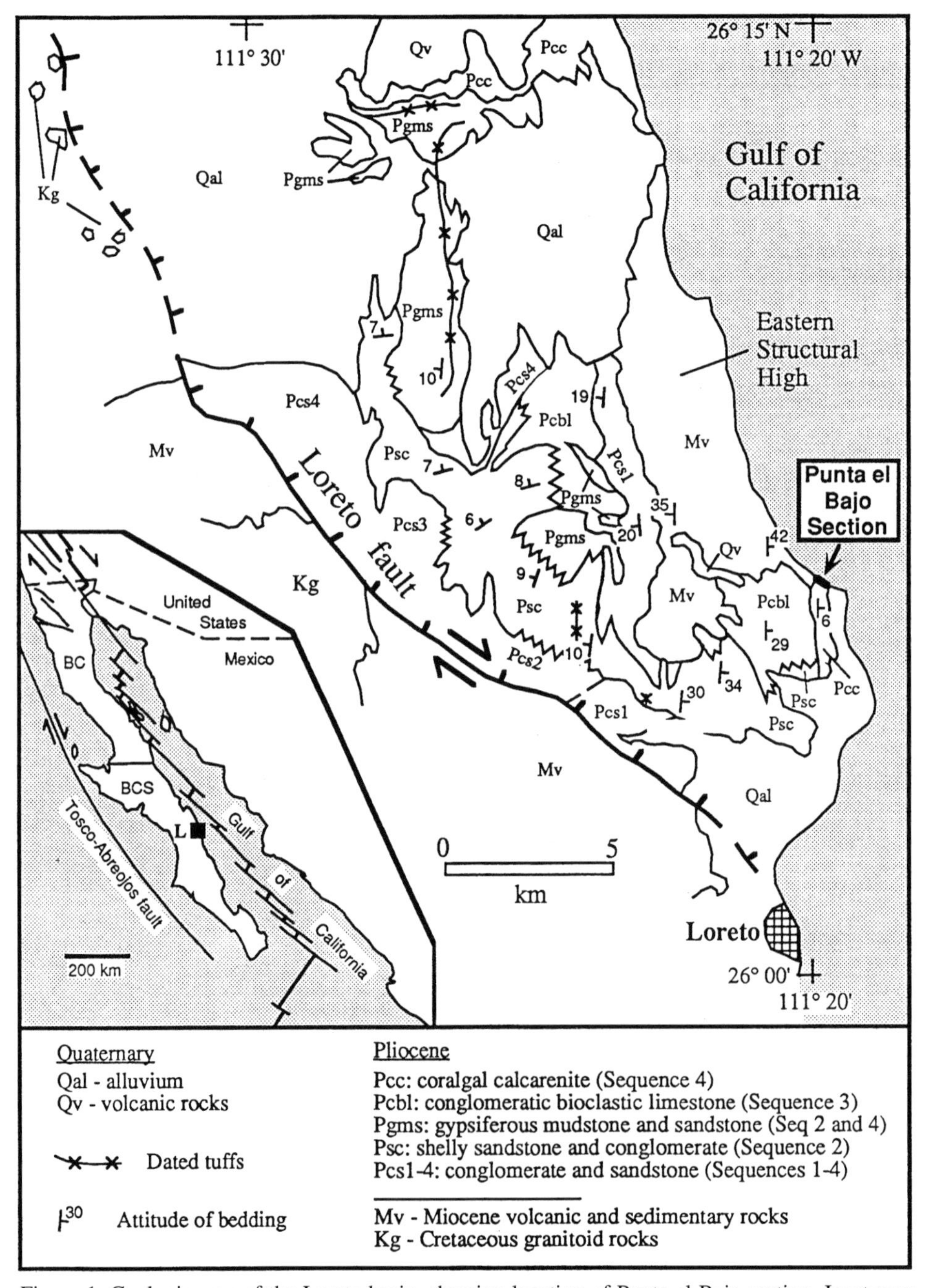

Figure 1. Geologic map of the Loreto basin, showing location of Punta el Baja section. Inset map shows position of the Loreto basin in the regional tectonic framework. BC = Baja California; BCS = Baja California Sur; L = Loreto area. Adapted from Umhoefer et al. (1994).

section belong to sequence 4, which is the youngest of four Pliocene sequences recognized throughout the basin (see Dorsey et al., this volume, for a summary of sequence stratigraphic terminology). In general, sequence 4 consists of rhodalgal and molluscan calcarenite with minor sandstone and conglomerate deposited during a phase of relatively slow basin subsidence and slow input of siliciclastic detritus to the basin. Dated tuffs and cross-cutting relationships in the southeastern Loreto basin show that normal and strike-slip faulting occurred within the southeastern basin during a brief time interval, between about 2.4 and 2.3 Ma, prior to deposition of sequence 4 (Umhoefer et al., 1994a, b). The section at Punta el Bajo accumulated at the south end of the eastern structural high, which represents the uplifted portion of the hanging-wall tilt block of the Loreto fault and probably formed a low-relief island during much of Loreto basin evolution (Fig. 1; Dorsey et al., this volume).

Rhodoliths are nodules and unattached branching growths of calcareous coralline red algae that are known to live and grow in a wide range of climates (tropical to arctic) and water depths (1 to >200 m) (Adey and MacIntyre, 1973; Bosence, 1983; Woekerling, 1988). Carbonate deposits that are dominated by rhodoliths represent a unique lithofacies type that is commonly found in regions transitional between tropical and temperate climatic zones (Lees, 1975; Nelson, 1988; Carannante et al., 1988). Rhodolith-rich carbonate facies in the Punta el Bajo section therefore are valuable for helping us to better understand paleoenvironmental conditions and carbonate sedimentation in this semitropical carbonate shelf system. The Punta el Bajo section is also significant because it contains a record of intrabasinal tilting late in the history of syntectonic sedimentation in the Loreto basin and thus provides important constraints on the style and timing of extensional deformation in the area.

DESCRIPTION OF SECTION

Physical stratigraphy

The Punta el Bajo section is well exposed along a modern beach cliff (Fig. 2). Here, the basal conglomerate of sequence 4 lies unconformably on Miocene volcanic rocks of the Comondú Formation. The section displays a distinctive fanning-dip geometry in which bedding dips decrease progressively from ~24° at the base of the section to ~11° higher in the section (Fig. 2). Farther to the southeast, to the left of the area shown in Figure 2, bedding dips continue to decrease up section into subhorizontal Pliocene strata. The total thickness of the Pliocene section at Punta el Bajo is about 50 to 60 m. Dipping Pliocene strata in the studied section are overlain by horizontally bedded Quaternary gravels along an angular unconformity (Fig. 2). Within the dipping Pliocene section there is a low-angle unconformity that consists of a calcite-cemented erosional surface colonized by numerous boring bivalves called lithofagids (Figs. 2, 3; Black, 1970; Frey and Pemberton, 1984). Wedging of units and thickening to the east is also evident at this locality (Fig. 2). The up-section and eastward decrease in bedding dips is thus accommodated geometrically by both an internal angular unconformity and by several zones of apparently nonerosive stratal wedging. The Pliocene section continues to the southeast, beyond and stratigraphically above the measured section (Fig. 4); these younger strata consist dominantly of sandy rhodolith-rich calcarenite with rare thin interbeds of calcareous granule to fine-pebble conglomerate.

Sedimentology

The physical sedimentology of the Punta el Bajo section is revealed in a measured section (Fig. 4) and photographs (Figs. 3, 5). The measured section contains four conglomerate beds 3 to 7 m thick that have sharp erosional bases and abrupt but nonerosive, conformable tops (Fig. 4). Some of these upper contacts show small-scale normal grading not visible at the scale of the measured section. Conglomerates are ungraded and massive, and the matrix consists of white mollusc- and rhodolith-bearing calcarenite similar to interbedded calcarenite units. In addition to the four numbered thick beds of conglomerate, two intervals of thin-bedded and planar-stratified pebble conglomerate are present in 2- to 3-m-thick intervals below conglomerate beds 3 and 4 (Fig. 4). Conglomerate clasts do not contain encrusting oysters or barnacles. The basal conglomerate contains rounded to subrounded cobbles and boulders up to 1.5 m long. Matrix in the lower part of the basal conglomerate consists of sand-rich, calcareous granule-pebble conglomerate. Matrix in the upper ~2.5 m of this bed is a distinctive white, pebbly calcarenitic shell hash that contains abundant molluscan shell debris and minor rhodolith material (Fig. 5A). Above this, tabular-bedded pebble-cobble conglomerate is interbedded with massive sandy calcarenite that contains a mix of rhodolith fragments and broken molluscan shell pieces (Fig. 5B).

Calcarenite beds typically are 1 to 4 m thick and contain a mix of crushed rhodoliths and mollusc fragments, with the relative abundance of rhodoliths increasing up section. Rhodolith and mollusc fragments occur in a matrix of well-sorted fine-grained lithic and/or bioclastic sand and are mixed with sparsely distributed coarse sand to granules in pervasively bioturbated structureless beds (Fig. 5C). Rhodoliths occur in three forms: (1) irregular, quasi-spherical, concentrically laminated calcareous nodules 1 to 3 mm in diameter; (2) cylindrical, elongate fragments 1 to 2 mm wide and 3 to 5 mm long, some displaying rounded, relict branches; and (3) broken to nearly complete ball-shaped algae plants displaying foliose and possibly fruticose (?) morphologies, ranging from about 0.5 to 4 cm in diameter (Fig. 5C; see also Foster et al., this volume, for examples of living rhodolith morphologies). Crushed rhodolith fragments (forms 1 and 2) consist of unidentifiable pieces and are by far the most abundant forms in the section. Form 3 was only rarely observed in this study, beginning stratigraphically ~5 to 10 m above the top of the section shown in Figure 4. The siliciclastic:carbonate ratio in the fine-sand matrix varies from about 9:1 to 1:9. Molluscan debris includes fragments of pectens, oysters, gastropods, and bivalves

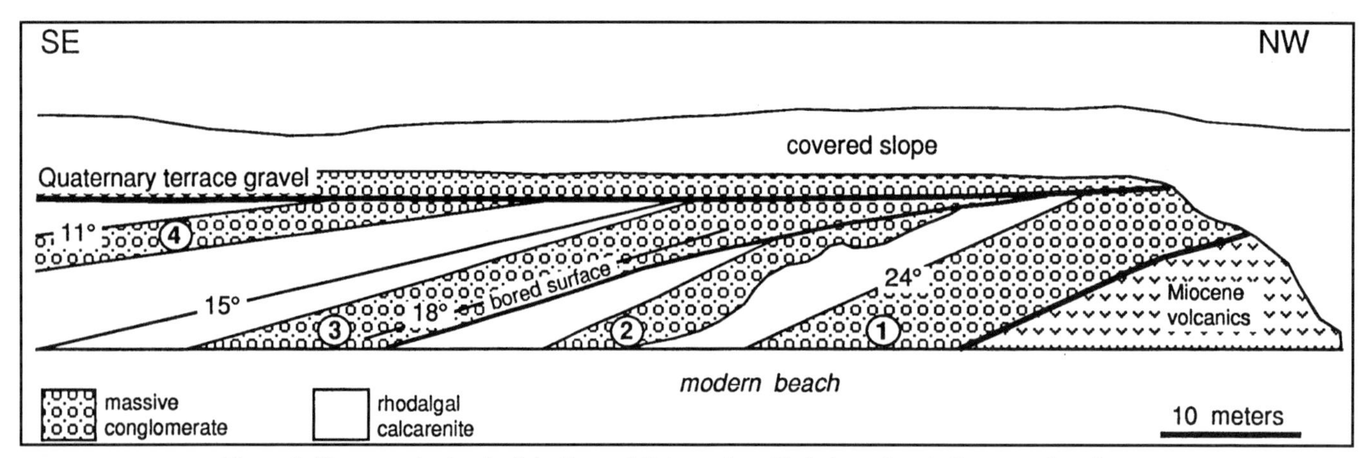

Figure 2. Photomosaic sketch of the Punta el Bajo section. Circled numbers indicate numbered conglomerate beds in the measured section (Fig. 4). Note fanning-dip geometry and angular truncation of strata below bored unconformity surface. Sediments in this part of the section consist dominantly of rocky-shore conglomerates and interbedded shoreface calcarenites. Deeper water (tidal channel?) rhodalgal calcarenites are found higher in the section, to the left of this view.

(e.g., Fig. 5A), as described in the Loreto area by Durham (1950), Smith (1991), and Piazza and Robba (1994). Rare thin horizons and small clusters of whole articulated pectens and oysters are also present.

INTERPRETATION

Depositional environments

Conglomerates. Sedimentary facies examined in this study are interpreted to be the interbedded deposits of rocky shoreline (conglomerates 1–4; Fig. 4), shoreface (rhodalgal calcarenites, lower ~30 m), and inner-shelf environments (rhodalgal calcarenites above ~30 m). A rocky shoreline setting (e.g., Johnson, 1988, 1992; Johnson and Hayes, 1993; Hayes et al., 1993) is inferred for conglomerate beds in this section based on the following observations and reasoning: (1) The basal cobble-boulder conglomerate sits directly on a major unconformity and contains abundant mollusc-rhodolith calcarenite matrix that likely formed during conglomerate deposition; (2) boulders in the basal conglomerate are too large (up to 1.5 m long) to be transported any significant distance by even the largest storms and therefore must have accumulated at or near the base of a steep coastal cliff; (3) rounding of boulders must have taken place by in situ abrasion in a high-energy environment such as a rocky shore; and (4) lithofagids seen on the bored internal unconformity (Figs. 2, 3) are diagnostic of unconformities that form in rocky shore environments (Frey and Pemberton, 1984). It is likely that cobble and cobble-boulder conglomerates formed as talus deposits at the base of steep coastal cliffs. Thin-bedded and stratified conglomerates appear to be gravelly deposits of the shoreface sequence; their occurrence below conglomerate beds 3 and 4 (Fig. 4) may record progradation of gravelly shoreline facies over shoreface calcarenites (see below).

Figure 3. Outcrop photograph of lithofagid-bored unconformity surface (see Figs. 2 and 4 for stratigraphic position). Lithofagids are environment-specific bivalves that live on and bore into hard substrates along rocky shorelines.

In spite of the above reliable criteria for a rocky-shore setting, no encrusting oysters or barnacles were observed on clasts in any of the four massive conglomerates. The lack of encrusting oysters may be due to the high energy of deposition, which can be inferred from the presence of thin pebble horizons and abundance of crushed rhodolith debris (below). This possibility is supported by recent observations of a modern carbonate setting in nearby Bahía Concepción, about 80 km north-northwest of the Punta el Bajo section, in which encrusting oysters are restricted primarily to the low-energy leeward side of Isla Requesón (Hayes et al., 1993). The lack of barnacles is more difficult to explain, especially considering the abundance of encrusting barnacles in other Pliocene carbonate deposits in the southeastern Loreto basin

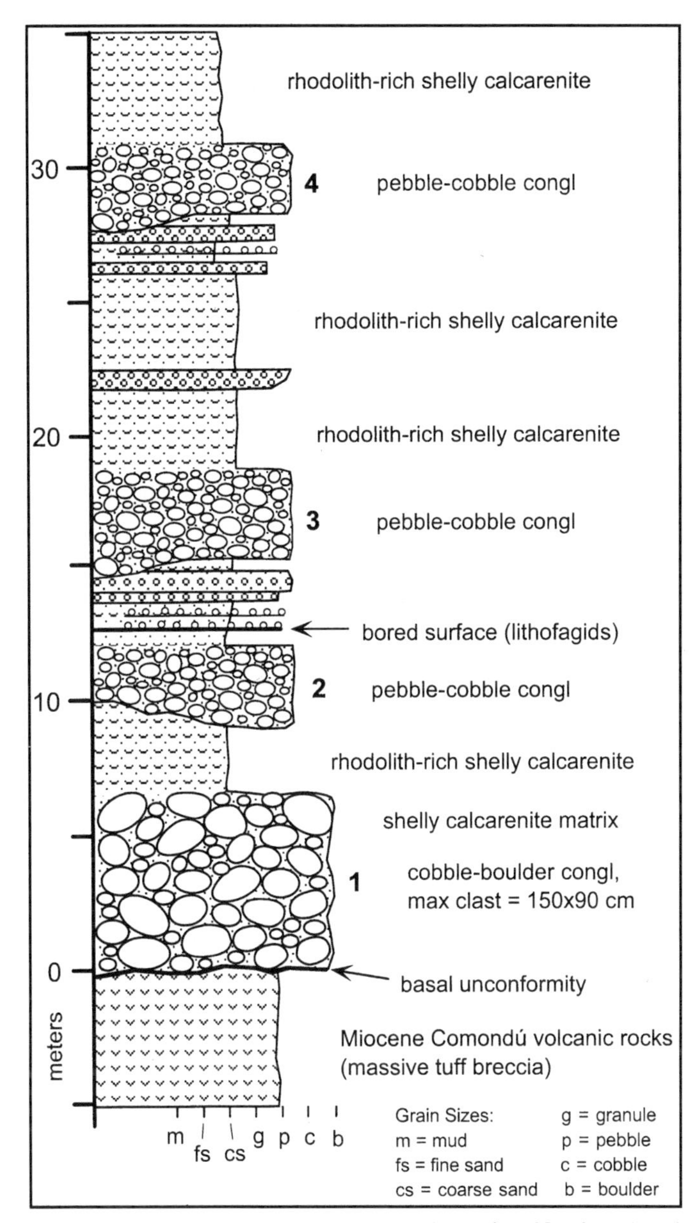

Figure 4. Measured log of the Punta el Bajo section. Numbers 1 to 4 correspond to numbered conglomerate beds in Figure 2. See Figure 1 for location of section.

Figure 5. Photographs of lithofacies in the Punta el Bajo section. A, mollusc-rich shelly calcarenite matrix found in the upper 2.5 m of the basal boulder conglomerate (cgl 1). Scale in cm. B, conglomerate bed 2 interbedded with massive calcarenite below and above. Note massive ungraded nature of conglomerate and abundance of calcarenite matrix. Calcarenite beds contain abundant nodular rhodolith fragments. C, detail of sandy rhodolith calcarenite. Contrast between rhodoliths and fine-grained lithic sand matrix in this example enhances fragmental and cylindrical rhodolith morphologies. Pen is 13.5 cm long.

(Dorsey et al., this volume; Cuffey and Johnson, this volume). Perhaps the large volume of crushed rhodolith debris acted as an abrasive in this high-energy setting, inhibiting the settlement of barnacles onto clasts or removing them by scouring and erosion during each transgression (M. E. Johnson, personal communication, 1995). This explanation is tentative, and the problem requires further study.

Calcarenites. Interpretation of calcarenite interbeds is aided by their association with conglomeratic rocky-shore deposits and the fragmental condition of rhodolith material. In a simple application of Walther's law, the conformable upper contacts of conglomerate beds require the calcarenite units to have been deposited in a marine environment that existed laterally adjacent to (contiguous with) the headlands and rocky beaches that produced the conglomerate beds. Thus, each overlying calcarenite interbed is interpreted to have been deposited in a high-energy, shallow-marine shoreface setting immediately offshore from the rocky coastline. The highly fragmental nature of rhodoliths (forms 1 and 2), the lack of mud, and the overall high degree of sorting in fine-grained sand matrix support this interpretation. Modern rhodoliths in the Loreto area live and grow in slightly offshore, inner-shelf beds that contain abundant whole plants (Foster et al., this volume). Mechanical breakdown, abrasion, and shoreward transport by waves are apparently required to produce the observed fragmentation, sorting, and concentric precipitation of calcium carbonate on crushed rhodolith fragments (e.g., Bosence, 1983). These processes are also important in deposition of foramol carbonate facies, which are dominated by bryozoans, molluscs, barnacles, and related biotas, in wave-dominated temperate shelves (Nelson et al., 1988; Scoffin, 1988; see also discussion below). The overall lack of stratification features in calcarenites of the Punta el Bajo section probably is the result of intensive bioturbation.

Starting at ~30 to 35 m in the section (Fig. 4), calcarenites reveal an upward transition to slightly deeper-water shelf deposits, possibly representing offshore tidal channels. This interpretation is based on the lack of thick, erosional-based, cobble-bearing conglomerates and the appearance of rare whole foliose rhodoliths. In a recent study of modern coastal and nearshore environments in Baja California Sur, Foster et al. (this volume) found that foliose-form rhodoliths are rare or absent in wave beds (2 to 12 m water depth) and are relatively abundant in tidal current deposits (deeper than 12 m). Because whole rhodoliths of any form are very rare in the Pliocene section at Punta El Bajo, the presence of several whole foliose plants high in the section is significant and suggests an upward transition to a deeper-water, offshore setting. This is a tentative conclusion, and a detailed analysis of stratigraphic variations in rhodolith morphologies in this section is clearly needed. Uncommon, thin, fine-pebble conglomerate beds high in the section are probably storm deposits.

Syndepositional tilting

The Pliocene section at Punta el Bajo preserves a detailed record of syndepositional tilting, subsidence, uplift, and erosion that was controlled by intrabasinal normal faulting in late Pliocene time. Prior to deposition of this section, older Miocene volcanic rocks in this area experienced uplift and erosion in the eastern structural high, which is the uplifted portion of the hanging-wall tilt block of the Loreto fault (Fig. 1; Dorsey et al., this volume). Beginning in late Pliocene time, between about 2.3 and 2.0 Ma (dates from Umhoefer et al., 1994a; extrapolated to this section using unpublished sequence stratigraphic relations), subsidence of the hanging-wall initiated deposition of rocky-shore and nearshore deposits at the Punta el Bajo locality. A simple transgression likely would have produced a simple deepening-up sequence. Instead, we see evidence for four episodes of shoreline erosion and deposition of strand-line conglomerates followed by transgression and deposition of shoreface deposits. Periods of alternating erosion and deposition are recorded in the four erosionally based rocky-shore conglomerate beds and conformably overlying rhodalgal calcarenites (Figs. 2, 4). The fanning-dip geometries seen through this stratigraphic interval suggest that the lateral migration of cliff, beach, and nearshore environments was controlled by the same processes that controlled progressive, syndepositional tilting of strata toward the east or southeast. These phenomena are readily explained as the product of fault-block tilting along down-to-the-west normal faults during deposition of the section. Alternatively, the up-section decrease in bedding dips could be the result of depositional infilling of preexisting steep topography, with no syndepositional tilting. However, detailed mapping (unpublished field data) shows that the stratigraphic interval exposed here coincides with an angular unconformity located immediately south of the studied section (Fig. 1) at the base of sequence 4, indicating that this section records the tilting episode that created that unconformity. Fanning-dip sections produced by syndepositional block tilting are common in tectonically active basins that experience regional extension and normal faulting (Gans et al., 1989; Hodges et al., 1989). Several north-striking normal faults that could have produced eastward tilting have been mapped nearby, to the west and southwest of the measured section (P. J. Umhoefer, unpublished field data).

The two observed geometries of fanning dips in this section are: zones of nonerosional wedging and a low-angle erosional unconformity (Fig. 2). These different geometries probably reflect different lateral positions of the fulcrum, or horizontal line of no vertical displacement, about which the section was tilted toward the east (e.g., Leeder and Gawthorpe, 1987). Zones of nonerosive wedging suggest a fulcrum that was located slightly to the west (right) of the facies panel in Figure 2. As long as deposits seen in this panel were being deposited in the zone of negative displacements (subsidence), and with eastward fault-controlled tilting, eastward stratal thickening and nonerosional wedging would be produced. In contrast, the erosional low-angle unconformity (Fig. 2) indicates that strata exposed at the Punta el Bajo section experienced a short period of uplift and erosion followed by submergence and drowning. I suggest that this unconformity was produced when the fulcrum shifted to the southeast, into the position of the measured section, causing uplift and erosion for a limited time. Submergence and deposition of younger strata occurred when the fulcrum of tilting stepped back to the west. This interpretation is supported by the presence of nonerosive wedging in strata above the bored unconformity surface (Fig. 2). Thus, the detailed geometries of fanning-dip sections, when integrated with vertical facies trends through the same sections, can be used to infer detailed aspects of intrabasinal faulting and tilting histories of active extensional basins.

DISCUSSION

Rhodolith-rich carbonate deposits in the Punta el Bajo section appear to occupy a transitional setting between tropical and temperate shelf environments along the climatically zoned western margin of the Gulf of California. Hermatypic (reef-building) corals, calcareous green algae, and accessory skeletal faunas (known collectively as coralgal or *chlorozoan* lithofacies) have been the dominant carbonate-producing organisms in tropical marine shelves and platforms since early Cenozoic time (Bathurst, 1975; James, 1983, 1984; Carannante et al., 1988). Temperate to polar carbonate shelf sediments (termed foramol, or *molechor* lithofacies) are constructed from bryozoans, bivalves, calcareous red algae, echinoderms, benthic foraminifers, and barnacles (Nelson, 1988; Nelson et al., 1988; Scoffin, 1988; Gostin et al., 1988). The boundary between (sub)tropical and temperate regions is generally defined as the 20° mean annual surface-water isotherm, which typically corresponds to about 30° latitude north and south (Nelson, 1988). Although reef-building corals and associated organisms typically are the dominant carbonate producers at latitudes less than 30°, carbonates of cooler-water affinity may extend into warm-water regions in response to anomalous environmental conditions (Lees, 1975; Nelson, 1988; Carannante et al., 1988). Carannante et al. (1988) defined an additional category of carbonate shelf deposits, *rhodalgal* lithofacies, which contains abundant encrusting coralline red algae in the form of algal bindstones and/or rhodoliths. Rhodalgal lithofacies form under environmental conditions (determined by temperature, salinity, and light intensity) that are transitional between the tropical and temperate-polar end members described above. A modern example of along-shelf climatic zonation of carbonate facies is found on the Atlantic continental shelf of Brazil, where carbonates containing abundant encrusting calcareous red algae make up a transition zone (from 15 to 23° S), between tropical coral reef–dominated facies in the north and cold temperate (molechor lithofacies) in the south (Carannante et al., 1988).

The occurrence of rhodolith-dominated deposits at Punta el Bajo (this study) appears consistent with the Carannante et al. (1988) model for carbonate deposition on climatically zoned continental shelves. The Gulf of California spans a climatic transition from sandy and rocky coastlines inhabited by echinoderms, bivalves, bryozoans, and barnacles (molechor lithofacies) in the north (32° N), to Cabo San Lucas (23°N) just south of the Tropic of Cancer, where abundant coral reefs and associated tropical faunas (chlorozoan lithofacies) are present. Located at about 26°N, the Loreto basin is presently situated near the northern limit of the tropical-subtropical climate zone, and it appears to fall within the zone of transition between tropical and warm-temperate climates. Assuming only minor changes in regional paleogeography due to post–2.0 Ma rifting, it is not surprising to find rhodolith-rich carbonate deposits (rhodalgal lithofacies) in the late Pliocene Punta el Bajo section. What is important is that the modern shelf just offshore of Punta el Bajo is also dominated by rhodoliths (Foster et al., this volume). It is interesting to note that the postulated rhodolith-rich "transition zone" in the Gulf of California extends at least up to Bahía Concepción and Punta Chivato (Foster et al., this volume; Meldahl et al., this volume), which is about 4° (450 km) poleward of the equivalent latitudinal limit of rhodalgal lithofacies on the Brazilian shelf (Carannante et al., 1988). This difference may be due to the fact that the Gulf of California is a restricted sea that experiences some anomalous warming of surface waters, unlike the Brazilian shelf, which faces the open Atlantic Ocean.

CONCLUSIONS

The Pliocene section at Punta el Bajo provides an intriguing glimpse into the history of intrabasinal faulting, tilting, erosion, and deposition of conglomerates and rhodolith-rich calcarenites in the southeastern Loreto basin. Massive conglomerates were deposited in rocky shoreline environments, and interbedded rhodalgal calcarenites accumulated in nearby shoreface to slightly deeper-water, offshore settings. Interbedding of conglomerate and calcarenite and one conspicuous internal unconformity provide evidence for four episodes of fault-controlled uplift and erosion followed by subsidence and submergence of the coastline. Detailed variations in fanning-dip geometries appear to be related to lateral changes in the position of the fulcrum about which syndepositional tilting of the section took place.

The Punta el Bajo section helps us understand carbonate sedimentation in this subtropical setting, which is located at about 26° N on the western margin of the Gulf of California. Although rhodoliths (calcareous red algae) are known from a wide variety of climatic settings, a search of the literature suggests that carbonate sediments consisting dominantly of rhodoliths, representing the rhodalgal facies, are characteristic of climate zones transitional between tropical and temperate regions (Carannante et al., 1988). The findings of this study are consistent with the Carannante et al. (1988) model for climatically controlled latitudinal zonation of carbonate shelves. In spite of the difficulties involved in reconstructing climate from carbonate facies, recognition of rhodalgal facies in ancient limestones may be helpful for placing stratigraphic sequences into a general context of climatic transition zones such as this.

ACKNOWLEDGMENTS

This research was supported by the National Science Foundation (EAR-9117269). Markes Johnson, Jorge Ledesma-Vázquez, and Keith Meldahl helped me learn to observe and appreciate the significance of rocky shoreline deposits and rhodolith-rich carbonate facies. Without their input and encouragement this paper would not have been written. Keith Meldahl kindly supplied a photograph that was used in constructing the facies panel (Fig. 2). Markes Johnson and Mike Foster are thanked for supplying me with early drafts of their papers and many valuable references. Jim Ingle, Joanne Borgois, and Markes Johnson are thanked for insightful and constructive reviews of the manuscript.

REFERENCES CITED

Adey, W. H., and MacIntyre, I. G., 1973, Crustose coralline algae: A reevaluation: Geological Society of America Bulletin, v. 84, p. 883–904.

Angelier, J., Colletta, B., Chorowicz, J., Ortlieb, L. and Rangin, C., 1981, Fault tectonics of the Baja California Peninsula and the opening of the Sea of Cortez, Mexico: Journal of Structural Geology, v. 3, p. 347–357.

Bathurst, R. G. C., 1975, Carbonate sediments and their diagenesis (second edition): Developments in Sedimentology, v. 12, Elsevier, Amsterdam, 658 p.

Black, R. M., 1970, The elements of paleontology: Cambridge, Cambridge University Press, 340 p.

Bosence, D. W. L., 1983, The occurrence and ecology of recent rhodoliths—a review, *in* Peryt, T. T., ed., Coated grains: Berlin, Springer-Verlag, p. 225–242.

Carannante, G., Esteban, M., Milliman, J. D., and Simone, L., 1988, Carbonate lithofacies as paleolatitude indicators: Problems and limitations: Sedimentary Geology, v. 60, p. 333–346.

Dorsey, R. J., Umhoefer, P. J., and Renne, P. R., 1995, Rapid subsidence and stacked Gilbert-type fan deltas, Pliocene Loreto basin, Baja California Sur, Mexico: Sedimentary Geology, v. 98, p. 181–204.

Durham, J. W., 1950, 1940 E. W. Scripps cruise to the Gulf of California. Part II: Megascopic paleontology and marine stratigraphy: Geological Society of America Memoir 43, 216 p.

Frey, R. W., and Pemberton, S. G., 1984, Trace fossil facies models, *in* Walker, R. G., ed., Facies models (second edition): Toronto, Geological Association of Canada, p. 189–211.

Gans, P. B., Mahood, G. A., and Schermer, E., 1989, Synextensional magmatism in the Basin and Range Province; A case study from the Eastern Great Basin: Geological Society of America Special Paper 233, 53 p.

Gostin, V. A., Belperio, A. P., and Cann, J. H., 1988, The Holocene non-tropical coastal and shelf carbonate province of southern Australia: Sedimentary Geology, v. 60, p. 51–70.

Hayes, M. L., Johnson, M. E., and Fox, W. T., 1993, Rocky-shore biotic associations and their fossilization potential: Isla Requeson (Baja California Sur, Mexico): Journal of Coastal Research, v. 9, p. 944–957.

Hodges, K. V., and eight others, 1989, Evolution of extensional basins and basin and range topography west of Death Valley: Tectonics, v. 8, p. 453–467.

James, N. P., 1983, Reef environment, *in* Scholle, P. A., Bebout, D. G., and Moore, C. H., eds., Carbonate depositional environments: American Association of Petroleum Geologists Memoir 33, p. 346–440.

James, N. P., 1984, Reefs, *in* Walker, R. G., ed., Facies models: Geological Association of Canada, Geoscience Canada Reprint Series 1, p. 229–244.

Johnson, M. E., 1988, Why are ancient rocky shorelines so uncommon?: Journal of Geology, v. 96, p. 469–480.

Johnson, M. E., 1992, Studies on ancient rocky shores: A brief history and annotated bibliography: Journal of Coastal Research, v. 8, p. 797–812.

Johnson, M. E., and Hayes, M. L., 1993, Dichotomous facies on a Late Cretaceous rocky island as related to wind and wave patterns (Baja California, Mexico): Palaios, v. 8, p. 385–395.

Leeder, M. R., and Gawthorpe, R. L., 1987, Sedimentary models for extensional tilt-block/half-graben basins, *in* Coward, M. P., Dewey, J. F., and Hancock, P. L., eds., Continental extensional tectonics: Geological Society of London Special Publication 28, p. 139–152.

Lees, A., 1975, Possible influences of salinity and temperature on modern shelf carbonate sedimentation: Marine Geology, v. 19, p. 159–198.

McLean, H., 1988, Reconnaissance geologic map of the Loreto and part of the San Javier quadrangles, Baja California Sur, Mexico: U.S. Geological Survey Miscellaneous Field Studies Map MF-2000, scale 1:50,000.

McLean, H., 1989, Reconnaissance geology of a Pliocene marine embayment near Loreto, Baja California Sur, Mexico: Pacific Section, Society Economic Paleontologists and Mineralogists Field Trip Guide, v. 63, p. 17–25.

Moore, D. G., and and Buffington, E. C., 1968, Transform faulting and growth of the Gulf of California since late Pliocene: Science, v. 161, p. 1238–1241.

Nelson, C. S., 1988, An introductory perspective on non-tropical shelf carbonates: Sedimentary Geology, v. 60, p. 3–12.

Nelson, C. S., Keane, S. L., and Head, P. S., 1988, Non-tropical carbonates on the modern New Zealand shelf: Sedimentary Geology, v. 60, p. 71–94.

Piazza, M., and Robba, E., 1994, Pectinids and oysters from the Pliocene Loreto basin (Baja California Sur, Mexico): Rivista Italiana di Paleontologia e Stratigrafia, v. 100, p. 33–69.

Scoffin, T. P., 1988, The environments of production and deposition of calcareous sediments on the shelf west of Scotland: Sedimentary Geology, v. 60, p. 107–124.

Smith, J. T., 1991, Cenozoic marine mollusks and paleogeography of the Gulf of California, *in* Dauphin, J. P., and Simoneit, B. R. T, eds., The Gulf and Peninsular Province of the Californias: American Association of Petroleum Geologists Memoir 47, p. 637–666.

Stock, J. M., and Hodges, K. V., 1989, Pre-Pliocene extension around the Gulf of California and the transfer of Baja California to the Pacific plate: Tectonics, v. 8, p. 99–115.

Umhoefer, P. J., Dorsey, R. J., and Renne, P., 1994a, Tectonics of the Pliocene Loreto basin, Baja California Sur, Mexico, and the evolution of the Gulf of California: Geology, v. 22, p. 649–652.

Umhoefer, P. J., Dorsey, R. J., and Stone, K. A., 1994b, Timing of transtensional deformation in the southern Loreto Basin, Baja California Sur: Geological Society of America Abstracts with Programs, v. 26, p. 100.

Woekerling, W. J., 1988, The coralline red algae: An analysis of the genera and subfamilies of nongeniculate Corallinaceae: London, Oxford University Press, 268 p.

Zanchi, A., 1994, The opening of the Gulf of California near Loreto, Baja California, Mexico: From basin and range extension to transtentional tectonics: Journal of Structural Geology, v. 16, p. 1619–1639.

Manuscript Accepted by the Society December 2, 1996

Printed in U.S.A.

Geological Society of America
Special Paper 318
1997

Living rhodolith beds in the Gulf of California and their implications for paleoenvironmental interpretation

Michael S. Foster
Moss Landing Marine Laboratories, P.O. Box 450, Moss Landing, California 95039
Rafael Riosmena-Rodriguez
Herbario Ficologico, Universidad Autónoma de Baja California Sur, Ap. Postal 19-B, La Paz, Baja California Sur, Mexico 23080
Diana L. Steller*
Hopkins Marine Station, Pacific Grove, California 93950
Wm. J. Woelkerling
Department of Botany, La Trobe University, Bundoora, Victoria, Australia

ABSTRACT

Subtidal surveys in the southwestern Gulf of California indicate that rhodolith beds are widely distributed, major sources of carbonate sediments, and habitats of high biodiversity. Beds with abundant branched rhodoliths ranging in size from 2 to 10 cm (longest dimension) have been found in two main types of environments: (1) gently sloping, subtidal soft bottoms with moderate wave action (wave beds; 2 to 12 m deep), and (2) relatively level bottoms in channels with tidal currents (current beds; below 12 m). Large individuals (to 11 cm) with up to 1 cm thick, densely packed branches are also found dispersed among sand and cobbles on more wave exposed shores. The relative abundance of fruticose forms and the sphericity and branch density of individual thalli are generally higher in wave beds than in current beds. Morphology within wave beds varies along gradients of water motion, with higher branch densities, more apical branching, and more branch fusions present as water motion increases. The extent to which these morphological differences represent different taxa is being evaluated. The abundance of rhodoliths in Pliocene and Pleistocene carbonate deposits and modern sediments indicates that this community has long been an important feature of nearshore environments in the Gulf. Our results suggest that measurements of a combination of morphological characters in populations of fossil rhodoliths, combined with detailed, small-scale stratigraphic analyses, may provide good estimates of paleoenvironmental conditions.

INTRODUCTION

"A thorough study of rhodoliths and their environments is required before critical paleoecological interpretations should be based on their morphology" (Adey and MacIntyre, 1973, p. 900).

Plants and animals with calcareous frameworks attached to the substratum are a major feature of the world's oceans. Calcareous algae and corals are particularly important in shallow water where suitable water chemistry and available light favor the use of $CaCO_3$ as a structural material (review in Barnes and Chalker, 1990). Extensive calcification in coralline and coral reefs has numerous consequences for the calcified organisms, the ecology of the community in which they are found, and geological processes and structures. Calcification presumably enhances survivorship by increasing resistance to disturbance by wave action and grazing and boring animals. The calcareous framework also

*Present address: Biology at Natural Sciences IV, University of California, Santa Cruz, California 95064.

Foster, M. S., Riosmena-Rodriguez, R., Steller, D. L., and Woelkerling, Wm. J., 1997, Living rhodolith beds in the Gulf of California and their implications for paleoenvironmental interpretation, *in* Johnson, M. E., and Ledesma-Vázquez, J., eds., Pliocene Carbonates and Related Facies Flanking the Gulf of California, Baja California, Mexico: Boulder, Colorado, Geological Society of America Special Paper 318.

provides a relatively stable habitat for a variety of other organisms (Younge, 1963).

These calcified frameworks preserve well and may, after death, be incorporated into sediments. Depending on conditions during and after preservation (Bosence, 1991; Scoffin, 1992), fossil assemblages of calcareous organisms can provide insight into geological processes and, combined with information on species composition and morphological attributes of the frameworks, may be used as indicators of paleoenvironmental conditions.

Rhodoliths, unattached nongeniculate coralline algae (Corallinales, Rhodophyta), represent another possible way to combine growth and morphological attributes with calcification to presumably enhance survival in dynamic environments. Their calcified thalli move with water motion, reducing damage from abrasion and impact and probably reducing the abundance of fouling organisms. Growth can occur from any part of the thallus surface, so growth is independent of thallus orientation. In many (all?) species, fragments resulting from breakage have the ability to grow. Damage can therefore result in vegetative reproduction. These red algae range in form from thin crusts covering pebbles and cobbles to entirely calcareous laminar or branched individuals, and vary in abundance from occasional individuals in surge channels to extensive beds many individuals thick (e.g., Bosellini and Ginsburg, 1971; Bosence, 1983a, b; Scoffin et al., 1985; Scoffin, 1988; Freiwald et al., 1991; Littler et al., 1991; Steller and Foster, 1995). Beds occur throughout the world from the intertidal zone to depths of over 200 m (Bosence, 1983b).

The contribution of rhodoliths to sediments and fossil assemblages is increasingly being assessed by geologists and used to establish paleoenvironmental conditions (e.g., Bosence and Pedley, 1982; Toomey, 1985; Braga and Martín, 1988; Freiwald et al., 1991; Johnson and Hayes, 1993; Dorsey, this volume). Carannante et al. (1988) point out that rhodolith-forming algae can be the most important carbonate producers in shallow subtropical/temperate waters. However, studies of the distribution and ecology of living beds are few (Bossellini and Ginsburg, 1971; Bosence, 1976; Scoffin et al., 1985; Scoffin, 1988; Prager and Ginsburg, 1989; Littler et al., 1991; Steller and Foster, 1995), and generalizations concerning environmental effects on growth, form, and composition used in paleoenvironmetal interpretations are still debated (Adey and MacIntyre, 1973; Reid and MacIntyre, 1988; Steller and Foster, 1995). Taxonomic difficulties have also hampered these interpretations (Woelkerling, 1988).

Living rhodoliths in the Gulf of California have been known to science since the late 1800s when Hariot (1895) described them from collections of Diguet. As part of his extensive natural history explorations in Mexico (Bois, 1928), Diguet (1911) obtained rhodoliths and described the general location of a number of beds while investigating pearl and other fisheries in the region. Pearl oysters were commonly associated with rhodoliths in the southwestern Gulf (Diguet, 1911; Dawson, 1960; Carino-Olvera and Caceres-Martinez, 1990), and pearl fishermen called the rhodolith beds "chicharrones" because their surfaces resemble fried pig skins. Dawson (1960) summarized the rhodolith species known from the Gulf based on prior information and extensive intertidal collecting and subtidal dredging in the 1940s and 1950s, and speculated that large subtidal beds might be present, particularly around La Paz. Schlanger and Johnson (1969) described an extensive bed in Canal de San Lorenzo near La Paz from dredge samples and diving observations. Steller (1993) discovered a large bed in Bahía Concepción, and Steller and Foster (1995) provided the first in situ observations of this and other beds in this bay. The report in Hayes et al. (1993) of rhodoliths and modern rhodolith sediments at Isla El Requéson in Bahía Concepción was based on Steller and Foster's (1995) observations. Extensive Pliocene and Pleistocene deposits occur in the Gulf (Anderson, 1950; Ortlieb, 1987, 1991; Meldahl, 1993), and entire rhodoliths and fragments can be abundant in these deposits (Dorsey, this volume; Simian and Johnson, this volume; M. Johnson and K. Meldahl, 1993, personal communication; M. Foster and R. Riosmena-Rodriguez, personal observation).

This information indicates that living rhodolith beds and fossil deposits are widespread along the southwestern shore of the Gulf of California (Fig. 1). The primary objective of our work is to determine the extent of living beds at selected locations between Punta Chivato and La Paz and to understand the environmental conditions that may affect their distribution. We also describe the morphological variation of rhodoliths within and among beds and discuss preliminary work on their taxonomy. Pleistocene deposits at four locations were qualitatively examined for rhodolith occurrence and distribution, and fossils were collected for taxonomic studies. Finally, we compare our information from living beds with that from fossil deposits to determine whether and how the former may be used to interpret the latter.

METHODS

Searches for living beds were first done using low-level aerial reconnaissance in Bahía Concepción (Steller and Foster, 1995) and more recently with dredging and scuba in areas where prior observations indicated beds occur (e.g., Canal de San Lorenzo; Fig. 1). Based on these findings, the environmental conditions in shallow beds influenced by moderate wave action reported in Steller and Foster (1995), the presence of Pleistocene deposits on shore, and information on the location of "chicharrones" provided by local fishermen, we made additional surveys in other areas (Fig. 1). The unexpected discovery of shallow-water rhodoliths in the more wave exposed locations at Calerita and Punta Bajo was based both on finding beach cast specimens and on luck.

Surveys were done with a dredge to 20-m depths, with a grab from 20 to 90 m, and with scuba from 1 to 25 m. Locations were established based on navigation charts and a portable Global Positioning System (GPS). Dredging was done with a small, conical (17-cm diameter, 27-cm long) rock dredge. The depth of each dredge haul was determined with a weighted line, and contents after a 1- to 2-min dredge were noted as entire living rhodoliths (>2-cm longest dimension, good purple/pink pigmentation),

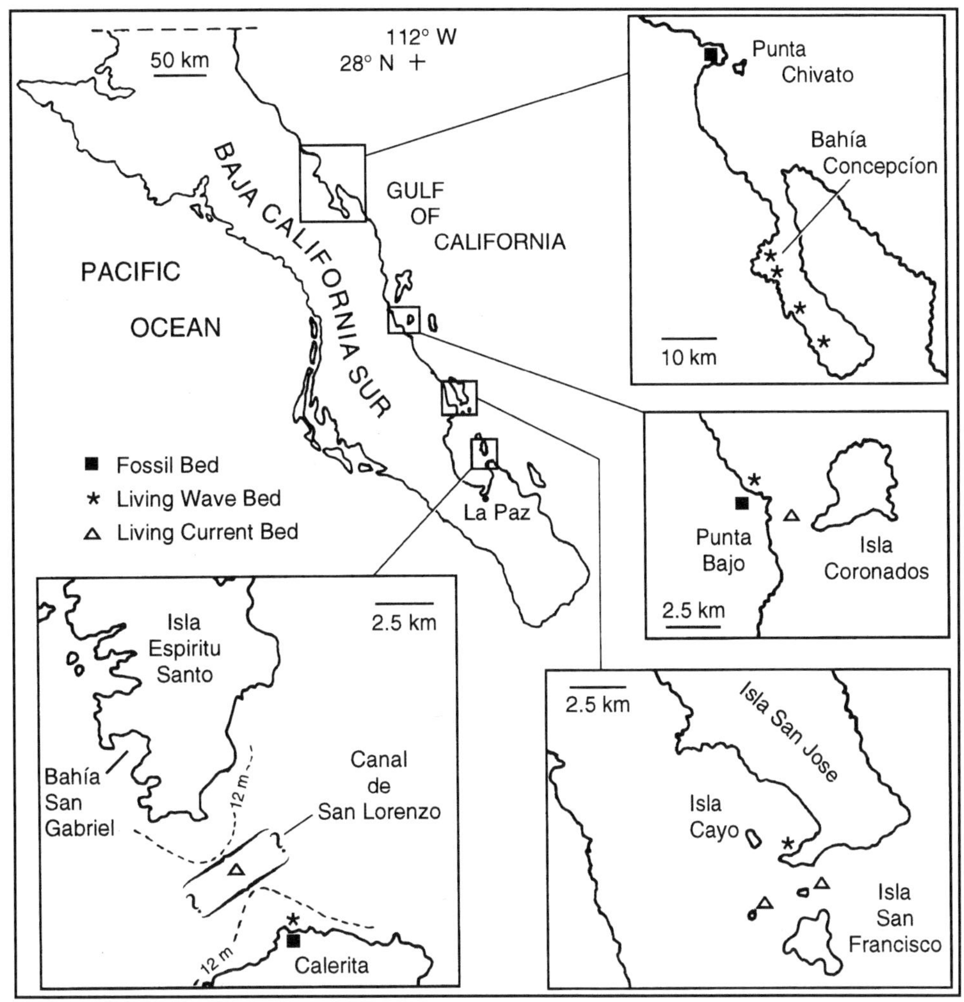

Figure 1. Study areas in the southwestern Gulf of California. Only a few living wave beds in Bahía Concepción are indicated (see Steller and Foster, 1995, for all those presently known). Current bed in Canal de San Lorenzo occurred at depths below12 m; the ? indicates east-west extent of this bed not sufficiently surveyed.

rhodolith fragments (~0.3- to 2-cm white or pigmented pieces), or rhodolith sand (<0.3-cm white fragments identifiable as derived from rhodoliths based on shape). Grab samples were obtained with a Van Veen grab (30 × 40 cm opening, 18-cm depth) and categorized as for dredge samples.

Scuba was used when dredging; skin diving or the environment indicated abundant rhodoliths were, or might be, present. The depth was noted with a calibrated depth gauge, and rhodoliths were usually collected for morphological analyses within a 20- to 40-m radius of the boat anchor. Collection was done by removing the largest individual within a 10-cm-diameter circle centered at a random point along a tape. The largest rhodoliths were sampled to reduce variance in morphological features and to ensure that these features were fully developed. Random points were selected with the restriction that successive points be separated by at least 1 m to avoid multiple samples within small aggregations. Ten or more individuals were collected at each site in current beds or at each of three depths in wave beds (see "Results" for differences in bed types). The three depths were just below the upper (shallow) limit of distribution of the bed, in the middle of the bed, and just above the lower limit of distribution (details in Steller and Foster, 1995). Rhodoliths collected at Steller and Foster's (1995) site 5 were used as representative of Bahía Concepción for all morphological characterizations except branch origin. The latter was determined for three widely separated (~200 m) areas around site 5 described in Steller (1993). Other haphazard collections were made for taxonomic analyses and associated species, and qualitative observations on rhodolith distribution and environmental conditions were recorded.

To determine if external morphology varied among locations and depths, randomly collected rhodoliths were sorted into fruticose and foliose morphologies as defined in Woelkerling et al. (1993). Three measurements of each individual were made to characterize variation in shape: the longest dimension (L), the intermediate dimension measured at the widest point 90° to the midpoint of the axis of L (I), and the shortest dimension as measured at 90° to the midpoint of I (S). These measurements were used to calculate sphericity (Sneed and Folk, 1958; Bosence, 1983a) and the coefficient of variation (sample standard deviation/mean, n = 3) for sphericity.

It has also been suggested that variation in branch density as well as the number of apical versus lateral branches is related to the frequency of rhodolith movement (Bosence 1976, 1991). Branch densities of fruticose rhodoliths were determined as the number of branches/cm^2 by taking the mean of the number of branch tips counted in four, 1 cm^2 quadrats placed haphazardly on the surface. The origin of the outermost branches was assessed by randomly picking four rhodoliths/depth, haphazardly removing the upper 1 cm of 10 branches/rhodolith (subsamples), and counting the number of lateral and apical branches above the point of removal. A branch was considered lateral if, at a branch point, the primary axis continued in the same direction and the branch from it diverged at 45° or more or, if the primary axis diverged at the branch point, the branch from it formed an angle of 45° or more relative to the imaginary continuation of the primary axis (i.e., where the primary axis would be if it continued in a straight line). All other branches of the primary axis were considered apical. The mean of the 10 subsamples was used to characterize an individual.

Statistically significant differences in these morphological variables among locations and depths were assessed with balanced, Model I Analyses of Variance (ANOVA), with the exception that any nested factors were random. If more than 10 rhodoliths were collected at a particular site, 10 were selected at random for analyses (four in analyses of branch origin). Variances that were not homogeneous according to Cochran's Test became so with transformation as noted in "Results." Multiple comparisons were done using the Tukey test and judged significant at $p \leq 0.05$.

Well-preserved, entire fossil specimens were collected from strata exposed along shores and coastal arroyos. Notes were made on the abundance of entire specimens and their relationship to other material among and below them.

RESULTS

Distribution of living beds

Living beds were found in all areas investigated, occurring at depths from 1 to 25 m depending on the site (Fig. 1). Our surveys and prior work in the regions suggest that these beds can be extensive, ranging from 100% or greater (layers of large individuals) cover of rhodoliths in approximately meter-square patches among rhodolith fragments and sand to complete cover of large individuals over the entire bottom. Such beds are commonly found in two environments: (1) gently sloping, shallow, subtidal soft bottoms with moderate wave action (wave beds, 2 to 12 m deep) and (2) relatively level bottoms in channels with tidal currents (current beds, below 12 m; Fig. 1).

The most abundant wave beds so far discovered occur along the mainland shore and around the offshore islands on the western side of Bahía Concepción (see also Meldahl et al., this volume). The shallow, upper margins of these beds, and sometimes the deeper, lower margins, were visible from the air on a calm day in May 1992 (Fig. 2). Some beds are more than 100 m wide and extend for more than 1 km along the shore. Observations and experiments indicate that they occur in areas where rhodoliths are moved by wave action associated with seasonal northeast winds moderated by the limited fetch within the bay. Rhodoliths placed above 3 m in this area were destroyed by wave action and those below 12 m covered by sediment (Steller and Foster, 1995). A bed like those in Bahía Concepción was found at 2- to 7-m depth in a bay at the south end of Isla San Jose where wave action is similarly controlled by northeast winds and bay fetch (Fig. 1).

Excavations to 1 m made by hand and with small, clear cores within these beds found that the 5- to 10-cm-thick surface layer of living rhodoliths and fragments rapidly grades into a mixture of white fragments and fine, dark sediment (Fig. 3). Fragments declined in abundance within a few centimeters and the sediment becomes anoxic. Anoxia occurred at 15- to 20-cm

Figure 2. Aerial photo (polarized light) of rhodolith beds in vicinity of Isla El Requéson in Bahía Concepción. Southern end of island in upper center (island ~100 m wide at widest point in photo). Beds are the dark scalloped and striated areas in the lower half of the photo (A) and dark area immediately below the white band of calcareous sand on offshore side of the island (B). Bar in lower right is airplane wing strut.

Figure 3. Clear core from middle of wave bed in Bahía Concepción. Core is 9-cm diameter.

depth in the center of beds and within a few centimeters of the surface at the deep margins.

We have also found large (commonly >5-cm diameter), lumpy, dense rhodoliths at 1- to 5-m water depths at Punta Bajo and Calerita (Fig. 1). Individuals at these sites, exposed to longer period waves from the Gulf, occur as widely separated (~ meters) individuals on sand/cobble bottoms.

A diverse assemblage of invertebrates, especially crustaceans, brittle stars, sea anemones, and micromolluscs, occurs as cryptofauna among the branches within the rhodoliths in wave beds. There is also an extensive associated infauna and epifauna (Steller, 1993). Rhodolith surfaces can be fouled by a variety of sponges, tunicates, and macroalgae. Beds in Bahía Concepción were extensively covered with blue-green algal/diatom mats in October 1991. Dense growths of the brown alga *Hydroclathrus clathratus* (C. Agardh) Howe can occur in the summer and early autumn, and large areas (~10^2 m) of the bed at Isla San Jose were also overgrown by the green alga *Caulerpa sertularioides* (Gmelin) Howe during late summer and early fall of 1994. Fouling was much less common in the more wave-exposed beds at Punta Bajo and Calerita.

Current beds were found in channels in three regions (Fig. 1). We found that the large bed in Canal de San Lorenzo covers most of the channel (>5 km^2) at depths from 12 to at least 19 m. We did not observe extensive bed development at 3 to 7 m as reported by Schlanger and Johnson (1969). Ten grab samples taken within, east of, and west of the channel revealed rhodoliths fragments grading into white calcareous sand from 20 to 35 m and fine, brown sediment in deeper water. Numerous fragments and occasional entire rhodoliths were found on soft bottoms in dredge hauls south of Calerita near the shipping channel in Bahía de La Paz, but extensive beds were not observed. Similarly, fragments and occasional entire, living individuals were also found in most of the 31 dredge samples (to 18 m) along the western side of Isla Espirtu Santo. Entire, living rhodoliths were abundant in Bahía San Gabriel (Fig. 1), but extensive beds were not found.

The current beds at Isla San Jose and Punta Bajo (Fig. 1) have only been partially surveyed, but reconnaissance dredging and diving indicated these beds also occur at 12 m, with fragments grading into sand below 18 m. Two beds were found at Isla San Jose, one in the northeast portion of the channel and an exceptionally deep bed (25 m) to the west. Unlike the wave beds, live, entire rhodoliths in the current beds are more clumped with approximately meter-square patches interspersed with similar sized areas of white fragments and sand. Excavations to 1 m deep through these clumps showed that the live rhodoliths occur on top of a layer of white fragments and sand with no dark, fine sediment and no indication of anoxia. Numerous fragments and occasional intact individuals were also found in dredge hauls and scuba surveys along the western shore of Isla Cayo near Isla San Jose (Fig. 1).

Numerous other plants and animals also occur in, on, and around the rhodoliths in current beds. Tiger snake eels (*Myrichthys maculosus* Cuvier) and bullseye electric rays (*Diplobatis ommata* [Jordan and Gilbert]) were observed within the beds, and cortez garden eels (*Taeniconger digueti* Pellegrin) can be common at the upper edges. These and other animals that burrow or excavate are no doubt responsible for the 0.1- to 1-m diameter pits and mounds commonly observed within the beds. The sea urchin *Toxpneustes roseus* Agassiz often formed large aggregations covered with rhodoliths. The green algae *Codium magnum* Dawson and *Caulerpa sertularioides* and the red alga *Gracilaria textorii* (Suringer) J. Agardh occurred from September through May of 1994-1995 but were most common in late summer and fall. A previously undescribed fleshy red algal flora (e.g., *Halymenia californica* Smith & Hollenberg, *H. megaspora*

Dawson, *Scinaia johnstoniae* Setchell, *S. latifrons* Howe, *Fauchea mollis* Howe, *F. sefferi* Howe) occurred as epiphytes or free living plants entangled with rhodoliths in winter and spring.

General morphology

Living (pigmented) rhodoliths are found in a variety of shapes and sizes in the Gulf of California (Dawson, 1960), and a range of forms was found at our sites (Fig. 4). Although rhodoliths can have a core of rock or hard material or, in the extreme, form only a thin veneer around cobbles or pebbles (commonly referred to as coralline crusts or "pink paint"), those from the beds in Bahía Concepción (Steller and Foster, 1995) and other sites we have sampled are generally coralline algal throughout (cross sections in Fig. 4) and probably grow vegetatively from

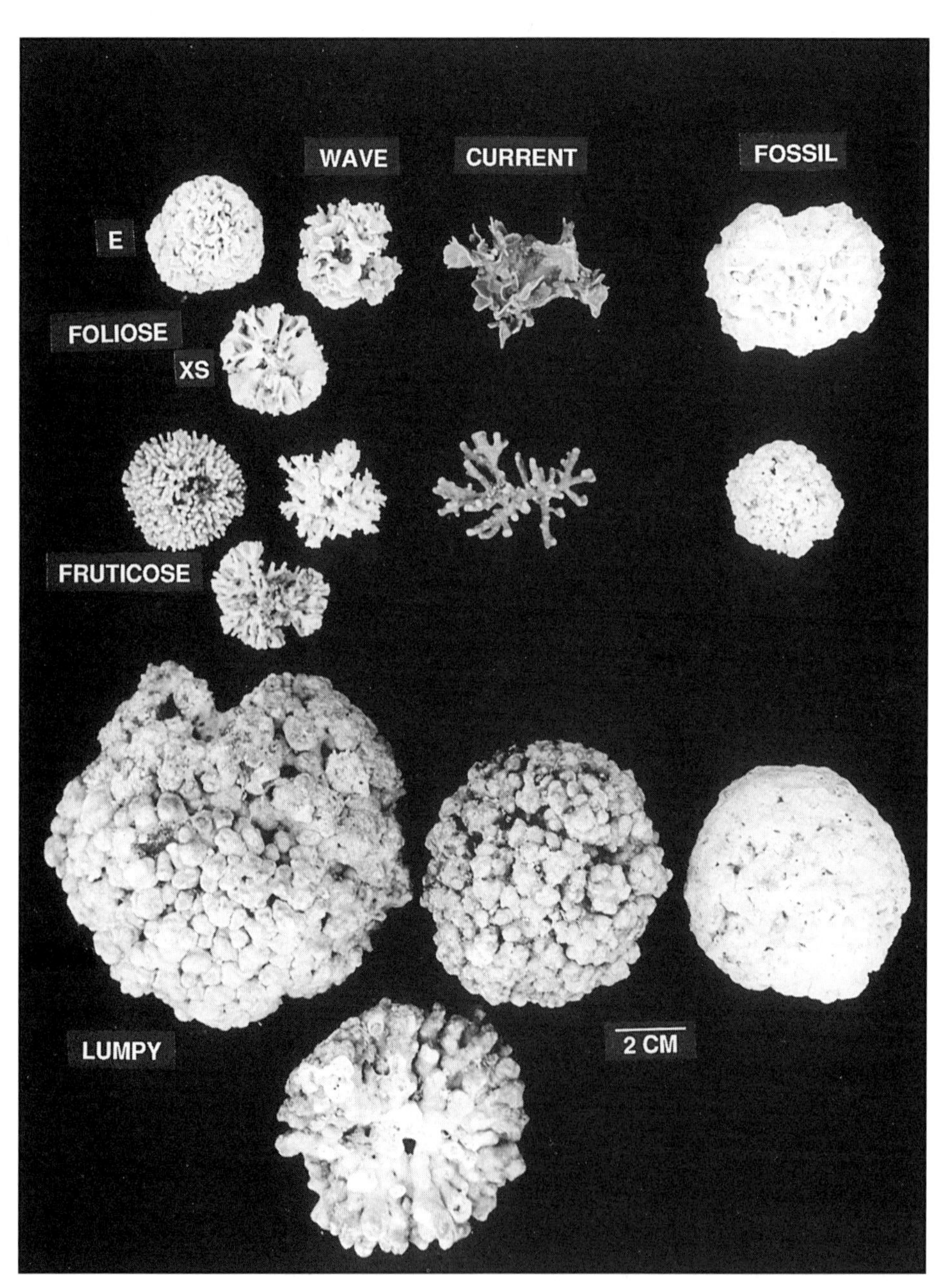

Figure 4. Representative rhodoliths from study sites in southwestern Gulf of California. Individuals from wave beds (WAVE, on the left) are grouped according to form (FOLIOSE, FRUTICOSE, LUMPY). In each group of three on the left side, the upper left rhodolith is entire (E) and typical of shallow water, the upper right is entire and typical of deep water, and the rhodolith below these two has been broken in half (XS) to show internal structure. Note that lumpy forms have not been found in current beds (CURRENT) and that the three nonfossil individuals identified as LUMPY are from wave beds. Pleistocene fossil examples (FOSSIL) are shown for each form.

fragments. Any particular bed may contain a variety of sizes from pigmented fragments less than 1 cm long to individuals over 10 cm in longest dimension.

Three basic forms were found: fruticose, foliose, and lumpy (Fig. 4). The first two forms were found together in beds exposed to moderate wave action and in current beds, whereas the last form was found only in wave-exposed beds. The morphology of fruticose and foliose forms (size, shape, branch density, branch origin) varies, in part, with habitat (see below).

Taxonomy

There have been no modern taxonomic studies of rhodolith-forming species in the Gulf of California, and consequently the number of species and genera present is uncertain and the reliability of characters used to delimit and identify them requires reassessment.

Our understanding of the taxonomy of rhodoliths in the Gulf of California began with the account of Hariot (1895), who reported three species he placed in the genus *Lithothamnion*, two of which were newly described. Heydrich (1901) transferred two of these three into *Lithophyllum* and described an additional *Lithophyllum* species, and Dawson (1944) described a fifth new species of *Lithophyllum* based on material from the Gulf. Subsequently, Dawson (1960) summarized existing taxonomic information on rhodoliths and other nongeniculate coralline red algae in the Gulf in his floristic treatment of the marine red algae of Pacific Mexico; no new detailed taxonomic studies have appeared since.

According to Dawson (1960), rhodoliths in the Gulf of California are thought to include two genera and seven species: *Lithophyllum diguetii* (Hariot) Heydrich, *L. lithophylloides* Heydrich, *L. margaritae* (Hariot) Heydrich, *L. pallescens* (Foslie) Heydrich, *L. veleroae* Dawson, *Lithothamnion australe* (Foslie) Foslie, and *L. fruticulosum* (Kützing) Foslie. All five species of *Lithophyllum* are based on type collections from the Gulf. Dawson (1960, p. 44), however, questioned whether *Lithophyllum diguetii*, *L. lithophylloides*, *L. margaritae*, and *L. veleroae* represented a single polymorphic species but left the matter unresolved.

Dawson (1960) adopted the concepts of *Lithothamnion* and *Lithophyllum* used by Mason (1953), but generic concepts in the Corallinaceae have undergone considerable change since 1953 (see Woelkerling, 1988; Penrose and Chamberlain, 1993), and the generic placement of all species dealt with by Dawson (1960) requires reassessment. Within *Lithothamnion* and *Lithophyllum*, Dawson (1960) delimited rhodolith-forming species from one another using characters relating to external morphology. Recent studies of *Lithophyllum* and *Lithothamnion* elsewhere (Woelkerling and Campbell, 1992; Wilks and Woelkerling, 1995), however, have shown that most such characters vary to the extent that they are unreliable for delimiting species and that characters relating to tetrasporangial/bisporangial conceptacle anatomy provide a much more stable basis for species delimitation. Information on conceptacle anatomy is presently lacking or scant for all rhodolith-forming species reported from the Gulf of California. This information is required to properly reassess the status and disposition of these taxa.

There also are problems with the application of the species names used by Dawson (1960). H. Johansen (*in* Abbott and Hollenberg, 1976, p. 384) concluded that specimens referred by Dawson (1960) to *Lithothamnion fruticulosum* really belong to *L. crassiusculum* (Foslie) Mason. This requires confirmation. More important, none of the type collections of rhodolith-forming species reported from the Gulf have been reexamined in a modern context, so the application of species names is based largely on tradition and not on firm scientific evidence.

We have initiated new taxonomic studies of rhodolith-forming species in the Gulf to determine how many species and genera are represented, what their correct names are based on the examination of type specimens, how the species present can be reliably delimited from one another, and whether and how features relating to external morphology can be used to help identify specimens or reflect the environmental conditions under which particular specimens are found. These studies involve the examination of newly collected *populations* of plants from a number of localities, along with the reexamination of older herbarium material and relevant type collections. Once these studies have been completed, it should be possible to correctly identify rhodolith material from the Gulf to species level in a modern context and to confidently and consistently use these names in ecological and geological studies of rhodoliths in this region. For these reasons we have not given species names to our material but have used forms (discussed previously) and measurements (discussed below) to describe individuals.

Morphological variation

Rhodolith size, shape, branch density, branch origin, and relative composition of fruticose versus foliose forms were examined to determine if these attributes differed among sites and depths, perhaps reflecting environmental differences that would be useful in ecological and paleoecological interpretation. Steller and Foster (1995) found that rhodolith size (longest dimension) ranged from between 3 and 7 cm in nine wave beds sampled in Bahía Concepción. There were significant differences between beds and among depths within beds, but the interaction of these factors was highly significant. Differences among beds were not associated with any known environmental differences. Variation in wave exposure and direction may be important but remains to be quantified. Within a bed, the largest individuals were generally found in the shallower parts of the depth distribution. This pattern was also found for the wave bed at San Jose Island, where rhodoliths at 3 m (mean = 8.1 cm) and 5 m (mean = 6.0 cm) were significantly larger than those from the lower margin at 7 m (mean = 4.7 cm; one-way ANOVA on ln transformed data, $F_{2,27} = 12.31$, $p < 0.0005$; 3 m = 5 m > 7 m). Although rhodoliths in two of the current beds were the smallest of all sites sampled (Canal de San Lorenzo, mean = 2.9 cm; Punta Bajo, mean = 2.8 cm), individuals from the eastern current bed at Isla San Jose had a mean longest dimension of 4.8 cm. Overall, comparisons among means between all individuals from all depths combined at Bahía Concepción versus Isla San Jose versus current beds indicated that rhodoliths from current beds were significantly smaller than those from the wave bed at Isla San Jose but not significantly different from those from the wave bed at Bahía Concepción (one-way ANOVA on ln transformed data, $F_{2,87} = 26.8$, $p < 0.0005$; Isla San Jose > current = Bahía Concepción; Fig. 5).

Rhodoliths were generally spherical at all depths in the wave beds at Bahía Concepción and Isla San Jose. Irregular shapes (between discoidal and ellipsoidal) increased with depth at the former site (Fig. 6). Irregular shapes were common in the current beds. The coefficient of variation reflected these trends in sphericity, with higher coefficients for the current beds and the lower margin of the wave bed at Isla San Jose (Fig. 6). Differences among all sites and depths (depths within wave sites treated as separate sites) were significant (one-way ANOVA, $F_{8,81} = 4.61$, $p < 0.005$). Multiple comparisons could not distinguish significant differences among individual means.

Branch density was usually highest near the upper (five of nine sites) or middle (four of nine sites) of the wave beds in Bahía Concepción (Steller and Foster, 1995) and was highest in the middle of the wave bed at Isla San Jose (one-way ANOVA, $F_{2,27} = 17.9$, $p < 0.0005$; middle > upper = lower). Branch densities of rhodoliths from the current beds were among the lowest, and similar to those from the lower limits of wave beds (Fig. 5). Densities were significantly different among all sites and depths (depths treated as separate sites; one-way ANOVA, $F_{8,81} = 41.4$, $p < 0.0005$; Bahía Concepción Upper > Isla San Jose Middle > all other sites).

The number of apical branches in rhodoliths from wave beds tended to decrease with increasing depth, whereas the number of lateral branches varied little (Fig. 7). Although not quantified, lateral fusion of branches increased as depth decreased. This fusion was so extensive in thalli from the upper margins of sites in Bahía Concepción that it was impossible to distinguish apical and lateral branches. Main effects for Bahía Concepción were analyzed by treatment level because of a significant site × depth interaction. There were significantly more apical branches on rhodoliths from the middle versus lower margins of the beds at two sites but not at the third (t tests adjusted for unequal variances, $t_{5,(2)} = 5.0$, 4.96, and 1.9 respectively, with p's of 0.004, 0.0043, and 0.12). The number of apical branches was significantly different among depths in the wave bed at Isla San Jose (one-way ANOVA, $F_{2,9} = 19.05$, $p = 0.001$; middle > shallow > deep). The number of lateral branches in Bahía Concepción differed significantly among sites but not depths (two-way ANOVA; Site, $F_{2,12} = 9.93$, $p = 0.003$; Depth, $F_{1,12} = 0.32$, $p = 0.585$; interaction not significant). The number of lateral branches differed significantly among depths at the Isla San Jose wave bed (one-way ANOVA, $F_{2,9} = 4.38$, $p = 0.05$), but multiple comparisons did not distinguish significant differences among individual depths.

Observations and data from collections indicate that fruticose rhodoliths are more common in wave beds (Fig. 5), and there was a significantly higher percentage of fruticose individuals in samples from wave versus current beds (depths within wave beds treated as separate sites; one-sided Mann Whitney U test, $W_{6,3} = 37$, $p = 0.047$). A few foliose rhodoliths were observed in the wave bed at Isla San Jose, but they were rare and did not occur in our samples. We have not sampled the current bed on the western side of Isla San Jose area, but fruticose rhodoliths were rarely observed in this 25-m-deep bed.

The morphology of rhodoliths from the wave beds exposed to the Gulf at Punta Bajo and Calerita have not been quantitatively sampled, and qualitative observations and collections along depth gradients have only been made at Calerita. All indi-

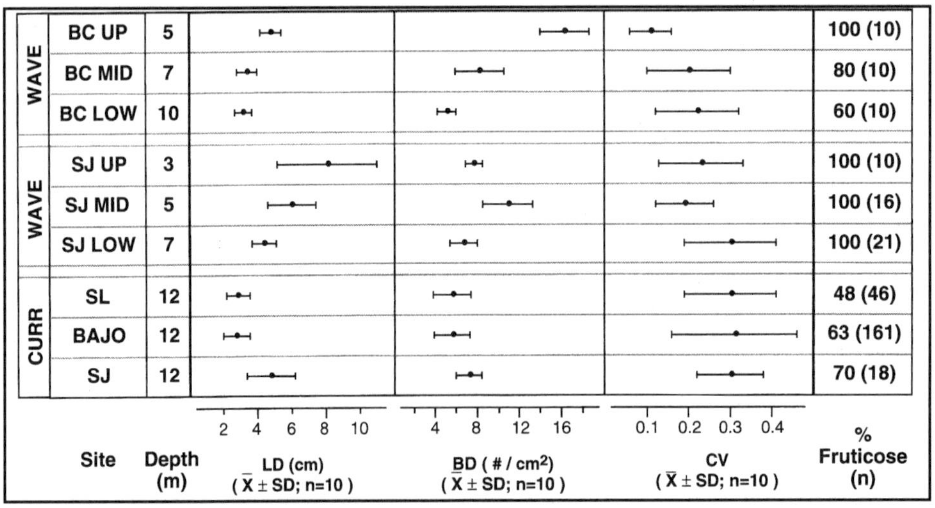

Figure 5. Longest dimension (LD), branch density (BD), coefficient of variation for sphericity (CV), and % fruticose for rhodoliths from wave and current (CURR) beds. BC, Bahía Concepción; SJ, Isla San Jose; SL, Canal de San Lorenzo; BAJO, Punta Bajo; UP, upper margin of wave bed; MID, middle of wave bed; LOW, lower margin of wave bed; m, depth in meters; SD, standard deviation; n, number of samples.

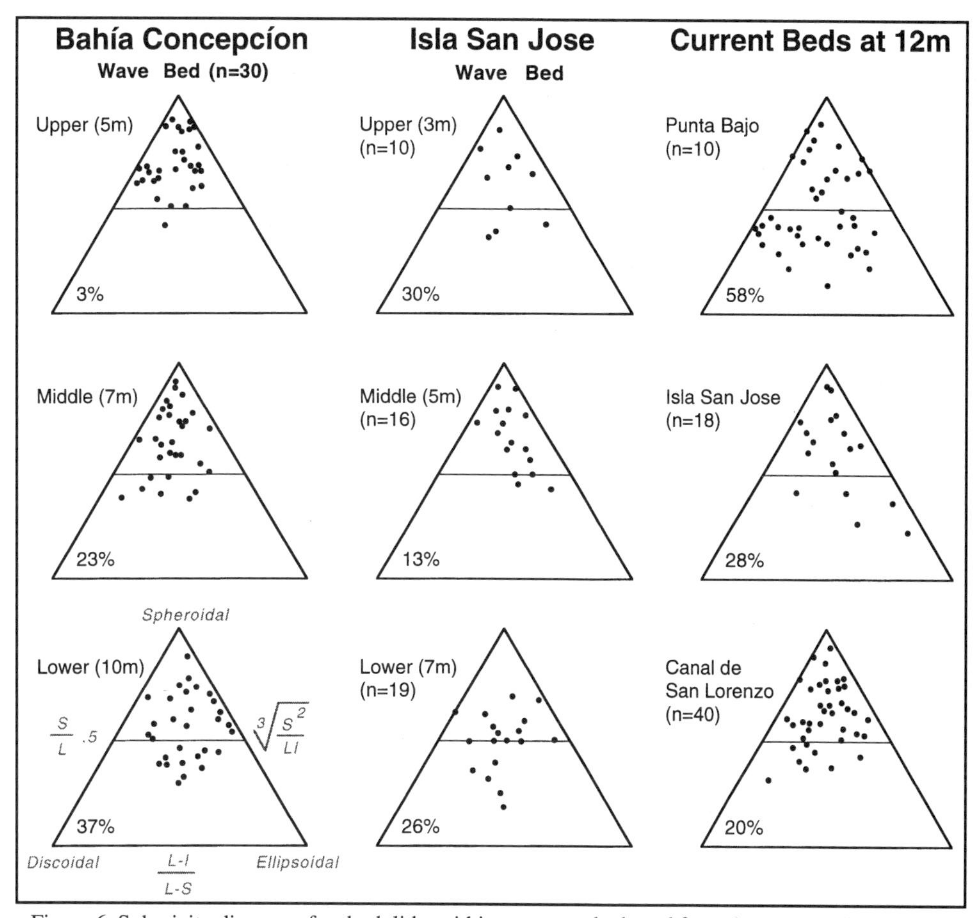

Figure 6. Sphericity diagrams for rhodoliths within two wave beds and from three current beds. Points falling above the internal horizontal line are more spheroidal. Percent of individuals falling below this line is given in lower left of each diagram. m, depth in meters; n, number of samples.

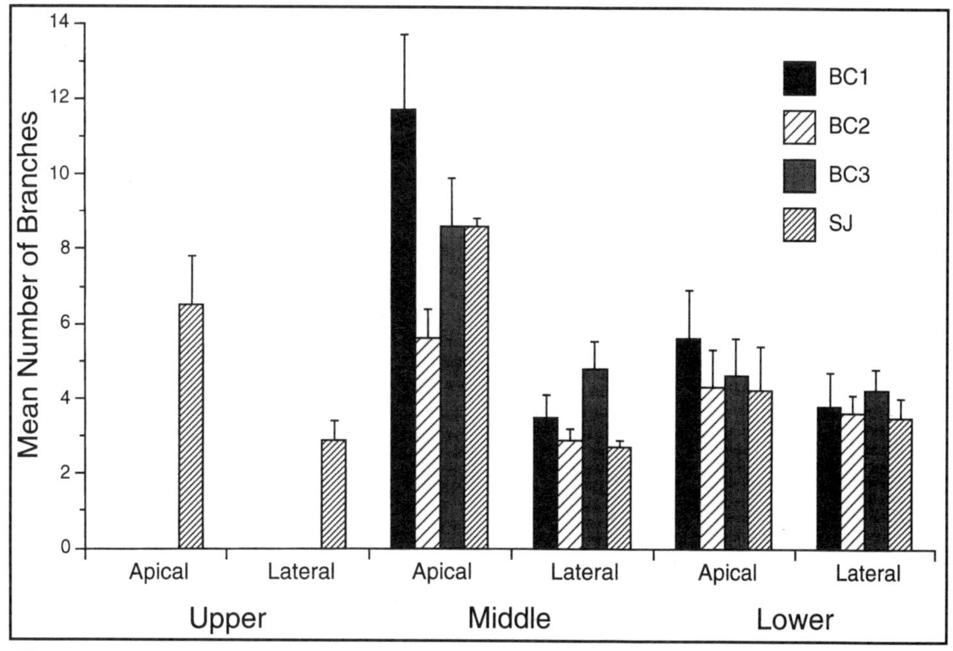

Figure 7. Number of apical and lateral branches per main branch on rhodoliths from three depths (Upper, Middle, Lower) in the wave bed at Isla San Jose (SJ) and two depths (Middle, Lower) within three wave beds in Bahía Concepción (BC1, BC2, BC3). Error bars are + 1 SD; n = 4.

viduals observed and collected at these sites have a lumpy surface and are very spherical (Fig. 4). The branches have diameters up to 1 cm, much larger than those of fruticose rhodoliths from other sites, and the primary axes are sparsely branched and tightly packed throughout the thallus (cross section in Fig. 4). Banding patterns are evident in longitudinal and cross sections of branches, and numerous buried conceptacles, sometimes occurring in bands, are also common in longitudinal sections.

Fossils

Well-preserved, complete fossil rhodoliths were found in all Pleistocene fossil sites (Fig. 1). Fruticose forms are especially abundant, but foliose forms are common in the Pleistocene deposit at Punta Bajo. Pliocene and Pleistocene deposits at Punta Chivato and Punto Bajo also contain abundant rhodolith fragments, sometimes overlain by intact individuals (Dorsey, this volume; Simian and Johnson, this volume; M. Foster, personal observation). Many of the Pleistocene sections exposed in coastal arroyos at Punta Chivato are suggestive of the vertical stratification in living current beds. One section located ~1 km from the tip of Punta Bajo along the western shore had an ~5-cm layer of intact rhodoliths overlying ~30 cm of fragments mixed with sediment that was very similar to the vertical stratification observed in the middle of living wave beds. Other sections with mixtures of rhodoliths and abundant disarticulated shells may represent sites of deposition. The ~1-m-thick deposit exposed along the shore at Calerita is a matrix of intact rhodoliths and clams and cobbles and may be representative of the living beds in this wave exposed area (K. H. Meldahl, personal communication, 1995).

DISCUSSION

Distribution

We located large beds in most regions where prior observations indicated such beds occur. Only a few of the collection areas listed in Dawson (1960) were investigated, suggesting that numerous other beds remain to be located. Recently (August 1995) we found a number of wave beds around the southern end of Isla Coronados, including one of the "exposed type" like those at Punta Bajo and Calerita. Collection records summarized in Dawson (1960) show that rhodoliths occur along the western shore and around offshore islands at least from Bahía de Los Angeles in the central Gulf south to its opening and along the eastern shore as well. In addition to our in situ studies, Hall (1992) filmed large beds at Isla San Pedro Martir southeast of Bahía de Los Angeles, J. Arreola (personal communication, 1994) photographed beds at 10 to 12 m on the northeast end of Isla San Jose, and E. Ochoa (personal communication, 1995) reports an extensive bed growing on a flat, rocky platform at a 25-m depth near Ensenada de Los Muertos 50 km south of La Paz. Beds like those at Punta Bajo and Calerita may also turn out to be abundant when additional gently sloping, cobble/sand habitats more fully exposed to Gulf waves are explored.

These observations and our results suggest that perhaps the only habitats in the photic zone (~30 m?) where rhodoliths do not occur in the Gulf are shallow rocky bottoms (individuals are destroyed by impact when moved by waves), steeply sloping shores (individuals cannot maintain position), calm areas where water motion is insufficient to prevent deposition of fine sediment, and perhaps soft bottoms exposed to large waves or tidal currents that destroy rhodoliths or transport them into other areas. Additional subtidal surveys are needed to test these predictions. High resolution side-scan sonar might be a particularly useful and rapid survey method. More aerial surveys for shallow beds would also be useful, especially if done in the spring when mats of diatoms and large, fleshy algal are least abundant.

Steller and Foster (1995) found that large, shallow beds in Bahía Concepción composed primarily of fruticose forms were best developed in areas with short-period waves produced by episodic winds of around 10 m/sec blowing over fetches on the order of 10 km. The resulting oscillatory water motion on the bottom at depths from 3 to 13 m appears sufficient to move individuals but not destroy them by impact, and to decrease deposition of fine sediment. Our continued observations in this bay and in the wave bed at Isla San Jose support this model. Moreover, that only occasional individuals, small fragments, or calcareous sand were found at depths below 12 m (with the exception of tidal channels) further suggests that wave action in the Gulf is generally insufficient for beds to develop below this depth.

Very large beds of foliose and fruticose forms below 12 m were found in tidal channels. The size of the bed in Canal de San Lorenzo and the distribution of rhodolith fragments and sand suggest this bed may be the primary source of carbonate sediments within the Bahía de La Paz. Fragments, perhaps transported by turbidity currents during hurricanes, have also been found to the east of Canal de San Lorenzo in a core taken from a water depth of 660 m (Schlanger and Johnson, 1969). Rhodoliths in these beds may be moved by tidal currents; current velocities can exceed 60 cm/sec at 5-m depth in Canal de San Lorenzo (Obeso-Nieblas et al., 1993). However, water velocity measurements are not yet available for the bottom of this channel or for any depths where other current beds were found. Accurate measurements of water motion on the channel bottoms with simultaneous determination of rhodolith response are needed. Such measurements would also help determine why rhodoliths are rare above 12 m in these channels. It is possible that, as suggested for some other rhodolith beds (e.g., Prager and Ginsburg, 1989) and by the numerous pits and mounds, bioturbation by associated animals is an important source of rhodolith movement in these current beds.

Environmental variables

Gulf waters are generally clear, so light may only be a controlling factor if currents are sufficiently strong to move rhodoliths in deeper water. Secchi disc depths of 9 m occur over the rhodolith beds in Bahía Concepción (Steller, 1993), up to 15 m in the middle of the bay (Lechuga-Deveze, 1994), and 14 m over the beds in Canal de San Lorenzo (Flores-Ramirez, 1994). According to Holmes (1970), this indicates euphotic zone depths (1% of surface illumination) of 18 to 30 m. Assuming 30 m is a maximum in the regions we studied, then the only beds observed with a high potential for light limitation are the deep current bed west of Isla San Jose (Fig. 1) and the deep bed south of La Paz.

Temperature/light/nutrient relations may affect bed distribution. We have observed rhodoliths in shallow water to become bleached and brittle when the water warmed to near 30° C (Steller, 1993; Lechuga-Deveze, 1994) during early fall in Bahía Concepción. It is interesting that during this time much of each bed was covered with diatom/blue-green algal mats and that rhodoliths beneath these mats were more densely pigmented. These mats and those of larger macroalgae may provide protection from high light at these temperatures, and the former might provide a source of fixed nitrogen.

Growth forms

Pending resolution of taxonomic uncertainties, we, like Steller and Foster (1995), compared morphological variation among beds according to forms and not species. The three forms are generally distinguishable in field-collected specimens. We assumed that differences in the morphology within foliose and fruticose forms were primarily due to environment and not to unknown differences in species. This assumption must be checked when species are properly delimited. The lumpy form and its morphological attributes (other than size and shape) are very distinct from the other forms.

Size is not consistently related to bed type, but sphericity diagrams and the coefficient of variation in sphericity showed rhodoliths in current beds to be more irregular in shape and more similar to individuals from the lower margins of wave beds (Figs. 5, 6). Although sphericity is a useful character, the coefficient of variation is much easier to calculate and illustrate. Coefficient of variation does not distinguish between discoidal and ellipsoidal forms (two of the three measures are similar in both), but these shapes are obvious from inspection, and at our sites, rhodoliths that were not spherical tended to be irregular in shape.

Rhodoliths from current beds also tended to have lower branch densities similar to those of individuals from the lower margins of wave beds. Studies by Bosellini and Ginsburg (1971) and Bosence (1976) indicate that densely branched, spheroidal shapes are indicative of more frequent turning, but this has yet to be experimentally documented in the field.

We examined branch origin based on the suggestion by Bosence (1976, 1983b) that more frequent turning induces lateral branching and that this character can be used to infer the degree of water motion. Since water motion declined with depth at our wave sites, one might thus expect more lateral branching in shallow water. Surprisingly, lateral branching remained relatively constant along depth gradients, whereas apical branching (along with branch density) increased in shallow water (Figs. 5, 7). Whether this represents environmental differences in water motion between our sites and those of Bosence (1976) or, alternatively, reflects variation in branching patterns among species or results from a difference in methodology remains to be determined. We also noted that branch fusions increased in shallow water but were not able to quantitatively measure this feature.

We do not know why some morphological features presumably indicative of more frequent movement were often best expressed in the middle of wave beds rather than at the upper margins (Steller and Foster, 1995; Isla San Jose bed in Figs. 5 and 7). It may be that local wave direction and/or refraction patterns result in more movement in the middle of the beds in some areas. Again, field measurements of water and rhodolith movement are needed.

The most distinctive difference between wave beds and current beds was the higher proportion of fruticose rhodoliths in the former (Fig. 5). We suggest that this feature, used in concert with branch density and the coefficient of variation in sphericity, may be the most accurate indicator of whether a particular population of fossil rhodoliths is from a current bed or a wave bed. The differences in the number of apical branches along depth gradients in wave beds (Fig. 7) suggest that this may also help distinguish bed types, but we have not yet assessed branching patterns in fruticose rhodoliths from current beds.

Fossil beds

The abundance of fossil deposits with well preserved rhodoliths suggests that these morphological analyses can be applied to estimate paleoenvironmental conditions. Given the differences in stratification within living beds, careful stratigraphic analyses at different spatial scales would provide further insights, as would studies of the associated fauna. Central to all these interpretations is the ability to distinguish modern depositional environments from living beds and to apply these distinctions in interpreting the origin of fossil deposits. The transition from living beds to a modern depositional lagoon on the west side of Isla El Requéson in Bahía Concepción (Fig. 2), deeper areas adjacent to living beds, and beach deposits such as those at Calerita might be excellent locations for such studies. The effects of extreme water motion associated with tropical storms on deposition would also be of interest.

Anthropogenic effects

Humans are currently affecting living beds in the Gulf of California. Hall (1992) documented the extreme negative effects of bottom trawling for shrimp on rhodolith and other benthic communities in the Gulf. We also noted extensive degradation (reduced cover of large individuals, increase in dead fragments, sedimentation, increase in dead shells) of rhodolith beds around Isla El Requéson in May 1993 after commercial divers began fishing for the scallop *Argopecten circularis* (Sowerby) in shallow water. These divers disrupt beds with boat anchors, by dragging air hoses over the bottom and moving over the bottom while collecting scallops. Fishing in shallow water began after an apparent overharvest of scallops (over 1.5 million kg including shell; Scallop [*Argopectin circularis*] harvest . . ., 1995) of traditional fishing grounds in deeper water and at the southern end of the bay in June 1991. By June 1994 the harvest was only 1,080 kg (Scallop [*Argopectin circularis*] harvest . . ., 1995).

In addition to these human disturbances, the ecology of the beds suggests that they would be damaged by activities such as sewage discharge that would increase sedimentation or nutrients. The former would bury individuals, and the latter would probably stimulate increased growth of epiphytes. All these real and potential disturbances directly affect rhodoliths and the diverse assemblage of organisms associated with them. The disturbances may also indirectly affect the abundance of species; we suspect that rhodoliths beds may be sites of larval recruitment because we have found juveniles of a number of invertebrates, some of which are commercially important, in and around the rhodoliths.

Living rhodolith beds are clearly widespread in the southwestern Gulf of California. The extensive, well-preserved, and well-exposed nearshore Pleistocene and Pliocene carbonate deposits in the region contain abundant rhodolith fossils, indicating that this community has been an important part of the shallow marine environment for millions of years. The close association of living beds and fossil deposits, similar to that found on a smaller scale by Friewald et al. (1991) in northern Norway, provides an ideal context for studies of ecology, paleoecology, and the processes of carbonate production, transport, and deposition that are just beginning to be realized.

ACKNOWLEDGMENTS

We thank M. Rivera, M. Palmeros, E. Ochoa, R. Yabur, C. Armenta, G. Brabata, P. Villa, K. Pelaez, and especially M. Medina and the Cuevas family for their help. The grab samples were obtained with the able assistance of the captain and crew of the C *Zamora*, Armada de México. D. Lamy assisted with obtaining literature concerning Diguet. R. Dorsey, M. Johnson, and K. Meldahl generously shared information on the geology of the Gulf and reviewed the manuscript. Additional reviews by D. Bosence and S. Murray and editorial assistance by M. Johnson were very helpful. MSF thanks the faculty and staff of Universidad Autonoma de Baja California Sur (UABCS) for their hospitality and support during his sabbatical stay. The research was supported by Moss Landing Marine Laboratories, UABCS, a Fulbright–Garcia Robles scholarship (to MSF), and Mexican Government CONABIO-UABCS Grant #FB135/B086/94 (to RRR).

REFERENCES CITED

Abbott, I. A., and Hollenberg, G. J., 1976, Marine algae of California: Stanford, California, Stanford University Press, 827 p.

Adey, W. H., and MacIntyre, I. G., 1973, Crustose coralline algae: A re-evaluation in the geological sciences: Geological Society of America Bulletin, v. 84, p. 883–904.

Anderson, C. A., 1950, 1940 E. W. Scripps cruise to the Gulf of California. Part I: Geology of islands and neighboring land areas: Geological Society of America Memoir 43, 53 p. plus one plate and one table.

Barnes, D. J., and Chalker, B. E., 1990, Calcification and photosynthesis in reef-building corals and algae, *in* Dubinsky, Z., ed., Ecosystems of the world, vol. 25: Coral reefs: Amsterdam, Elsevier, p. 109–131.

Bois, D., 1928, Notice nécrologique sur Léon Diguet, *in* Diguet, L., Les cactacées utiles du Mexique: Paris, Archives d'Histoire Naturelle, p. 1–12.

Bosellini, A., and Ginsburg, R. N., 1971, Form and internal structure of recent algal nodules (rhodolites) from Bermuda: Journal of Geology, v. 79, p. 669–682.

Bosence, D. W. J., 1976, Ecological studies on two unattached coralline algae from western Ireland: Paleontology, v. 19, p. 365–395.

Bosence, D. W. J., 1983a, Description and classification of rhodoliths (rhodoids, rhodolites), *in* Peryt, T. M., ed., Coated grains: Berlin, Springer-Verlag, p. 217–224.

Bosence, D. W. J., 1983b, The occurrence and ecology of recent rhodoliths—a review, *in* Peryt, T. M., ed., Coated grains: Berlin, Springer-Verlag, p. 225–242.

Bosence, D. W. J., 1991, Coralline algae: Mineralization, taxonomy, and palaeoecology, *in* Riding, R., ed., Calcareous algae and stromatolites: Berlin, Springer-Verlag, p. 98–113.

Bosence, D. W. J., and Pedley, H. M., 1982, Sedimentology and palaeoecology of a miocene coralline algal biostrome from the Maltese Islands: Palaeogeography, Palaeoclimatology, Palaeoecology, v. 38, p. 9–43.

Braga, J. C., and Martín, J. M., 1988, Neogene coralline-algal growth-forms and their palaeoenvironments in the Almanzora River Valley (Almeria, S.E. Spain): Palaeogeography, Palaeoclimatology, Palaeoecology, v. 67, p. 285–303.

Carannante, G., Esteban, M., Milliman, J. D., and Simone, L., 1988, Carbonate lithofacies as paleolatitude indicators: Problems and limitations: Sedimentary Geology, v. 60, p. 333–346.

Carino-Olvera, M. M., and Caceres-Martinez, C., 1990, La perlicultura in la peninsula de Baja California a principios de siglo: Universidad Autonoma de Baja California Sur, Serie Cientifica, v. 1, p. 1–6.

Dawson, E. Y., 1944, The marine algae of the Gulf of California: Allan Hancock Pacific Expeditions, v. 3, p. 189–454.

Dawson, E. Y., 1960, Marine red algae of Pacific Mexico. Part 3: Cryptonemiales, Corallinaceae subf. Melobesioideae: Pacific Naturalist, v. 2, p. 3–124.

Diguet, L., 1911, Pecheries du Golfe de Californie: Bulletin de la Societe Centrale d'Aquiculture et de Peche, v. 23, p. 186–196.

Flores-Ramirez, S., 1994, Utilizacion ecologica de La Bahia de La Paz, Baja California Sur por el rorcual tropical *Balaenoptera edeni* (Cetacea: Balaenopteridae), 1988–1991 [Tesis de Maestro en Ciencias]: Ensenada, Universidad Autónoma de Baja California, 93 p.

Freiwald, A., Henrich, R., Schafer, P., and Willkomm, H., 1991, The significance of high-boreal to subarctic maerl deposits in northern Norway to reconstruct holocene climatic changes and sea level oscillations: Facies, v. 25, p. 315–340.

Hall, H., 1992, Shadows in a desert sea: Del Mar, California, Howard Hall Productions, video.

Hariot, P., 1895, Algues du Golfe de Californie recueillies par M. Diguet: Journal de Botanique, v. 9, p. 167–170.

Hayes, M. L., Johnson, M. E., and Fox, W. T., 1993, Rocky-shore biotic associations and their fossilization potential: Isla Requeson (Baja California Sur, Mexico): Journal of Coastal Research, v. 9, p. 944–957.

Heydrich, F., 1901, Die Lithothamnien des Museum d'histoire naturelle in Paris: Botanische Jahrbucher, v. 28, p. 529–545.

Holmes, R. W., 1970, The secchi disk in turbid coastal waters: Limnology and Oceanography, v. 15, p. 688–694.

Johnson, M. E., and Hayes, M. L., 1993, Dichotomous facies on a Late Cretaceous rocky island as related to wind and wave patterns (Baja California, Mexico): Palaios, v. 8, p. 385–395.

Lechuga-Deveze, C. H., 1994, Shift of acetone-extracted pigments due to unknown natural phytoplankton populations from Conception Bay, Gulf of California: Bulletin of Marine Science, v. 55, p. 248–255.

Littler, M. M., Littler, D. S., and Hanisak, M. D., 1991, Deep-water rhodolith distribution, productivity, and growth history at sites of formation and subsequent degradation: Journal of Experimental Marine Biology and Ecology, v. 150, p. 163–182.

Mason, L. R., 1953, The crustaceous coralline algae of the Pacific Coast of the United States, Canada and Alaska: University of California Publications in Botany, v. 26, p. 313–390.

Meldahl, K. H., 1993, Geographic gradients in the formation of shell concentrations: Plio-Pleistocene marine deposits, Gulf of California: Palaeogeography, Palaeoclimatology, Palaeoecology, v. 101, p. 1–25.

Obeso-Nieblas, M., Jimenez-Illescas, A., and Troyo-Dieguez, S., 1993, Modelacion de la marea en la Bahia de La Paz, B.C.S.: Investigationes Marina CICIMAR, v. 8, p. 13–22.

Ortlieb, L., 1987, Neotectonique et variations du niveau marin au Quaternaire dans la region du Golfe de Californie, Mexique: Paris, Institut Francais de Recherche Scientifique pour le Developpement en Cooperation Collection Etudes et Theses (2 volumes: 779 p. and 257 p.).

Ortlieb, L., 1991, Quaternary vertical movements along the coasts of Baja California and Sonora, *in* Dauphin, J. P., and Simoneit, B. R. T., eds., The Gulf and Peninsular Province of the Californias: American Association of Petroleum Geologists Memoir 47, p. 447–480.

Penrose, D., and Chamberlain, Y. M., 1993, *Hydrolithon farinosum* (Lamouroux) comb. nov.: Implications for generic concepts in the Mastophoroideae (Corallinaceae, Rhodophyta): Phycologia, v. 32, p. 295–303.

Prager, E. J., and Ginsburg, R. N., 1989, Carbonate nodule growth on Florida's outer shelf and its implications for fossil interpretations: Palaios, v. 4, p. 310–317.

Reid, R. P., and MacIntyre, I. G., 1988, Foraminiferal-algal nodules from the eastern Caribbean: Growth history and implications on the value of nodules as paleoenvironmental indicators: Palaios, v. 3, p. 424–435.

Scallop (*Argopectin circularis*) harvest from Bahia Concepcion, BCS, Mexico—1988-1994, 1995: La Paz, Secretaria del Medio Ambiente Recursos Naturales y Pesca (data compiled from original reports).

Schlanger, S. O., and Johnson, C. J., 1969, Algal banks near La Paz, Baja California—Modern analogues of source areas of transported shallow-water fossils in pre-alpine flysch deposits: Palaeogeography, Palaeoclimatology, Palaeoecology, v. 6, p. 141–157.

Scoffin, T. P., 1988, The environments of production and deposition of calcareous sediments on the shelf west of Scotland: Sedimentary Geology, v. 60, p. 107–124.

Scoffin, T. P., 1992, Taphonomy of coral reefs: A review: Coral Reefs, v. 11, p. 57–77.

Scoffin, T. P., Stoddart, D. R., Tudhope, A. W., and Woodroffe, C., 1985, Rhodoliths and coralliths of Muri Lagoon, Rarotonga, Cook Islands: Coral Reefs, v. 4, p.71–80.

Sneed, E. D., and Folk, R. L., 1958, Pebbles in the lower Colorado River, Texas, a study in particle morphogenesis: Journal of Geology, v. 66, p. 114–150.

Steller, D. L., 1993, Ecological studies of rhodoliths in Bahía Concepción, Baja California Sur, Mexico [MS thesis]: San Jose, California, San Jose State University, 89 p.

Steller, D. S., and Foster, M. S., 1995, Environmental factors influencing the distribution and morphology of rhodoliths in Bahía Concepción, B.C.S., Mexico: Journal of Experimental Marine Biology and Ecology, v. 194, p. 201–212.

Toomey, D. F., 1985, Paleodepositional setting of rhodoliths from the Upper Pennsylvanian (Virgil) Salem School limestone of northcentral Texas, *in* Toomey, D. F., and Nitecki, M. H., eds., Paleoalgology: Contemporary research and applications: Berlin, Springer-Verlag, p. 297–305.

Wilks, K. M., and Woelkerling, W. J., 1995, An account of southern Australian species of *Lithothamnion* (Corallinaceae, Rhodophyta): Australian Systematic Botany, v. 8, p. 549–583.

Woelkerling, W. J., 1988, The coralline red algae: an analysis of the genera and subfamilies of the nongeniculate Corallinaceae: New York, Oxford University Press, 268 p.

Woelkerling, W. J., and Campbell, S. J., 1992, An account of southern Australian species of *Lithophyllum:* Bulletin of the British Museum (Natural History) (Botany), v. 22, p. 1–107.

Woelkerling, Wm. J., Irvine, L. M., and Harvey, A. S., 1993, Growth-forms in non-geniculate coralline red algae (Corallinales, Rhodophyta): Australian Systematic Botany, v. 6, p. 277–293.

Younge, C .M., 1963, The biology of coral reefs: Advances in Marine Biology, v. 1, p. 209–260.

MANUSCRIPT ACCEPTED BY THE SOCIETY DECEMBER 2, 1996

Printed in U.S.A.

Geological Society of America
Special Paper 318
1997

Miocene-Pleistocene sediments within the San José del Cabo Basin, Baja California Sur, Mexico

Genaro Martínez-Gutiérrez
Departamento de Geología Marina, Universidad Autónoma de Baja California Sur, La Paz, Baja California Sur, Mexico 23080
Parvinder S. Sethi
Department of Geology, Box 6939, Radford University, Radford, Virginia 24142

ABSTRACT

The San José del Cabo Basin at the southern tip of the Baja California peninsula is considered a half-graben basin that formed in association with the opening of the Gulf of California. The basin consists of Tertiary-Quaternary sediments that range in age from middle Miocene to Pleistocene and accumulated in settings ranging from continental to marine. New formational names are proposed for the different sedimentary deposits. The formations studied include the La Calera, Trinidad, Refugio, Los Barriles, and El Chorro.

The La Calera Formation records the onset of sedimentation (terrestrial) within the basin as a result of middle Miocene block faulting. A marine transgression within the basin is recorded by the Trinidad Formation that conformably overlies the La Calera Formation. The transgression is attributed to subsidence that resulted from the development of a strike-slip zone at the mouth of the Gulf of California. A regressive phase was initiated within the San José del Cabo Basin during the lower Pliocene. Tectonic activity or slow subsidence within the basin marks the onset of a regression with shoaling and terrestrial deposition.

Late Pliocene listric normal faulting (San José del Cabo fault) affected the sedimentary sequence, producing a half-graben structure. The faulting produced terrestrial deposits of the Los Barriles Formation that consist of high-gradient, alluvial fan facies. Movement along the San José del Cabo fault continued by latest Pliocene through Pleistocene time.

During Pleistocene through Holocene time, alluvial sediments of the El Chorro Formation prevailed in the study area, reflecting the continued denudation of the La Victoria basement complex.

Lithologic evidence suggests that sediment accumulation in Pliocene-Pleistocene time was probably affected not only by sea-level change but also by local tectonism associated with syndepositional extension and transpression within the basin.

The objectives of this study are to interpret the depositional history of upper Miocene through Pleistocene sediments and to test the hypothesis that the basin developed through two tectonic stages (late Miocene extensional phase and Pliocene-Recent stage transform displacements).

INTRODUCTION

The San José del Cabo Basin contains continental and marine strata that range in age from early middle Miocene to Recent. The present work provides information to assume that the origin of the basin is associated with opening of the Gulf of California and formation of the Baja California peninsula. Development of both the gulf and the peninsula is related to a Tertiary-

Martínez-Gutiérrez, G., and Sethi, P. S., 1997, Miocene-Pleistocene sediments within the San José del Cabo Basin, Baja California Sur, Mexico, *in* Johnson, M. E., and Ledesma-Vázquez, J., eds., Pliocene Carbonates and Related Facies Flanking the Gulf of California, Baja California, Mexico: Boulder, Colorado, Geological Society of America Special Paper 318.

Quaternary rift system that originated about 10 Ma (Hausback, 1984; Stock and Hodges, 1989; Lonsdale, 1991). Sedimentary deposits of the San José del Cabo Basin contain distinctive tectonostratigraphic features associated with a rift-type basin.

One of the purposes of the present work is to characterize the sedimentary facies of the late Miocene to Pleistocene sediments and to relate their distribution to the unique structural setting associated with the Gulf of California. The overall objective of this study is to reconstruct the deposition of the late Miocene to Pleistocene formations and evaluate the effects of syndepositional extensional relative sea-level changes and basin subsidence.

The study area is located at the southern tip of the Baja California peninsula, Mexico (Fig. 1). The basin covers roughly 2,000 km^2, and includes the La Calera, Trinidad, Refugio, Los Barriles, and El Chorro Formations. The basin forms a north-south depression that is bounded on its western margin by the Sierra La Victoria and along the eastern margin by the Sierra La Trinidad.

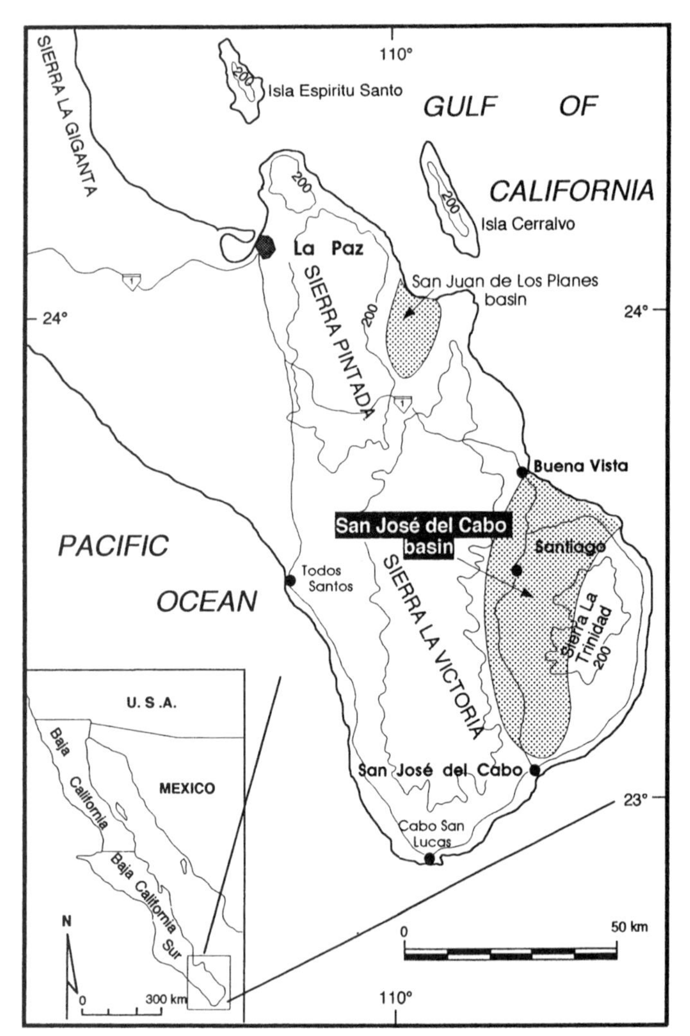

Figure 1. Location map of the study area.

PREVIOUS WORK

The basin was designated by different names: San José del Cabo region (Beal, 1948), Santiago–San José del Cabo region (Pantoja-Alor and Carrillo-Bravo, 1966), and San José del Cabo trough (McCloy, 1984). The study area is referred to as the San José del Cabo Basin in the present work because of its geologic features and geographic location.

The basin was the subject of only a few geologic reconnaissance investigations, most of which included it either as part of regional geologic summaries or focused on specific topics such as descriptive paleontology (Table 1).

Based on field reconnaissance, Beal (1948) proposed that the northern area of the San José del Cabo Basin was affected by a north-south–trending fault along the eastern front of the Sierra La Trinidad. He also recognized that the basin was depressed below sea level during Miocene and Pliocene time and that marine sediments may directly overlie granitic rocks. Mina (1957) assigned the name Salada formation to Pliocene marine deposits within the basin.

Hertlein (1925, 1966) provided some of the first detailed paleontological observations for the peninsular strata. He collected principally Tertiary fossils (mollusks), including some fossils from the "Salada formation" within the San José del Cabo Basin. Hertlein (1925) described some new species from the El Refugio locality (Fig. 2) to which he assigned late Miocene or early Pliocene ages.

Pantoja-Alor and Carrillo-Bravo (1966) mapped and described the formations that crop out in the San José del Cabo Basin. They correlated some volcanic rocks and continental deposits exposed in the study area with the volcanic deposits located at Sierra La Giganta (Comondú Formation) and the youngest marine deposits with the Salada Formation from the Purísima-Iray-Magdalena basin. They also proposed the informal name of Trinidad Formation for the oldest marine sedimentary strata that unconformably overlie what they named Comondú Formation (Table 1). They assigned a late Miocene age to the volcanic rocks and continental deposits on the basis of their similarity to the stratotype of the Comondú Formation. However, the Comondú Formation was not identified or observed.

McCloy (1984) studied the Tertiary sedimentary units that crop out in the study area. She included some stratigraphic sections, megafossils, and planktic foraminiferal data, concluding that sediments deposited in the basin correspond to alluvial and shelf deposits.

Subsequent workers utilized informal names proposed for the upper Miocene to Pliocene sediments (Mina, 1957; McCloy, 1984; Gaitán, 1986; Martínez-Gutiérrez, 1986). In the present work, new stratigraphic nomenclature is formally proposed for the San José del Cabo Basin, because names furnished by previous authors do not have type localities (Table 1). Among the new formation names are La Calera Formation, Refugio Formation, Los Barriles Formation, and El Chorro Formation. The name

TABLE 1. STRATIGRAPHIC TABLE OF UNIT NAMES OF EARLY WORKERS IN THE STUDY AREA*

	Epoch		Pantoja-Alor and Carrillo-Bravo, 1966	McCloy, 1984	Martínez-Gutiérrez, 1986	Smith, 1991	McCloy, in preparation†	This Study
Cenozoic	Pleistocene		Terraces and Coquina	Alluvium	Alluvium / Post-Salada	Alluvium	Alluvium	El Chorro
Cenozoic	Pliocene	Late	Salada	Salada	Salada	"Salada"	Refugio	Los Barriles
Cenozoic	Pliocene	Early	Salada / Trinidad	Salada	Salada	"Salada"	Refugio	Refugio / Trinidad
Cenozoic	Miocene	Late	Comondú	Trinidad	Trinidad	Trinidad	Trinidad	La Calera
Cenozoic	Miocene	Mid		Trinidad	Trinidad	Trinidad / ?	Trinidad	La Calera / Volcanic Rocks
Cenozoic	Miocene	Early		Coyote Red Beds	Coyote Red Beds	?	Coyote	Volcanic Rocks
Cenozoic	Ologocene					Coyote Red Beds		
Cenozoic	Eocene	L						
Cenozoic	Eocene	M						
Cenozoic	Eocene	E						
Cenozoic	Paleocene							
Mesozoic		Late	Granite	Granite	Granite	Granite	Granite	Granite-granodiorite
Mesozoic		Early	Granite	Granite	Granite	Granite	Granite	Granite-granodiorite
Mesozoic	Jurassic							

*Martínez-Gutiérrez, 1994, produced a geologic map of a 1:100,000 scale. The geographic names derive from the topographic maps of scale 1:50,000 as edited by the Instituto Nacional de Estadística, Geografía e Informática (1983a through f).
†McCloy, C., 1997, Reconnaissance geologic map of the San Jose del Cabo trough, Baja California Sur, Mexico, Plates I and II: U.S. Geological Survey Open File Report, in preparation.
[shaded] = No deposition. —·—·—·— = Unconformity

Trinidad Formation is formalized in this work by providing a neostratotype.

GEOLOGIC SETTING

The southern portion of the Baja California Sur state (south of about 24.5°N) is dominated by batholithic massifs that resemble northerly trending basins and ranges. The ranges are represented by the Sierra La Victoria and Sierra La Trinidad ridges, and the basins are represented by the San José del Cabo and San Juan de Los Planes Basins (Fig. 1). The basins contain Cenozoic nonmarine and marine sediments that range in age from late Miocene to Recent. Quaternary deposits consist of alluvial fans, braided river deposits, dunes, and beach deposits. The rocks that crop out at the southern tip of the Baja California peninsula are of igneous, sedimentary, and metamorphic. The igneous and associated metamorphic rocks are dated as pre-Tertiary and are thought to be of Mesozoic age (Gastil et al., 1978).

Two major crystalline complexes bound the San José del Cabo Basin: the La Victoria igneous-metamorphic complex on the west and the Trinidad igneous complex on the east. These two crystalline complexes are the sources of most of the sediments that filled the basin.

The La Victoria complex (identified as Kim, Fig. 2) consists of granite, granodiorite, tonalite, gneiss, schist, and mafic dikes (andesitic). Only regional geologic mapping (Lopéz-Ramos, 1973; Instituto Nacional de Estadística Geografía e Informática, 1987) and local geologic studies exist along the complex (Altamirano, 1972; Gastil et al., 1976; Aranda-Goméz and Pérez-Venzor, 1988; Ortega-Gutiérrez, 1982; Frizzell, 1984; Murillo-Muñetón, 1991; Carrillo-Chávez, 1991; and Sedlock et al., 1993). Gastil et al. (1976) estimated an age of Late Cretaceous (98.4 to 93.4 Ma) for the complex, using radiometric dating.

The La Trinidad complex (Kgr-grd, Fig. 2) has received minimal study and consists mainly of three major lithologic units: granite, granodiorite, and lava flows (rhyolitic). Gastil et al. (1976) assigned an age of Late Cretaceous–Early Tertiary (88.2 to 54.1 Ma) to the complex, based on radiometric dating.

Acid volcanic and volcaniclastic rocks of possible late mid-

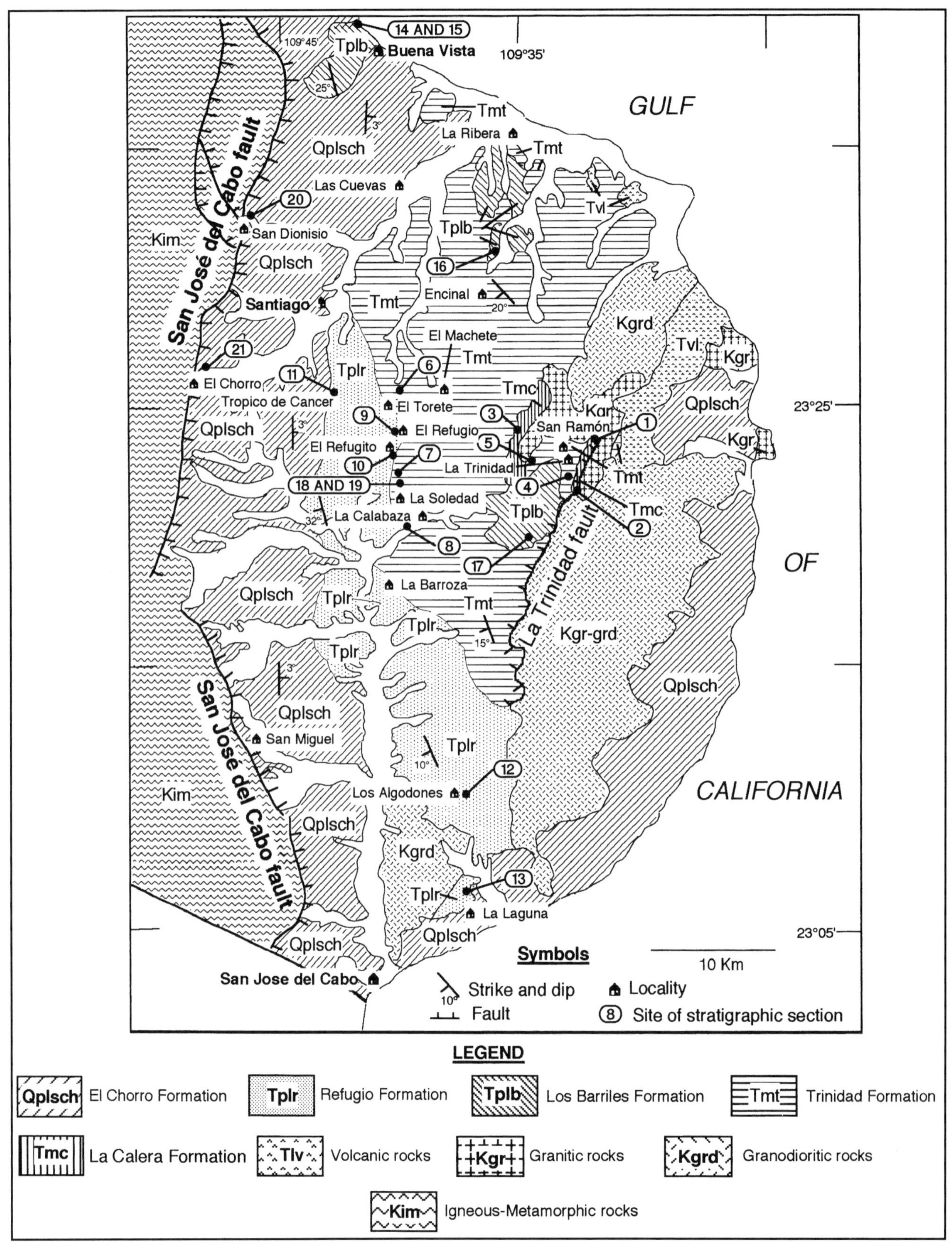

Figure 2. Generalized geologic map of the San José del Cabo Basin, adopted from Martínez-Gutiérrez (1994).

dle Miocene age also crop out at the northeastern margin of the San José del Cabo Basin.

METHODS

Twenty-one stratigraphic sections were measured within the San José del Cabo Basin (Fig. 2). Table 2 shows the characteristics of the sites as well as the purpose of measuring each section in terms of facies, paleocurrent indicators observed, and stratigraphic interval. The goal was to obtain more detailed information to characterize the facies of each formation, to infer the depositional processes associated with their accumulation, and to demonstrate the spatial relation among the sedimentary units.

Each stratigraphic section was measured, where possible, from the bottom to top. The thickness of individual beds in the stratigraphic sections shown is schematic. Composition, bedding, grain size, color, sorting, sedimentary structures (inorganic and organic), fossils, and paleotransport indicators were recorded in the field for each section (Fig. 3). Paleotransport indicators are not abundant. The indicators encountered with measurable transport azimuths include imbrication, cross-bedding (tabular and trough), channels, climbing-ripple marks, and groove marks. The azimuths of the paleocurrent indicators from each unit are plotted in rose diagrams.

RESULTS

La Calera Formation

The La Calera Formation (LCF) is the oldest sedimentary deposit and crops out at the eastern margin approximately 26 km east of Santiago. Pantoja-Alor and Carrillo-Bravo (1966) correlated this unit with the Comondú Formation cropping out near La Paz; however, the LCF lacks volcanogenetic rocks that characterize the Comundú Formation. McCloy (1984) proposed the informal name of Coyote Red Beds, yet the name of arroyo El Coyote does not appear on the topographic maps. The LCF is a distinctive red-colored deposit of conglomerate and sandstone.

The LCF is composed of about 25% conglomerate and 75% sandstone beds in a crudely "fining-upward" sequence. Conglomerate beds are concentrated mainly at the bottom of the sequence.

Exposures are restricted to a small downfaulted area of the La Trinidad complex. Based on its good exposures and easy access, we designate the stratigraphic section at arroyo La Calera, near La Trinidad, to be the type section (Fig. 2). Three columnar sections totaling 300 m were measured from this formation. The most complete and best-exposed interval is located at the intersection of arroyo La Trinidad and arroyo El Sauce (Fig. 4), where its thickness is about 130 m. The section of La Trinidad shows the sequence's general fining-upward trend.

The base of the LCF is characterized at all of these locations mainly by poorly sorted clast-supported conglomerate beds containing coarse pebbles and small cobbles.

The LCF grades from coarse to fine sandstone and from sandy conglomerate to conglomeratic sandstone (40 to 20% coarser clasts and 60 to 80% sand, respectively) at the locality of arroyo La Calera and La Trinidad. At the La Trinidad site, the LCF is composed of sandy conglomerate to pebbly sandstone with a conglomeratic content between 30 and 40% and a sandy content between 70 and 60%.

The top of the LCF is well exposed at the junction of arroyos La Calera and La Trinidad and is composed of medium to fine sandstone (pebbly arenite). Here, it grades upward into the fine-

TABLE 2. CRITERIA CONSIDERED FOR SELECTION OF SECTIONS

Site	Formation	Facies	Interval	Paleocurrent Indicators
1 San Ramón	La Calera	Alluvial	Bottom-middle	Fair
2 La Trinidad	La Calera	Alluvial	Middle-top	Good
3 Arroyo El Sauce	La Calera	Alluvial	Bottom-top	Fair
4 La Trinidad	Trinidad	Shallow marine	Bottom	Poor
5 Arroyo El Sauce	Trinidad	Shallow marine	Bottom-middle	Poor
6 El Torete	Trinidad	Outer shelf	Middle	Poor
7 El Rosarito	Trinidad	Inner shelf	Top	Good
8 La Calabaza	Trinidad	Shallow marine	Top	Poor
9 El Refugio	Refugio	Shallow marine	Bottom	Poor
10 El Refugito	Refugio	Shallow marine	Bottom-middle	Poor
11 Trópico de Cancer	Refugio	Shallow marine	Top	Good
12 Los Algodones	Refugio	Shallow marine	Bottom	Poor
13 Cerro La Laguna	Refugio	Shallow marine	Bottom-top	Fair
14 Arroyo Buenos Aires	Los Barriles	Alluvial	Bottom-top	Good
15 San Bartolo	Los Barriles	Alluvial	Bottom-top	Poor
16 El Encinal	Los Barriles	Alluvial	Bottom-top	Good
17 La Trinidad	Los Barriles	Alluvial	Middle	Poor
18 El Rosarito 1	Los Barriles	Alluvial	Bottom	Good
19 El Rosarito 2	Los Barriles	Alluvial	Bottom	Good
20 Arroyo San Dionisio	El Chorro	Alluvial	Bottom-middle	Poor
21 El Chorro	El Chorro	Alluvial	Bottom-top	Fair

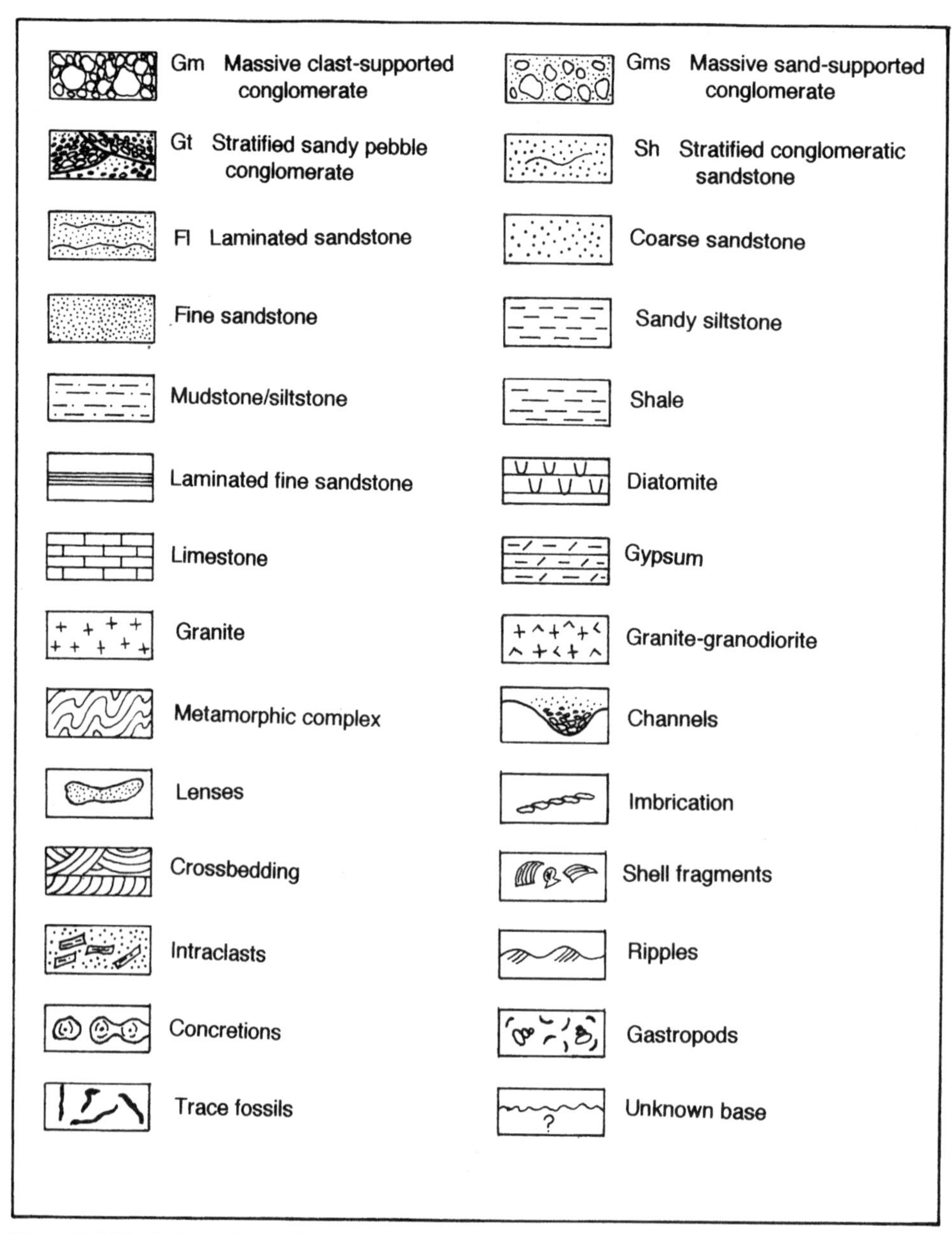

Figure 3. Lithofacies code and legend symbols for stratigraphic columns in Figures 4, 7, 8, 9, 11, 13, 14, 15, and 17.

grained sediments of the Trinidad Formation. The top of the LCF exhibits tabular stratification and massive stratification that ranges in thickness from 0.05 to ~2.0 m. Channels, scours, and clast imbrication are present in this part of the formation. At arroyo La Calera, channels and scours as well as slump structures in sandy beds are present. The clast composition of the La Calera Formation include granitic (69%), rhyolitic (20%), and pyroclastic (11%) rock fragments (Fig. 5) (Martínez-Gutiérrez, 1994).

Coarse-grained, poorly sorted, crudely stratified beds of sand-supported conglomerate punctuated by scour band lenses of clast-supported conglomerate indicate that the formation probably was deposited under alluvial conditions. Paleotransport indicators are scarce within the LCF; however, channels, trough cross-bedding, tabular cross-bedding, and imbrication were recorded from Sites 2 and 3 (Fig. 6). The indicators show that the flow patterns of these deposits had a west-northwest trend. Thus, clast types and limited cross-bedding converge to indicate that most of the La Calera clasts were probably derived from the adjacent La Trinidad complex and were transported generally northward during an initial depositional stage of the basin formation.

The age of the La Calera Formation remains undetermined. A reasonable age estimate is assigned using simple stratigraphic principles. The LCF unconformably overlies the La Trinidad basement complex and conformably underlies the Trinidad Formation. Rhyolitic lava flows of early Tertiary age

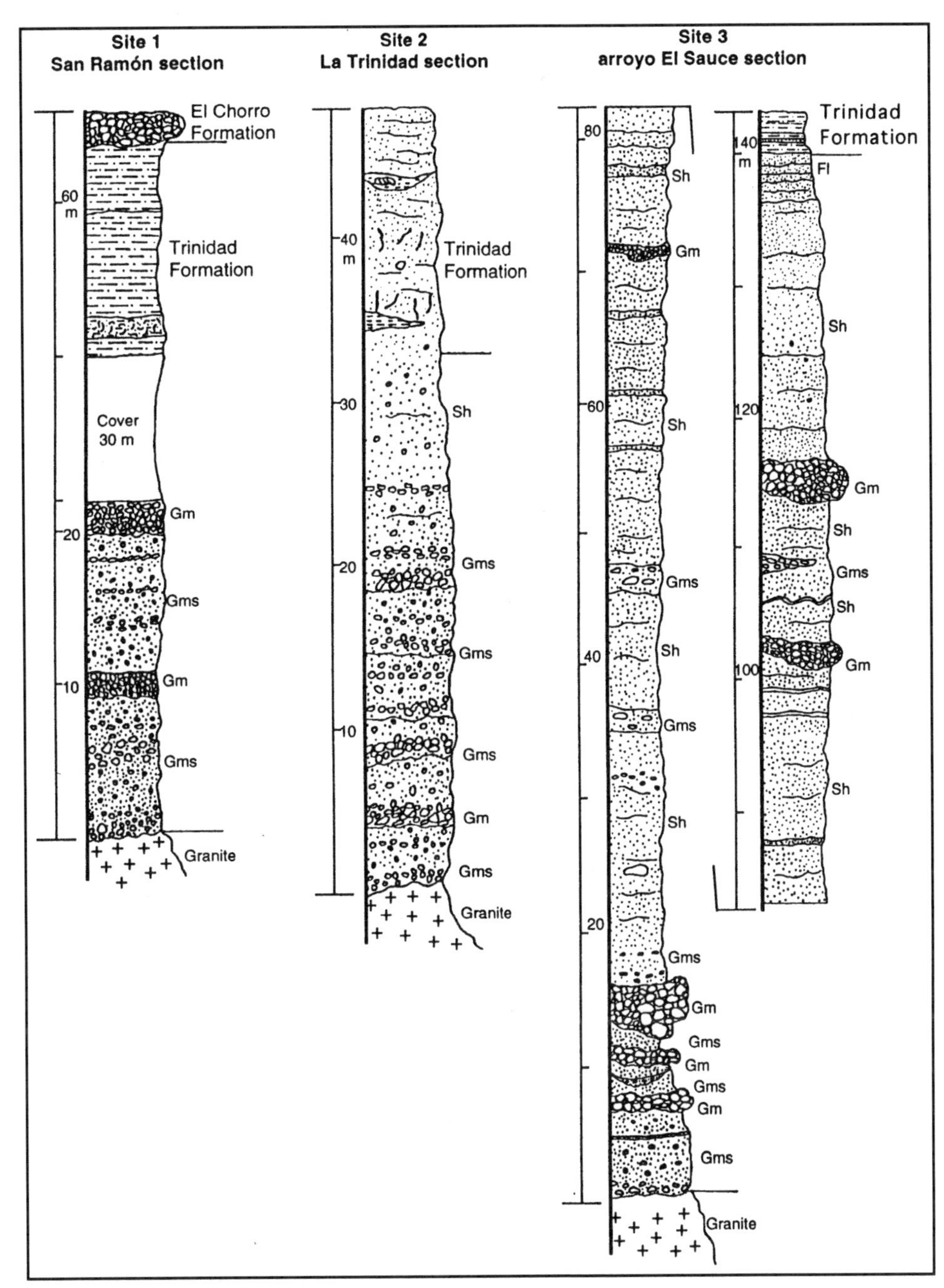

Figure 4. Stratigraphic columns of the La Calera Formation. See Figure 3 for legend.

overlie the Trinidad basement. An age of late Miocene based on megafossils and microfossils is assigned to the lower part of the Trinidad Formation (McCloy, 1984; Smith, 1989; Martínez-Gutiérrez, 1994). Therefore we estimate that the La Calera Formation could be middle-late Miocene in age. We propose the name of La Calera Formation and use it to designate the older, alluvial, basement-derived sequence within this basin.

Trinidad Formation

The Trinidad Formation (TF) was first described by Pantoja-Alor and Carrillo-Bravo (1966) and consists of greenish shale, mudstone, and sandstone that collectively represent the initial marine transgression. Marine fossils and marine trace fossils at the base of the unit provide evidence of marine deposition. The TF exhibits both vertical and lateral lithological changes from the northern to the southern region of the basin. No locality shows a complete stratigraphic section, but five stratigraphic columnar sections were measured (Fig. 2) to characterize typical facies and its lateral variations. The TF exhibits a prevalent northwest strike and dips southwest from 10 to 30°. Toward the eastern basin margin (the rancho La Trinidad area), the formation is slightly deformed into an antiform structure.

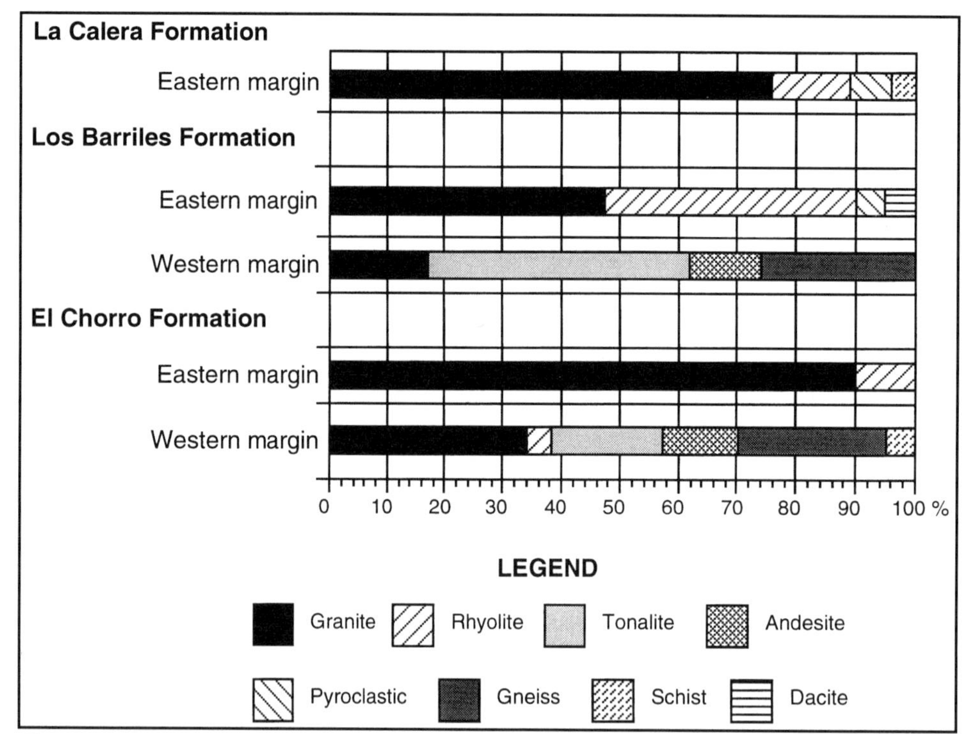

Figure 5. Clast composition of the alluvial formations: La Calera, Los Barriles, and El Chorro Formations. An average of 125 clast counts was done for each location.

The type section of the TF is located 28 km from the town of Santiago in the rancho La Trinidad area (Fig. 7). At rancho La Trinidad, the base of the TF conformably overlies the La Calera Formation; elsewhere it unconformably overlies the granitic-granodioritic basement. The TF is conformably overlain by the Refugio Formation, but toward the southern part of the basin the two formations are unconformable.

The TF is composed mainly of gray-greenish, laminated, fine- to medium-grained marine sandstone, shale, and siltstone and some diatomite laminae toward the center of the basin. Fine-grained sandstone, siltstone, and mudstone tend to dominate the formation. The sands of the Trinidad Formation are classified petrographically as medium-fine, calcareous, texturally mature arkose and consist mainly of angular to subrounded quartz, mica (mainly biotite), and K-feldspar grains and rare igneous and metamorphic rock fragments. An overall vertical trend is suggested for coarse-grained sandstone beds that grade upward into shale-siltstone with diatomite laminae, which grade near the top of the formation to coarse-to medium-grained sandstone beds. We estimate a total thickness of 400 m from the six stratigraphic sections.

To the west of the Sierra La Trinidad ridge (Fig. 1) toward the Sierra La Victoria ridge (Figure 1), the Trinidad Formation changes lithology. To better characterize the Trinidad Formation, it has been divided into three facies: lower, middle, and upper.

Lower facies. The lower facies is best exposed toward the eastern margin in the rancho La Trinidad area where it directly overlies the La Calera Formation. The grain size at the base ranges from coarse to medium sand, whereas at arroyo El Sauce the formation ranges from fine to medium sand that grades into clayey and silty sandstone beds. This basal sand facies is roughly equivalent to McCloy's (1984) subunit "A," which consists mainly of white coarse-grained quartzose sandstone, green

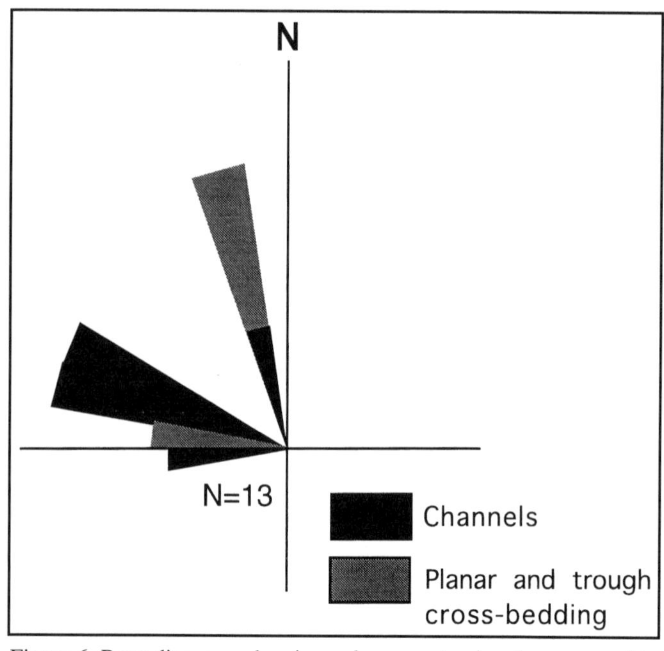

Figure 6. Rose diagrams showing paleocurrent azimuths measured in the La Calera Formation.

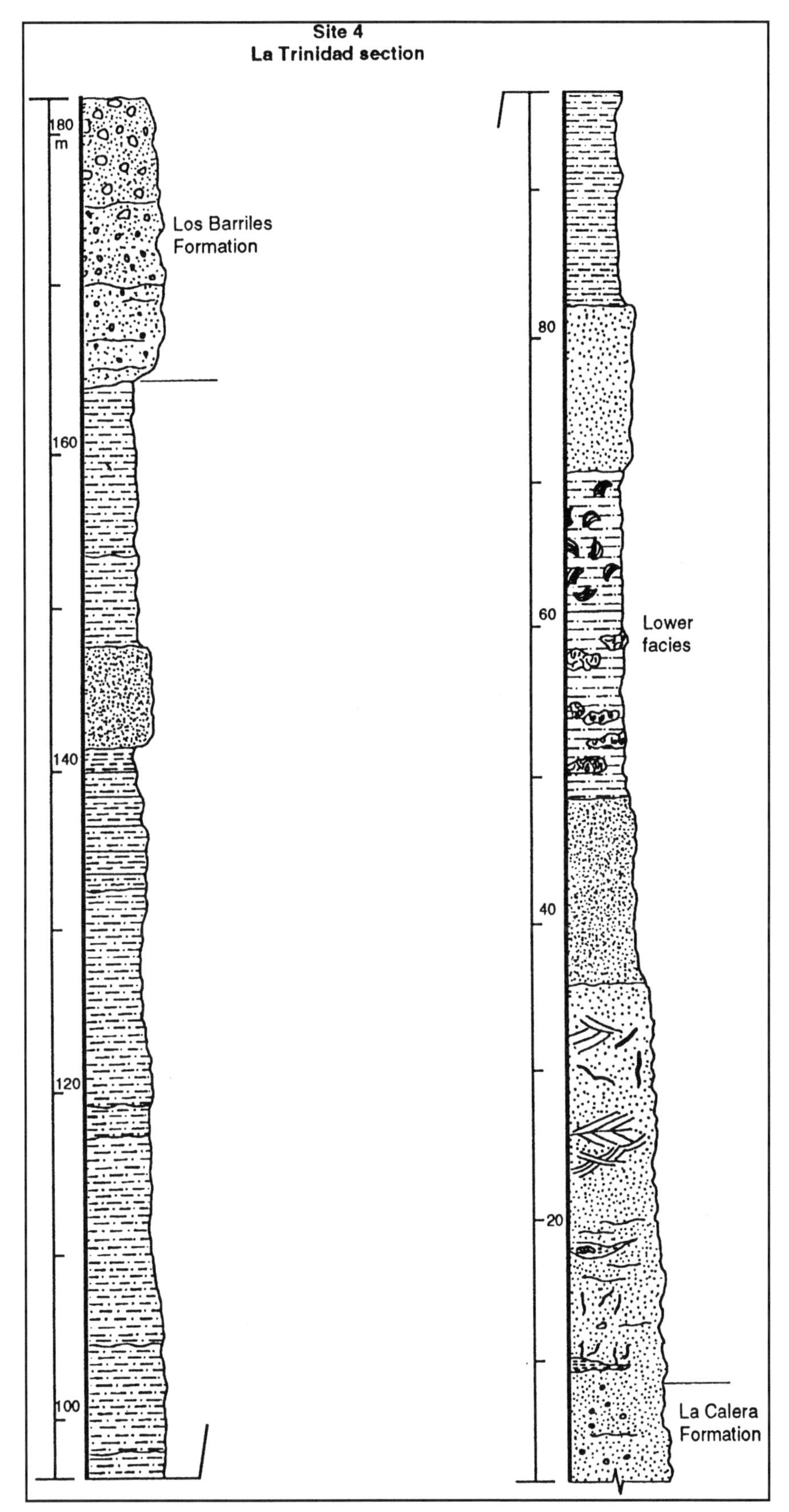

Figure 7. Stratigraphic columns of the Trinidad Formation, lower facies. See Figure 3 for legend.

siltstone, and shale beds. It appears to be restricted to the eastern portion of the basin (Fig. 7).

The basal quartz sandy facies contains a great variety of sedimentary structures. It is characterized by massive beds to thickly bedded sandstone. Herringbone cross-bedding, small-scale ripple marks, and planar bedding are present (McCloy, 1984). Burrows and fossil shell fragments are also present within these fine-grained, white, quartzose sandstone beds. Lenticular stratification within the shale-siltstone laminated beds is present. Lenses and thin beds of sandstone and marl are rich in fossil gastropods and pelecypods. These fossils suggest shallow, brackish-marine environments. Rodríguez-Quintana (1988), on the basis of the presence of *Cerithea* sp., *Anadara* (*Grandiarca*), *Strombus*, *Melongena*, *Murex*, *Conus*, and *Oliva*, inferred a lagoonal environment for the lower Trinidad. Impressions of rafted plant debris also are present at the locality of rancho La Trinidad, confirming nearby terrestrial habitats. Scattered lenses and thin beds of limestone containing small communities of oysters and scallops are present within the muddy sequence. Martínez-Gutiérrez (1991) reported the first horse jaw fossil found (*Merychippus-Pliohippus*, Miocene ?) in the state of Baja California Sur from the lower facies of the Trinidad Formation.

Middle facies. The middle facies, which conformably overlies the lower facies, is a sequence of alternating tan shale-siltstone laminae and fine-grained sandstone beds containing rare to abundant diatomite laminae. The middle facies grades from fine-grained sandstone to shale-siltstone laminae; occasionally muddy sandstone facies occur toward the top of the formation. Small outcrops of mudstone, diatomaceous shale, and diatomite occur at the central part of the basin (Site 6). This fine-grained facies is also equivalent to McCloy's (1984) subunits "B" and "C."

Massive to laminated fine-grained sandstones and siltstone beds contain rare concretions and large-scale slumps (~1.5-m width). These slumps are more common in the localities of rancho La Soledad and arroyo La Trinidad near rancho El Machete (Fig. 2). No other sedimentary structures were observed, but the remains of whale bones are present within mudstone beds near the rancho El Torete locality (Fig. 8).

Slumps of 1.5-m width were observed in the Trinidad Formation, generally occurring within the muddy sandstone facies. The slumps probably originated by overstepping of depositional slopes, possibly associated with tectonic activity.

Upper facies. This facies is characterized by a sequence of greenish to reddish, medium-to fine-grained quartzose sandstone and siltstone beds. The localities at rancho El Rosarito and rancho La Calabaza (Fig. 9) represent the best exposures of this facies. The top of the Trinidad Formation is truncated by coarse material (pebbles and cobbles) that belongs either to Recent alluvial fans (Sites 4 and 7) or to the Los Barriles Formation. The Trinidad Formation appears to grade both vertically and laterally southward into the overlying Refugio Formation.

The upper part of the TF exhibits massive (beds greater than 1.20 m in thickness) to thickly bedded coarse- to medium-grained sandstone that contains minor siltstone beds. Tabular cross-bedding, trough cross-bedding, and climbing ripple marks are common near El Rosarito and La Calabaza. In the area of Las Cuevas, basal scour, chevron, and groove marks are present in the sandy beds that represent shallow, sandy marine shoals.

The upper facies of the Trinidad Formation also yielded some paleocurrent data. Tabular and trough cross-bedding indicate transport toward the northeast and the south-southwest. Trinidad Formation paleocurrent indicators also included measured groove marks showing a different direction of flow in the Las Cuevas area (Fig. 10).

The combination of lithology and fossil assemblages indicates that the lower, middle, and upper facies of the Trinidad Formation were probably deposited in three different environments. The base of the formation (lower facies) probably corresponds to nearshore-lagoonal deposits, as indicated by the fossil assemblages containing *Cerithea* sp., *Anadara* (*Grandiarca*), *Strombus*, *Melongena*, *Murex*, *Conus*, and *Oliva*. The white, texturally mature, coarse-to medium-grained quartz-enriched, sandstone strata also suggest considerable reworking of shallow deposits by some combination of wave-and tide-generated currents or wind-generated currents. The alternating fossiliferous mudstone strata of the middle facies probably reflect deeper or protected bottom areas (McCloy, 1984; Martínez-Gutiérrez, 1994). The marine microfauna (*Globorotalia lenguaensis, Globigerina angustiumbilicata,* and *Globorotalia mayeri*) described by McCloy (1984) and the fine-grained sandstone, mudstone, and diatomite beds indicate that the middle facies of the TF was deposited in shelf depths slightly greater than that of normal wave base. Medium- to fine-grained sandstone with cross-bedding and shell fragments characterize the upper facies of the Trinidad Formation. These features indicate that the facies accumulated in high-energy, shallow marine waters affected by unidirectional currents, probably related to inner shelf shoals and bars. Outliers of the TF also suggest that the sea transgressed across significant topographically subdued areas of what are now La Trinidad basement exposures.

Different ages are assigned to the Trinidad Formation: middle? Miocene to upper Pliocene (McCloy, 1984), lower Pliocene (Pantoja-Alor and Carrillo-Bravo, 1966), and upper Miocene (Smith, 1991). Using the different fossil assemblages described and found in the Trinidad Formation, we estimate that the unit ranges from late Miocene to early Pliocene in age.

Refugio Formation

The present work uses the Refugio Formation (RF) to indicate the youngest marine unit that crops out throughout the basin. The RF forms a widespread outcrop belt and has a wedge shape in map view (Fig. 2). The RF strikes to the northwest and dips from 10 to 20° to the southwest. Toward the south, the RF unconformably overlies the Sierra La Trinidad basement complex. Different thicknesses were measured for the RF. Pantoja-Alor and Carrillo-Bravo (1966) measured a thickness of 80 m, whereas McCloy (1984) found a thickness of 280 m. Gaitán (1986) measured a thickness of about 360 m at the southern

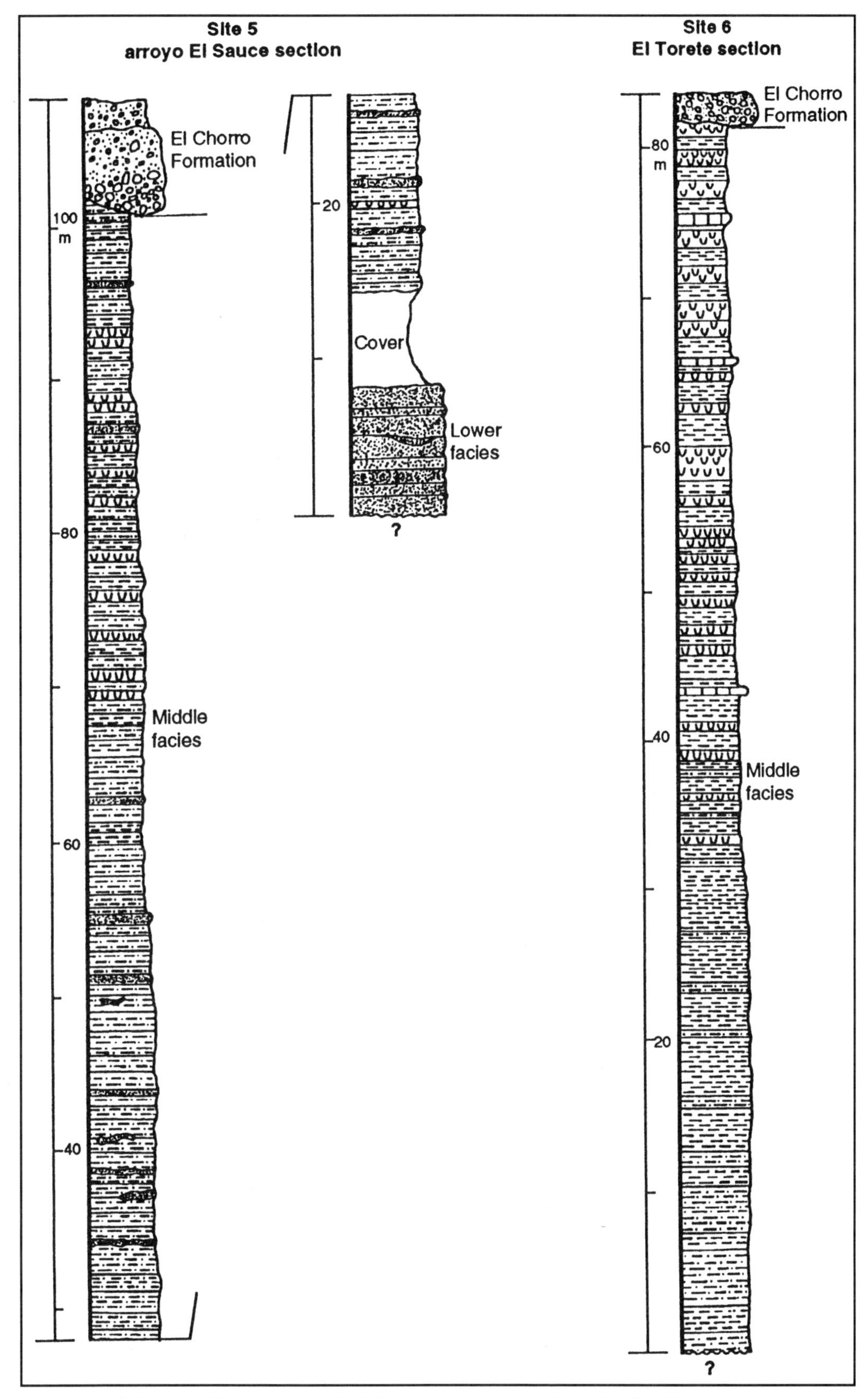

Figure 8. Stratigraphic columns of the Trinidad Formation, middle facies. See Figure 3 for legend.

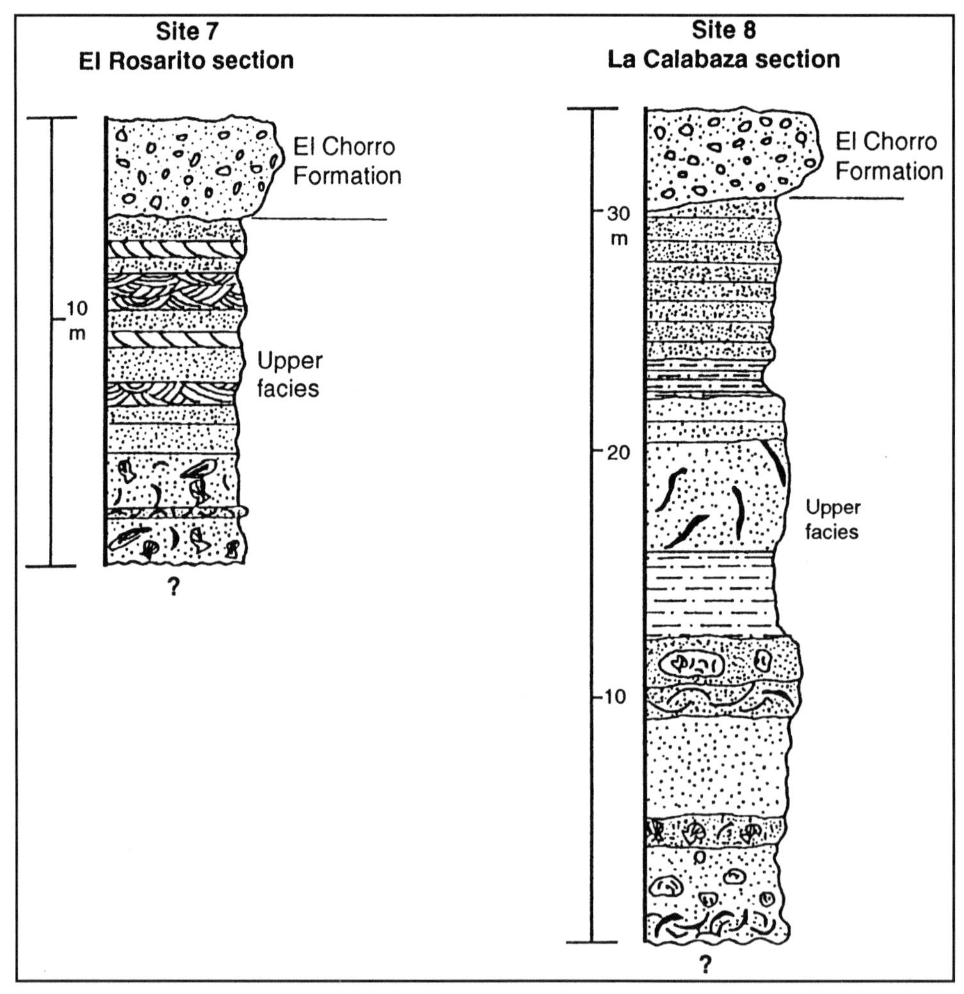

Figure 9. Stratigraphic columns of the Trinidad Formation, upper facies. See Figure 3 for legend.

portion of the basin. In the present work, we suggest that the formation has a maximum total thickness of 380 m, based on six measured sections.

Although none of the localities shows a complete stratigraphic section, the Refugio and Refugito areas (Fig. 2) show the best exposures of the formation. The Refugio area is considered the type section and lower boundary stratotype (lat. 23°24.1′N, long. 109°38.5′W). The nature of the basal contact is variable. In places, the formation unconformably overlies older rocks (i.e., La Trinidad basement complex); elsewhere it grades conformably into the underlying Trinidad Formation. Near cerro La Laguna (Figs. 2 and 11), the base of the formation unconformably overlies the granite-granodiorite rocks of the La Trinidad complex, and near the basin center (e.g., El Refugio), the Refugio Formation usually conformably overlies the Trinidad Formation.

The RF consists mainly of gray-white, coarse- to medium-grained sandstone beds, locally interbedded with limestone and shale. Lithologically, the RF is dominated by generally friable to weakly carbonate cemented, grayish-white, coarse- to medium-grained, poorly sorted, arkosic sandstone. Quartz, K-feldspar, and mica grains are the main components, with lesser amounts of volcanic lithic fragments. Brackish marine fossils are common throughout. In general, the unit is considered to be a regressive sequence, exhibiting a tendency for coarser sandstone grain size and decreased marine fossil content toward its top. Miller (1980) and Espinoza-Arrubarena (1979) reported terrestrial deposits at the rancho Los Algodones as part of the Refugio Formation; however, these deposits correspond to small outcrops of the alluvial facies of the overlying Los Barriles Formation. Parallel lamination and planar stratification, trough and planar cross-bedding, trace fossils, and concretions are present in the facies.

The section near El Refugio (Fig. 11) best represents the lower part of the RF, which consists of bioturbated fine-grained sandstone beds (0.3 to 1.00 m) that contain interbeds of laminated fine-grained arkosic sandstone (0.5 to 0.15 m). The bioturbated beds in the lower ~1.2 m pass upward into ~6.3 m of thickly bedded, medium-grained sandstone. At the adjacent section of El Refugito (Fig. 11) a higher part of the unit may be exposed. Here, the interval consists of a series of alternating

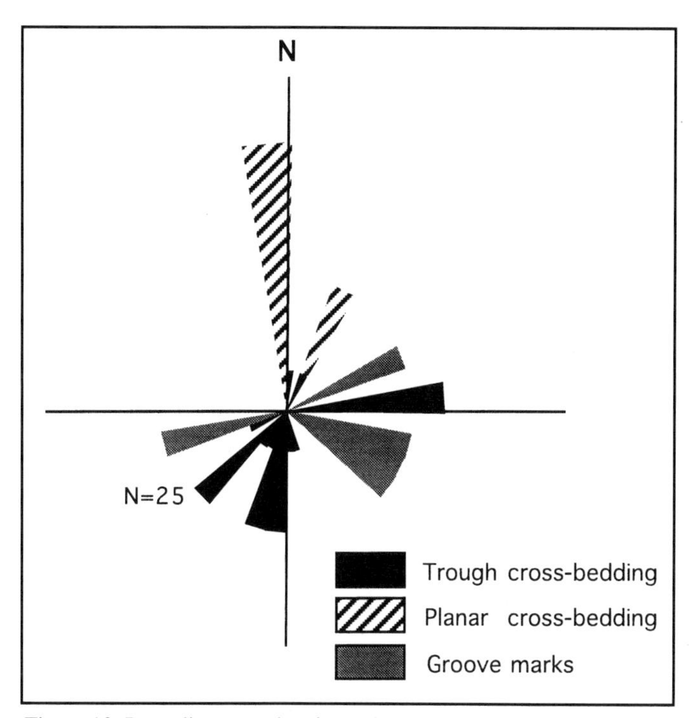

Figure 10. Rose diagrams showing paleocurrent azimuths measured in the Trinidad Formation, upper facies.

very coarse- and medium-grained white sandstone beds with thicknesses that range from 0.15 to 0.40 m. The medium-grained sandstone is moderately sorted and subrounded to rounded, whereas the very coarse-grained sandstone is poorly sorted and angular to subangular. The contact between beds is gradational, and some of the beds (both coarse and medium grained) fine upward. All sandstone beds have marine shell fragments, but no complete fossil shells are present. The top of the RF is exposed at Trópico de Cancer. It consists of commonly cross-bedded, texturally supermature, fine-grained sandstone.

The RF near rancho Los Algodones (Fig. 11) consists of ~30 m of massive, coarse-grained sandstone beds, unconformably overlying the Trinidad Formation. This sand is moderately sorted, with subrounded to rounded grains. It contains scattered whole and fragmented marine mollusks. Primary bedding is rarely preserved.

The arkosic sandstone is intercalated with siltstone and fossiliferous beds (coquina) at the Trópico de Cancer and cerro La Laguna localities. The sandstone beds are weakly cemented by $CaCO_3$, but in the cerro La Laguna area the base of the unit is strongly cemented by $CaCO_3$ and occasionally by SiO_2.

The southernmost section, cerro La Laguna, was originally measured by Gaitán (1986). Our unit description differs in terms of lithology from the northern exposures within the basin. Its base consists of white, carbonate- and silica-cemented, very coarse to coarse-grained quartzose sandstone that is moderately sorted and rounded to well rounded (similar to that at Site 12). The middle part of the RF consists of pinkish-gray limestone beds alternating with greenish-yellow siltstone beds that are overlain by fine-grained arkosic sandstone containing interbedded siltstone layers. The limestone beds contain shell fragments and lesser amounts of fine-grained sand. The upper part presents alternating deposits of yellowish-gray limestone and greenish-yellow siltstone beds. These alternating deposits are overlain by a white, coarse-grained, texturally mature to supermature, quartz-rich sandstone.

Tabular and trough cross-bedding measured in the RF at Trópico de Cancer (Fig. 12) suggests general southerly transport, probably in response to wave-and/or tidal-induced bottom currents. In contrast, older formations revealed paleotransport toward the west-northwest (Martínez-Gutiérrez, 1994).

The Refugio Formation is rich in fossils, including gastropods, pelecypods, arthropods, and marine vertebrates. Hertlein (1966) reports *Argopecten calli, Clhamys tamiamiensis grewingki, Euvola refugioensis, Pecten aletes, Striotsrea* sp., *Undulostrea megadon, Conus multiliratus, Ficus carbasea, Strombus obliteratus,* and *Amonia peruviana* at the rancho El Refugio locality. McCloy (1984) also includes *Pecten (Plagioctenium) calli* and *Pecten* (*Euvola*) *keepi.*

Smith (1991) states that the RF contains a Tertiary fauna of Caribbean affinity that is characterized by *Clementia dariena, Turritella abrupta fredeai, Florimetis tritinana, Cyathodonta gatunenis,* and *Raeta undulata.* Based on the fossil assemblage, Hertlein (1966) assigned an age of middle Pliocene to the Refugio Formation. Pantoja-Alor and Carrillo-Bravo (1966) designated a Pliocene age; McCloy (1984) assigned an upper Pliocene–Pleistocene age. Smith (1991), using the Caribbean affinities of Cenozoic marine mollusks from several Californian formations, suggests that "Salada Formation" (herein called the Refugio Formation) is a lower Pliocene unit. In this study an early Pliocene age is suggested for the Refugio Formation based on the fossil assemblage described by previous authors and by its conformable contact with the Trinidad formation.

The Refugio fossil assemblage described by Hertlein (1966) in the area of rancho El Refugio and rancho El Refugito suggests deposition in warm, shallow marine waters. Pantoja-Alor and Carrillo-Bravo (1966) interpreted the unit as a high-energy marine deposit with abundant turbidity currents. Based on megafossils (mollusks and gastropods), McCloy (1984) interpreted this formation as an inner shelf deposit in the El Refugio area that shoals upward into beach and lagoonal deposits in the Santiago area to the west. The presence of a coarsening-upward sequence, brackish marine fossils, and limestone beds suggests that the Refugio Formation was deposited in shallow marine waters as a regressive package in response to a sedimentary progradation.

Los Barriles Formation

The Los Barriles Formation (LBF) is typical of the conglomeratic deposits exposed along arroyo El Datilar at the western margin of the Sierra La Trinidad basement complex and

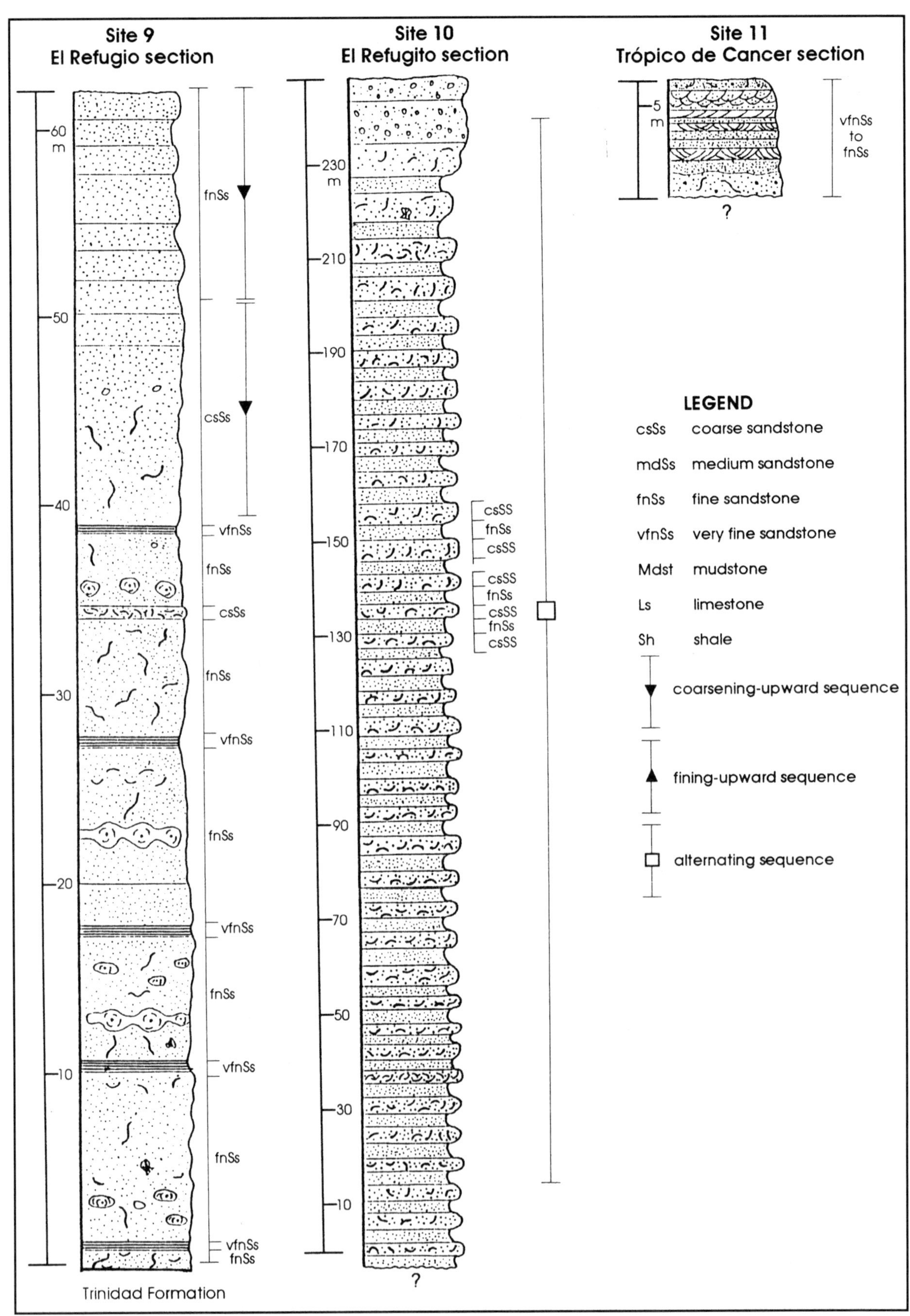

Figure 11. Stratigraphic columns of the Refugio Formation. See Figure 3 for legend.

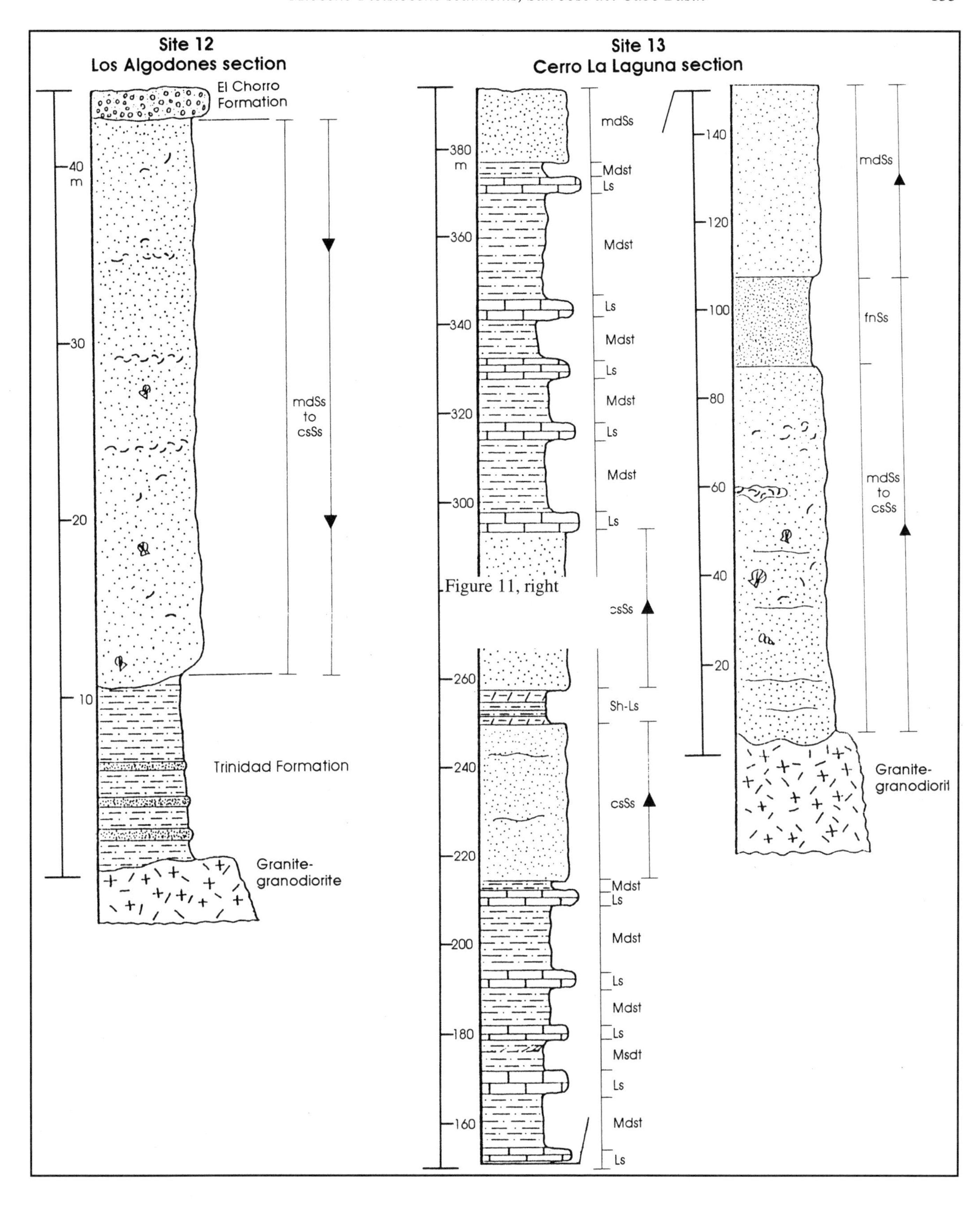
Site 12
Los Algodones section
El Chorro Formation
40 m
30
20
10
mdSs to csSs
Trinidad Formation
Granite-granodiorite
Site 13
Cerro La Laguna section
380 m
360
340
320
300
260
240
220
200
180
160
mdSs
Mdst
Ls
Mdst
Ls
Mdst
Ls
Mdst
Ls
Mdst
Ls
csSs
Sh-Ls
csSs
Mdst
Ls
Mdst
Ls
Mdst
Ls
Msdt
Ls
Mdst
Ls
Figure 11, right
140
120
100
80
60
40
20
mdSs
fnSs
mdSs to csSs
Granite-granodiorit

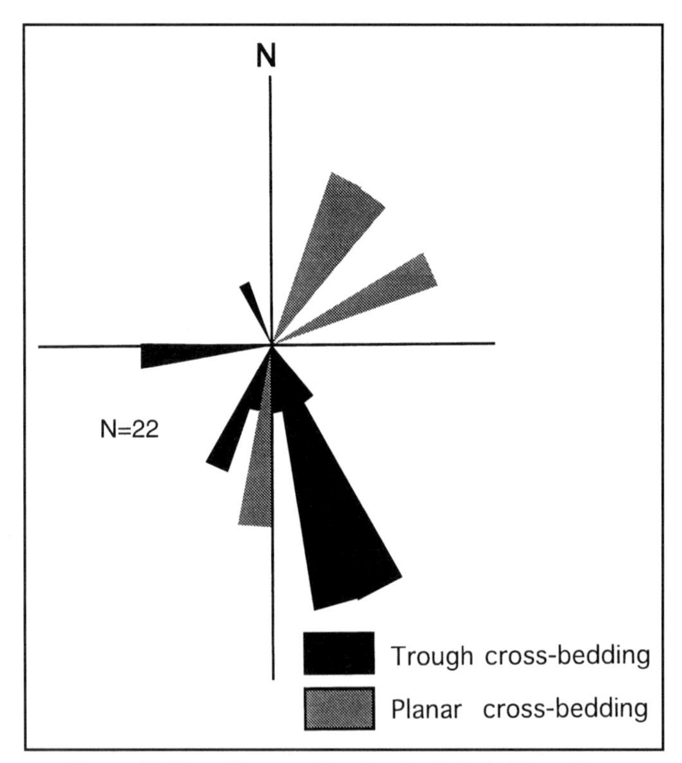

Figure 12. Rose diagrams showing the Refugio Formation.

along arroyo Buenos Aires at the eastern margin of the Sierra La Victoria basement complex. At arroyo Buenos Aires a thickness of ~1,650 m was measured. The type section is located in the small town of Los Barriles (lat. 23°40.5′N, long. 109°42′W). No exposures were observed in the southern part of the basin. The LBF consists of medium to coarse sandstone and conglomerate that ranges from coarse pebbles to boulders. The LBF occasionally is strongly dominated by conglomerate beds, with subordinate sandstone beds that are concentrated near the base (Fig. 13). Adjacent areas within about 2 km show a greater sand content with subordinate amounts of conglomerate (Fig. 14). Elsewhere, conglomerate is more abundant near the base of the formation.

At arroyo Buenos Aires (Fig. 13), the formation is characterized by a gradual upward transition from the coarse siltstone beds of the Refugio Formation to the medium-grained sandstone beds of the Los Barriles Formation. Large boulders of granite and tonalite are observed in this part of the section. In the basin center at El Rosarito (Fig. 15), the massive pebbly arenite that forms the base of the LBF interfingers with the coarse- to medium-grained sandstone of the upper Refugio Formation. The grain size of the pebble sandstone ranges from coarse sand to coarse pebbles, with few cobbles. Here, the formation thins along the east-central basin margin. Farther to the southeast the LBF unconformably overlies the Trinidad Formation (Fig. 15). In the north (Site 14), the formation conformably overlies Refugio strata; southward (Sites 18 and 19), they interfinger (fan-delta facies).

Larger clasts (pebbles and cobbles) of the LBF along the western margin consist of granite-granodiorite (17%), tonalite (45%), andesite (12%), and gneiss (26%) (Fig. 5). The deposits located at the eastern margin are mainly composed of granitic-granodioritic (47.5%), dacitic (4.5%), rhyolitic lava (43%), and pyroclastic (5%) rock fragments (Martínez-Gutiérrez, 1994). Clasts observed along both margins range from subrounded to rounded and exhibit a poorly to well-sorted texture. Clast-supported conglomerates become more abundant toward the top of the unit.

The deposits of the LBF vary in dip from 10 to 25° toward the southwest along both margins of the basin. The tilting of the LBF along the western margin is probably related to the uplift of the La Victoria basement complex and onset of the formation of the San José del Cabo fault, perhaps activated as a growth fault by sedimentary loading.

Sedimentary structures are uncommon in the LBF. Massive bedding generally characterizes the clast-supported conglomerate. The base of the formation at the arroyo Buenos Aires and arroyo San Bartolo areas also exhibits scour-based, lenticular channels infilled with fine-grained sandstone or siltstone. No fossils were found within the formation, except at El Rosarito, where reworked shell fragments are present within the lower pebbly conglomerate beds.

Paleotransport indicators are scarce in the Los Barriles Formation. Channels, imbrication, and trough cross-bedding were identified (Fig. 16). Indicators from the eastern margin (Sites 16, 18, and 19) reveal that the flow patterns had a west-northwest trend. In contrast, data collected from alongside the western margin reveal flow patterns with a south-southeast trend (Site 14).

The presence of coarse, poorly sorted, crudely layered units of sand-supported clasts eroded by scour, sand lenses of clast-supported conglomerate, and the absence of mud suggest that the unit was probably deposited under alluvial conditions as the product of debris or high-energy stream flows in a semi-arid regime. The LBF is interpreted as representing deposition in a high gradient alluvial fan environment. Deposition of the formation may be related to uplift and faulting of the La Victoria and La Trinidad complexes. The age of these deposits is unknown; however, stratigraphic relationships suggest an age of upper Pliocene–lower Pleistocene.

El Chorro Formation

The term El Chorro Formation (ECF) is used to designate the youngest continental deposits within the San José del Cabo Basin. The ECF represents a classic alluvial fan landform and overlies older strata along an angular unconformity.

The formation is best preserved along the western margin (Fig. 2). However, some deposits are present along the eastern margin. Gaitán (1986) measured a thickness of 80 m for the alluvial fan deposits on the western margin compared to a thickness of 2 to 4 m on the eastern margin. In this study, we measure a thickness of about 150 m at the localities described on the western margin (Fig. 17) and about 3 to 8 m for the deposits in the eastern

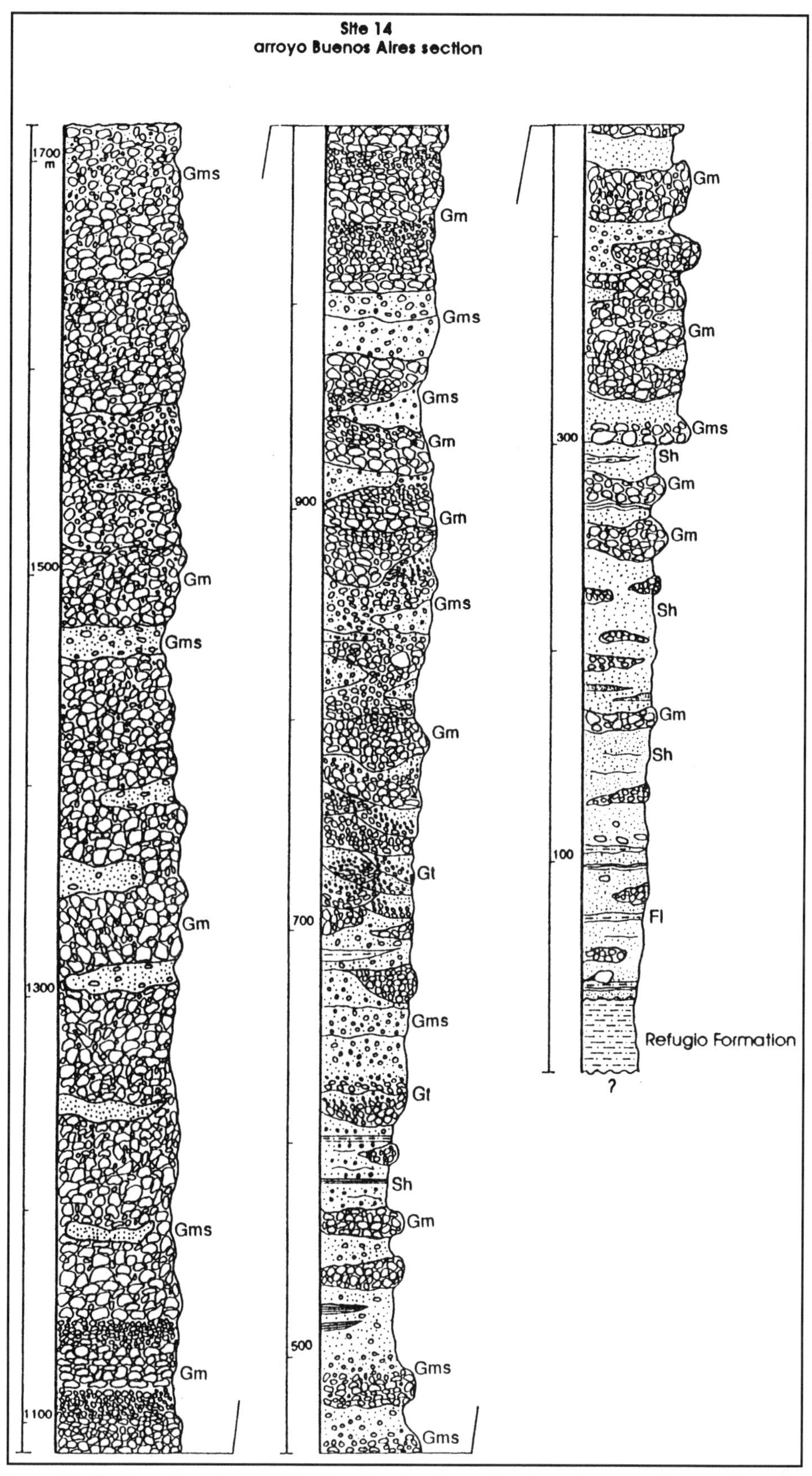

Figure 13. Stratigraphic columns of the Los Barriles Formation along the arroyo Buenos Aires. See Figure 3 for legend.

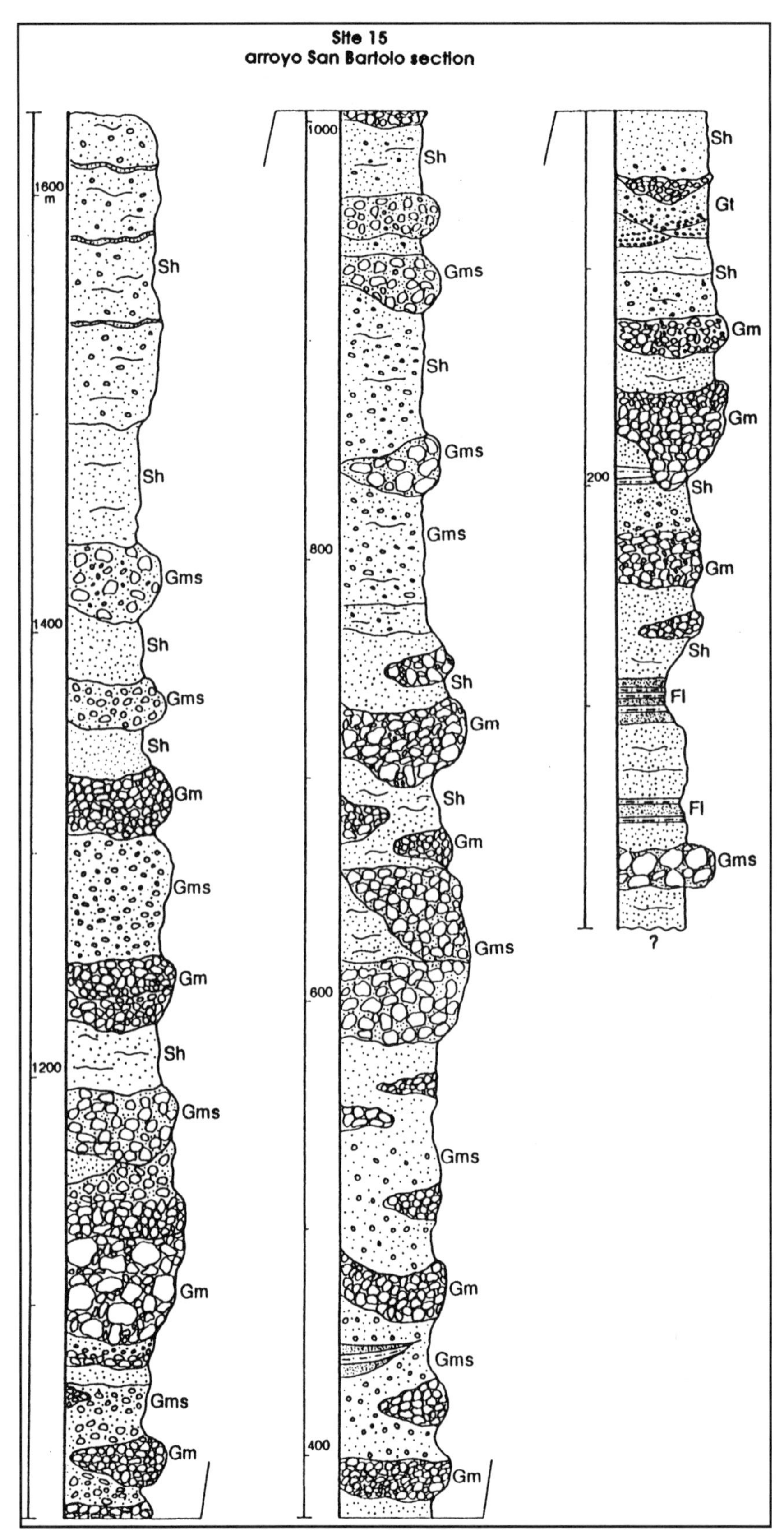

Figure 14. Stratigraphic columns of the Los Barriles Formation along the arroyo San Bartolo. See Figure 3 for legend.

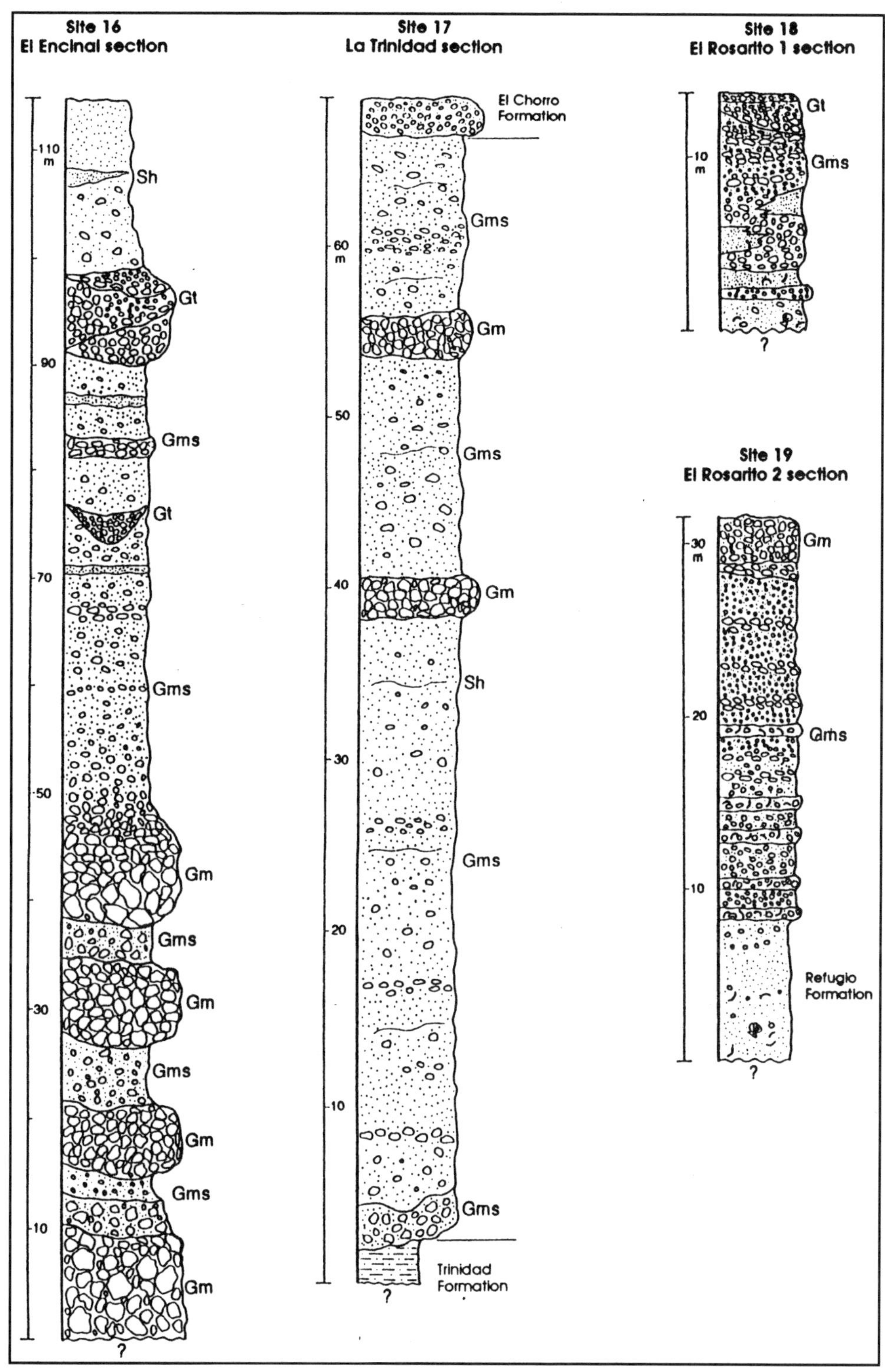

Figure 15. Stratigraphic columns of the Los Barriles Formation at the eastern margin of the basin. See Figure 3 for legend.

margin. The most complete stratigraphic sections measured are located near rancho El Chorro (section type, lat. 23°26.5′N, long. 109°48.5′W) and along arroyo San Dionisio (Fig. 17).

The El Chorro Formation is composed of nonmarine, coarse-grained sandstone and conglomerate. Lithic clasts range in size from coarse sand to boulders and vary from angular to subrounded. Clasts from the eastern margin and the western margin of the basin differ in composition. The eastern margin is composed of granitic (90%) and rhyolitic (10%) rock fragments (Fig. 5), whereas the western margin consists of granitic (23%),

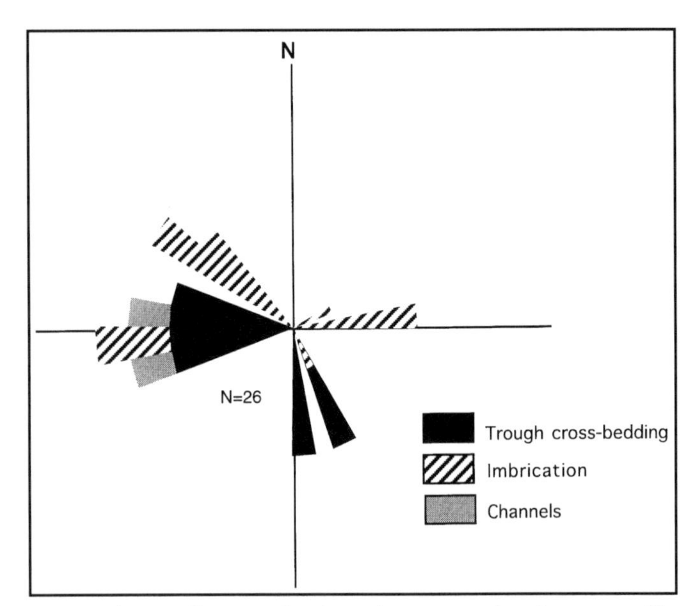

Figure 16. Rose diagrams showing paleocurrent azimuths measured in the Los Barriles Formation.

granodioritic (11%), tonalitic (19%), rhyolitic (4%), andesitic (13%), gneissic (25%), and schistic (5%) rock fragments (Martínez-Gutiérrez, 1994). Conglomerate beds are concentrated near the base of the formation (Site 20) and rarely interbedded with sandstone beds throughout the formation. The thickness of the conglomerate beds ranges from 0.5 to 5 m. Color is grayish orange or pale yellowish orange.

The lower part of the ECF is composed of conglomerate beds made of poorly sorted boulders and cobbles, with lesser amounts of sand beds. Locally the base of the unit is in fault contact with the Mesozoic rocks of the La Victoria basement complex (Fig. 17). The middle and upper part of the formation consists of massive, matrix-supported pebbly arenite, with local lens and channel fills of clast-supported conglomerate (Fig. 17). The median grain size in the middle and upper portion of the unit ranges from pebbles to coarse-medium sand. The upper part of the ECF is characterized by the presence of large-scale cross-bedding. Clast imbrication is scarce.

The alluvial deposits of the ECF along both margins gently dip between 5 and 10° toward the basin center. In the south-central region of the basin, a north-south–trending fault, parallel to arroyo San José, offsets both the Refugio and El Chorro Formations.

Indicators of transport direction are rare in the ECF. The few that were measured along Federal Highway 1 (Fig. 18) are consistent with the observable fan morphology and confirm that transport direction was predominantly eastward.

The El Chorro Formation is flat lying and most commonly lies in angular unconformity above older strata, including the La Calera, Trinidad, Refugio, and Los Barriles Formations. The ECF represents recent alluvial fan deposition with a probable age of upper Pleistocene–lower Holocene.

INTERPRETATION AND CONCLUSIONS

The San José del Cabo Basin contains Miocene to Quaternary sedimentary strata that range from alluvial to shallow marine facies. In an east-west cross section, the sedimentary deposits are dipping toward south-southwest, where the San José del Cabo Basin resembles a half graben that is bounded by the San José del Cabo fault on the western margin and the La Trinidad fault on the eastern margin (Fig. 2).

Basement rock types of the eastern margin (i.e., granite, granodiorite, and volcanic rocks) are considered to be a remnant of mainland tectonic activity. Normal fault traces in the Sierra La Trinidad basement complex show a north-northwest to northwest direction. These directions are correlated with the east tilted blocks of the southern Sinaloa coast (Stock and Hodges, 1989). The petrologic affinity between the La Trinidad basement complex and the Jalisco block (Gastil et al., 1978; Wallace et al. 1992; Ferrari, 1995) shows also that this portion of the peninsula was still adjacent to the mainland at least during lower Miocene time. The rhyolitic lava, pyroclastic, and dacitic lava rocks that crop out in the northeastern part of the La Trinidad complex accumulated as a result of calc-alkaline volcanism that resulted from subduction of the Guadalupe plate under the North American plate (Fig. 19) between ~16 to ~13 Ma (Lyle and Ness, 1991).

Block faulting of the La Trinidad batholithic basement and the volcanic sequence probably began during the middle Miocene. Faults and fractures in the Trinidad basement complex show that this faulting continued during the late Miocene when the peninsula was part of the mainland. The oldest sedimentary sequence (i.e., La Calera Formation) is not affected by this type of deformation. This tectonic activity presumably could be related to one of the main phases of Neogene rifting in the Gulf of California, a late Miocene phase of orthogonal east-west extension (Stock and Hodges, 1990; Wallace et al., 1992).

As a result of this faulting, the deposition of the massive boulder-pebble deposits of the La Calera Formation began to accumulate. A fining-upward sequence is evident in the La Calera Formation. The presence of this fining-upward sequence in the LCF implies that these deposits were associated with either high-energy stream to debris flows that grade to low-energy regime sheet flows or a gradient reduction as basins filled and differential relief was reduced.

An initial subsidence or relative sea-level rise within the San José del Cabo Basin began prior to or during upper Miocene time. The onset of a transgressive phase within the basin is recorded by marshy to shallow-marine deposits of the lower part of the Trinidad Formation (Fig. 20). The most prominent transgression is represented by the muddy and diatomaceous middle facies of the formation. The subsidence and transgression continued through the upper Miocene to lower Pliocene. The transgressive phase is explained in terms of a subsidence of the basin. A pre-8.2 Ma subsidence is also reported by McCloy (1987) where similar marine sediments at Isla María Madre indicate this event. The subsidence of the basin can be attributed to

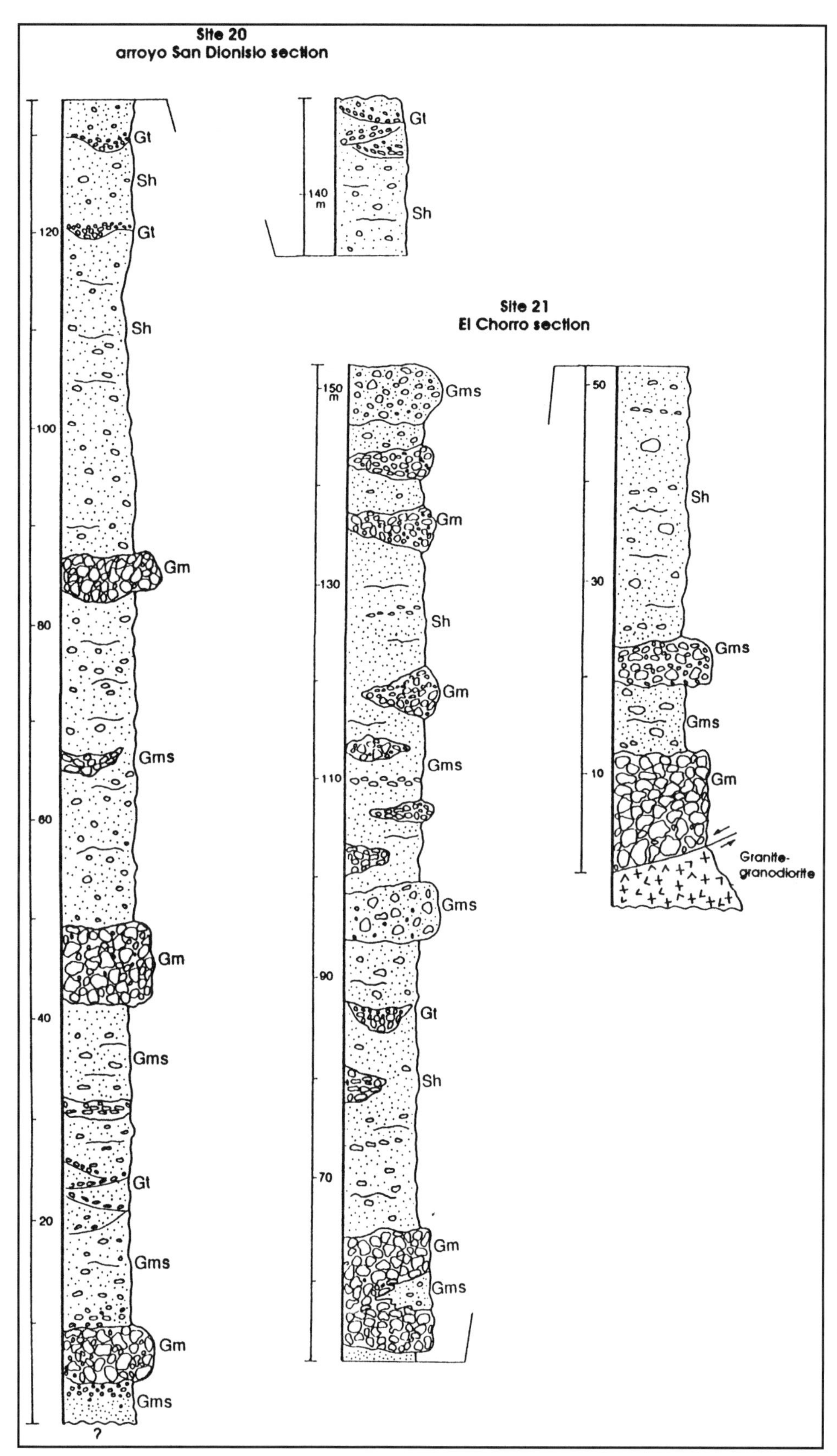

Figure 17. Stratigraphic columns of the El Chorro Formation. See Figure 3 for legend.

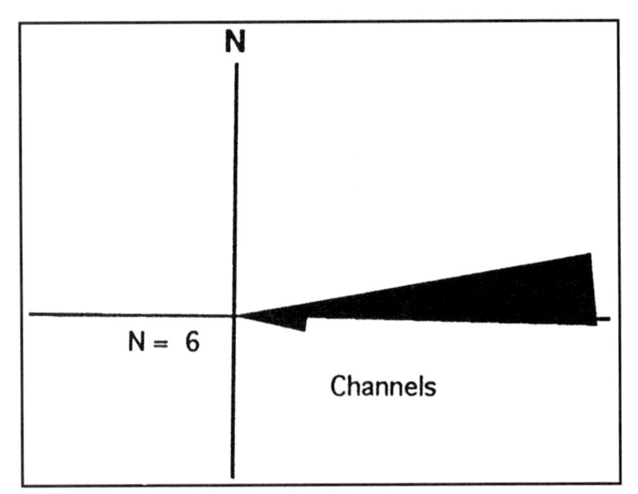

Figure 18. Rose diagrams showing paleocurrent azimuths measured in the El Chorro Formation.

development of the strike-slip fault zone (the Tosco-Abreojos fault) that initiated about 12 Ma (Spencer and Normak, 1979). However, strike-slip displacements were not observed in the Trinidad Formation.

During the lower Pliocene, a regressive phase initiated within the San José del Cabo Basin. The outer-shelf, marine, muddy-sandy facies and the inner-shelf, marine, sandy facies developed (i.e., the upper facies of the Trinidad Formation and lower facies of the Refugio Formation, respectively). Coarsening-upward sedimentation of the TF and a paleocurrent direction change indicate that depositional conditions were modified toward the south-southeast during the deposition of the RF. This variation is related to Pliocene tectonic activity within the basin. Although the strike-slip motion continued along the plate boundary, producing subsidence at the mouth of the Gulf of California, the geologic record within the basin shows an uplift or drop in sea level. As a result of the uplifting the Sierra La Victoria basement complex began to emerge. The uplifting is recorded by the coarsening-upward

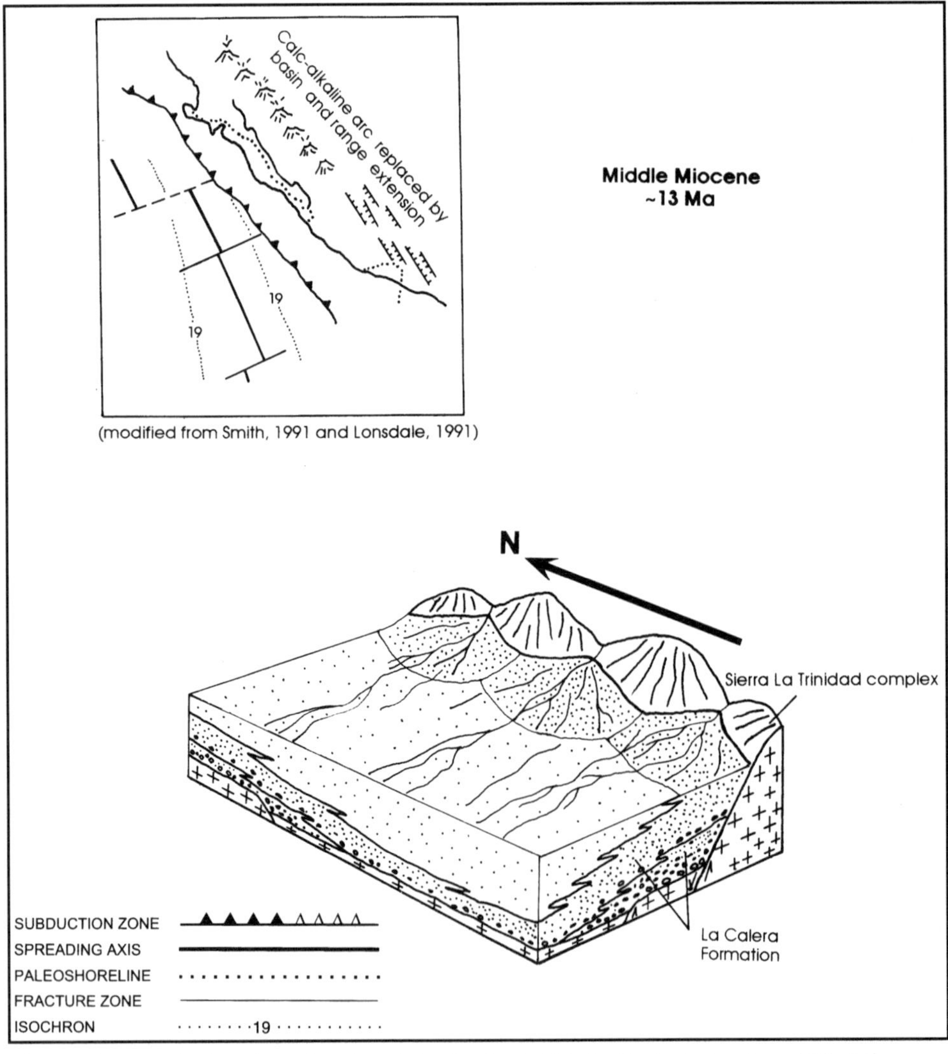

Figure 19. Tectonic-depositional setting of the San José del Cabo Basin during the middle Miocene. A block faulting of the Cretaceous-Tertiary batholithic basement and the volcanic sequence produced the alluvial sedimentation within the basin.

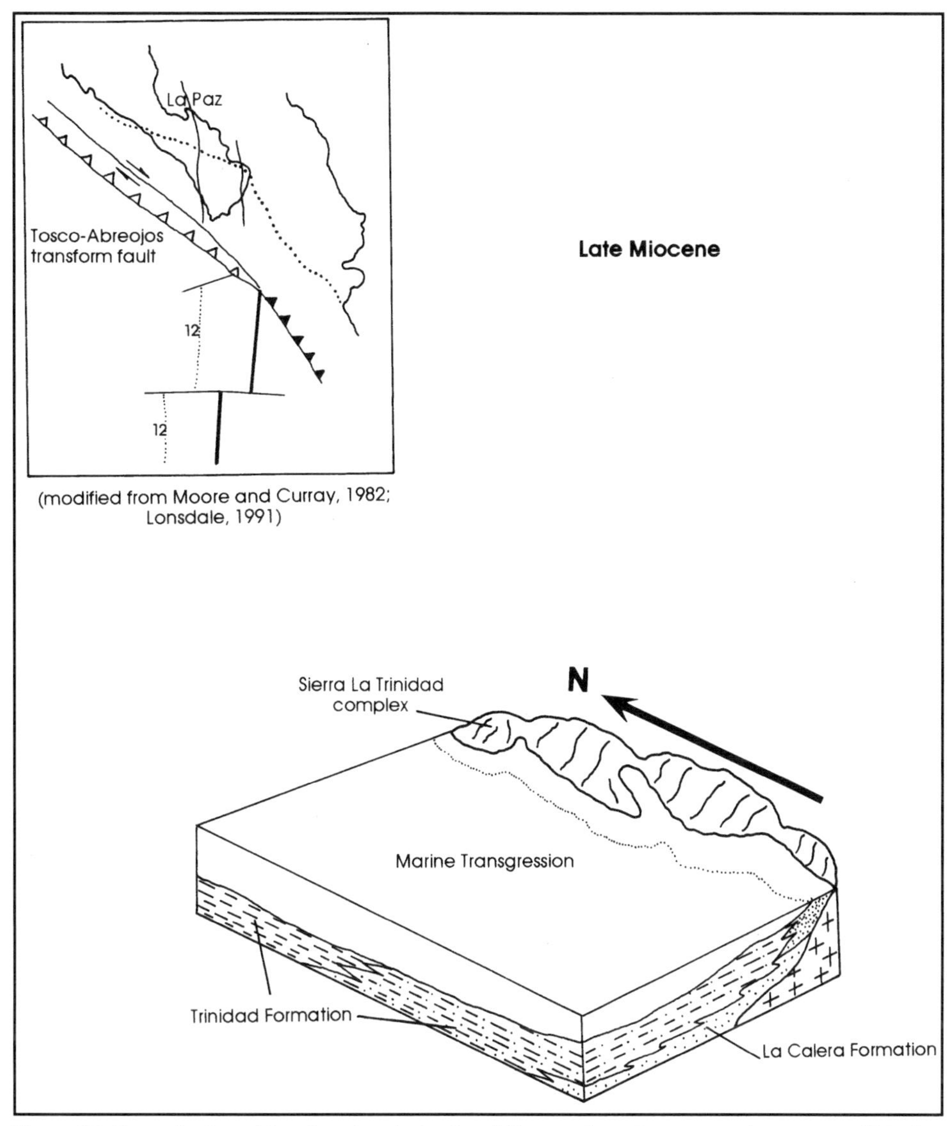

Figure 20. Tectonic-depositional setting during late Miocene time. A transgression occurs within the basin with the deposition of the Trinidad Formation. The transgressive phase is associated with a slip along basin-bounding faults. Key as on Figure 19.

nature of the Refugio Formation, which is only observable at the northern part of the basin; correlative strata in the southern part of the basin show evidence for deposition in a deeper water column (cerro La Laguna section).

Continued shoaling of the basin at the end of the late Pliocene, is evidenced by the shallow marine sandy facies of the Refugio Formation. This was perhaps due to tectonic uplift or a slowing subsidence at this time (i.e., the onset of deposition of the Los Barriles strata). The shoaling of the basin is recorded by the shallow marine sandy facies of the RF (Fig. 11). The shallow marine-lagoonal facies of the upper part of the RF that accumulated in the San José del Cabo area also suggests that a relative sea-level fall and/or tectonic uplift occurred. Paleocurrent data from the upper part of the RF indicate a general southward transport.

The transgressive and regressive phases that occurred within the basin do not coincide with the global sea-level curve (Haq et al., 1988). Therefore, we can assume that deposition within the basin was controlled by tectonism, presumably Miocene-Quaternary extension associated with rifting of the Gulf of California.

By late Pliocene time probably two tectonostratigraphic events occurred: (1) a continuation of the regressive phase that is represented by the change of the shallow marine deposits of the Refugio Formation to the terrestrial deposition of the Los Barriles Formation and (2) the continuation of uplift of the La Victoria and La Trinidad complexes that supplied sediments for deposition of the Los Barriles Formation. Both events apparently contributed to basin shoaling, marking the final phase of the marine sedimentation and heralding the onset of a terrestrial depositional environment on both sides of the basin (Fig. 21). Reactivation of the La

Trinidad fault followed in late Pliocene time, resulting in the deposition of the alluvial facies (Los Barriles Formation) along the eastern margin. The western margin, however, represents the formation of the San José del Cabo fault, a listric normal fault initiated as a result of the emergence of the La Victoria basement complex during the late Pliocene. The sediments supplied from the La Victoria basement complex were volumetrically more important than the sediments supplied from the La Trinidad basement complex. The La Victoria basement complex apparently was characterized by a higher rate of tectonic uplift than the La Trinidad basement complex. The San José del Cabo fault affects the older formations, producing a half-graben tilting toward west-southwest. The regressive phase shown by the clastic sedimentation of the Refugio and Los Barriles Formations can be related to the second Neogene rifting phase, which implies transform displacements and seafloor spreading in the Gulf (Stock and Hodges, 1989). As result of this tectonic activity, a reactivation of the basin-bounding faults occurred.

By latest Pliocene through Pleistocene time, movement along the San José del Cabo fault continued. During this time, continuous seafloor spreading had been initiated along the East Pacific Rise, and the crustal plates began to move apart, causing the widening of the Gulf of California (Lyle and Ness, 1991). Seafloor spreading and transform faulting associated with thermal expansion (Gaitán, 1986) presumably contributed to vertical uplift and deformation on the adjacent margins of the tip of the peninsula, recorded as an angular unconformity between the Los Barriles Formation and the overlying El Chorro Formation. Clastic sedimentation continued through the Pleistocene with the deposition of the alluvial lithofacies of the ECF alongside both margins of the San José del Cabo Basin (Fig. 22). These alluvial deposits derived from both complexes were deposited unconformably over both the older tilted formations and the crystalline basement. The alluvial facies shows a coarsening-upward sequence devoid of a muddy matrix, perhaps a response to arid climatic conditions and very episodic rainfall.

The San José del Cabo Basin exhibits a variety of depositional environments and tectonic settings that are characteristic of rift basins, including fluvial, brackish, and marine environments related to extensional phases. In addition, the San José del Cabo Basin presents evidence of Neogene plate motions at the mouth of the Gulf of California since ~12 Ma; we can therefore conclude that the San José del Cabo Basin is associated with the opening of the Gulf of California.

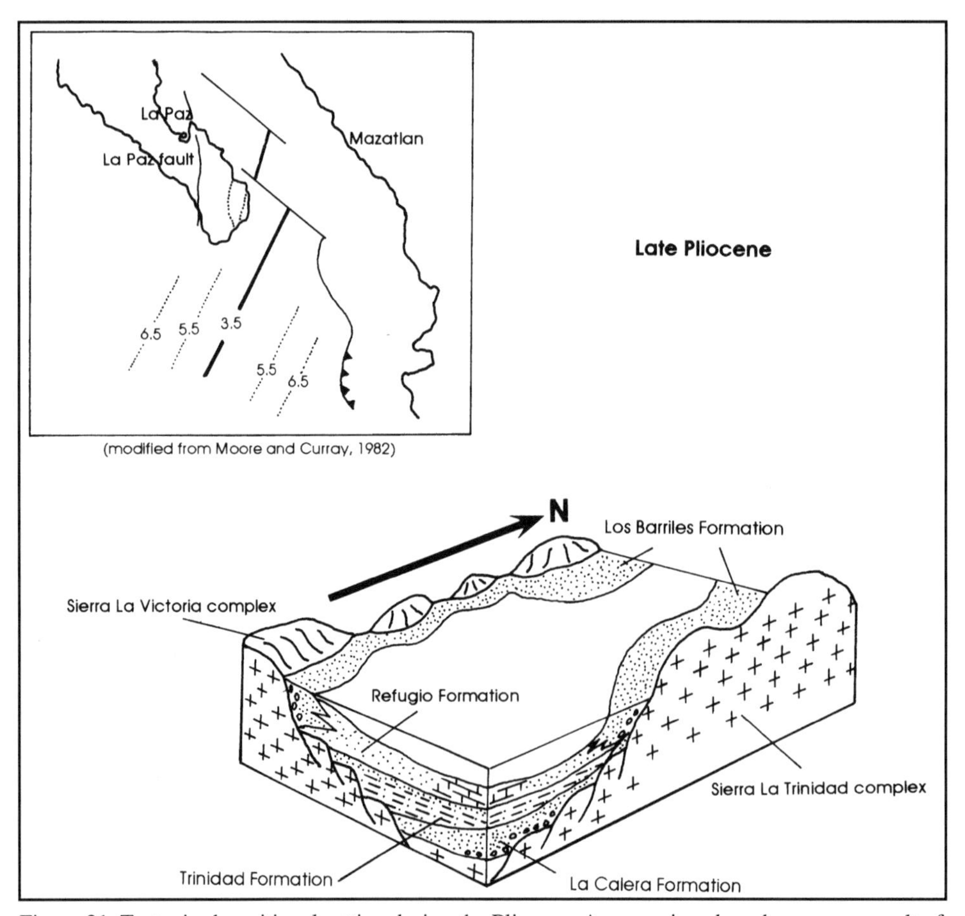

Figure 21. Tectonic-depositional setting during the Pliocene. A regressive phase began as a result of reactivation of the basin-bounding faults. The terrestrial sedimentation is represented by the Los Barriles Formation. Key as on Figure 19.

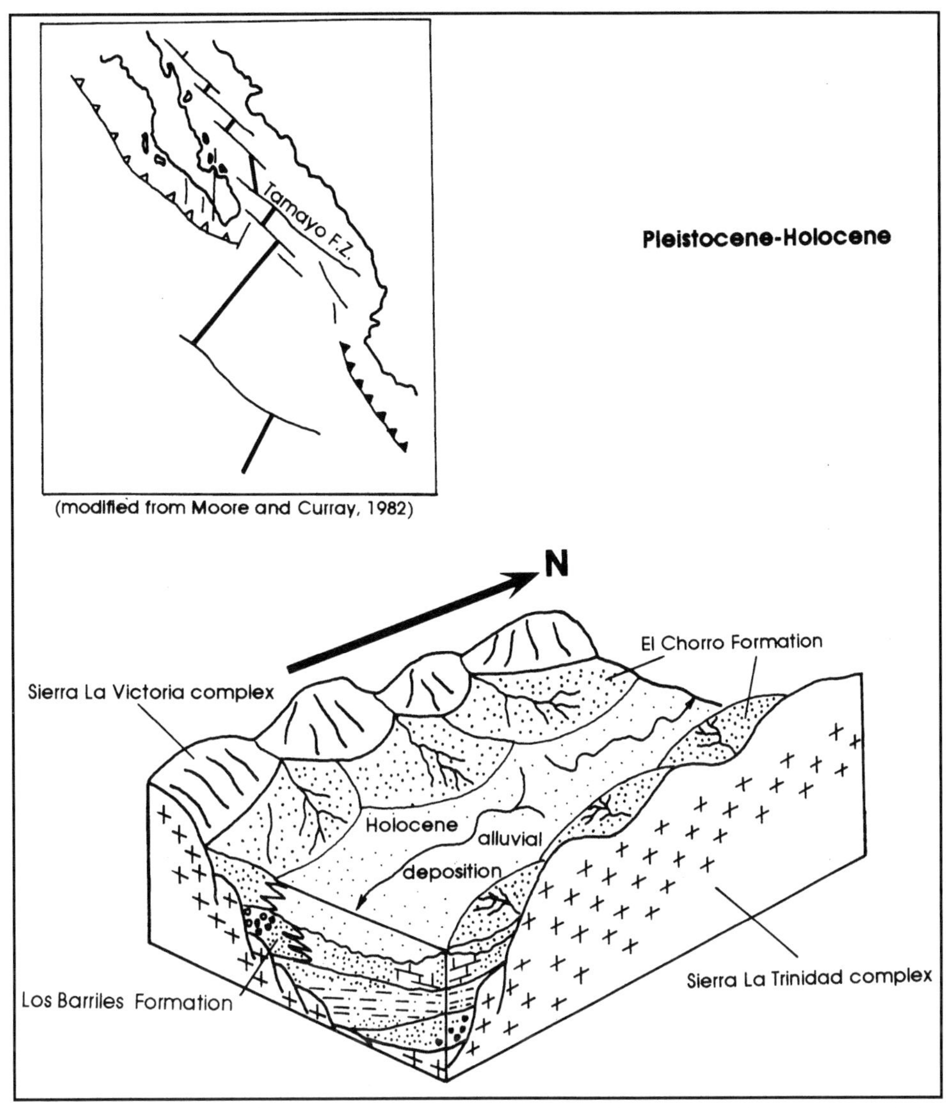

Figure 22. Tectonic-depositional setting during the Pleistocene. The spatial distribution of the alluvial units within the basin is like that of the present time. Key as on Figure 19.

In contrast to some other rift basins along the Gulf of California (e.g., Loreto basin and Santa Rosalia basin), the San José del Cabo Basin did not accumulate any volcanoclastic deposits (see Dorsey et al., 1995; Umhoefer et al., 1994). Moreover, the basin presents some sedimentologic and tectonic features that can be correlated with other Neogene basins along the Gulf of California, showing that they were formed by at least one late Miocene to Recent extensional phase (Stock and Hodges, 1989) that took place during the formation of the Gulf of California.

ACKNOWLEDGMENTS

We want to acknowledge Dr. Rebecca Dorsey (University of Northern Arizona), Dr. Karl W. Flessa (Arizona University), and Dr. Kurt Grimm (University of British Columbia) for their careful and constructive reviews, which helped to improve the quality of the final manuscript. We also appreciate the comments and suggestions provided by Dr. Larry Mayer (Miami University), which contributed to improvement of the manuscript.

REFERENCES CITED

Altamirano-R., F. J., 1972, Tectónica de la porción meridional de Baja California Sur: Resúmenes, Sociedad Geológica Mexicana Convención Geológica Nacional, 2nd, 1972, Mexico City: Mexico City, Sociedad Geológica Mexicana, p. 113–114.

Aranda-Goméz, J. J., and Pérez-Venzor, J. A., 1988, Estudio geológico de Punta Coyotes, Baja California Sur: Revista del Instituto de Geología, Universidad Nacional Autónoma de México, v. 7, p. 1–21.

Beal, C. H., 1948, Reconnaissance of the geology and oil possibilities of Baja California, Mexico: Geological Society of America Memoir 31, 138 p.

Carrillo-Chávez, A., 1991, Las alteraciones como guías mineralógicas en yacimientos de oro tipo dique falla. Los Uvares, B.C.S., ejemplo característico: Proceedings, First International Meeting on Geology of the Baja California Peninsula, La Paz, Baja California Sur, Universidad Autónoma de Baja California Sur, p. 14.

Dorsey, R. J., Umhoefer, P. J., and Renne, P. R., 1995, Rapid subsidence and stacked Gilbert-type fan deltas, Pliocene Loreto basin, Baja California Sur, Mexico, *in* Chough, S. K., and others, eds., Fan deltas, depositional styles and controls: Sedimentary Geology, v. 98, p. 181–204.

Espinoza-Arrubarena, L., 1979, Los tiburones fósiles (Lamniformes) del rancho Algodones, Baja California Sur, México [B.S. thesis]: Mexico City, Universidad Nacional Autónoma de México, Facultad de Ciencias, 60 p.

Ferrari, L., 1995, Miocene shearing along the northern boundary of the Jalisco block and the opening of the southern Gulf of California: Geology, v. 23, p. 751–754.

Frizzell, V. A., 1984, The geology of Baja California peninsula: An introduction, *in* Frizzell, V. A., Jr., ed., Geology of the Baja California peninsula, Field Trip Guidebook: San Diego, Society of Economic Paleontologists and Mineralogists, Pacific Section, v. 39, p. 1–8.

Gaitán, J., 1986, On neotectonic evidences in the southern peninsular region, Baja California, Mexico [M.S. thesis]: Amsterdam, The Netherlands, The International Institute of Aerospace Survey and Earth Sciences, 110 p.

Gastil, G., Krummenacher, D., Douppont, J., Bushee, J., Jensky, W., and Barthelmy, D., 1976, La zona batolítica del sur de California y el occidente de México: Boletín de la Sociedad Geológica Mexicana, v. 37, p. 84–90.

Gastil, G., Morgan, G. J., and Krummenacher, D., 1978, Mesozoic history of peninsular California and related areas east of the Gulf of California, *in* Howell, D. G., and McDougall, K. A., eds., Mesozoic paleography of the western United States: Society of Economic Paleontologists and Mineralogists, p. 107–115.

Haq, B. U., Hardenbol, J., and Vail, P., 1988, Mesozoic and Cenozoic chronostratigraphy and cycles of sea-level change, *in* Wilgus, C. K., Hastings, B. K., Posamentier, H., Wagoner, J. V., Ross, C. A., and Kendall, C. G. St. C., eds., Sea-level changes—An integrated approach: Society of Economic Paleontologists and Mineralogists Special Publication 42, p. 71–108.

Hausback, B., 1984, Cenozoic volcanic and tectonic evolution of Baja California Sur, Mexico, *in* Frizzell, V. A., Jr., ed., Geology of the Baja California peninsula, Field Trip Guidebook: Society of Economic Paleontologists and Mineralogists, Pacific Section, v. 39, p. 219–235.

Hertlein, L. G., 1925, Pectens from the Tertiary of lower California, California: Academy of Science Proceedings, v. 14, no.1, p. 1–35.

Hertlein, L. G., 1966, Pliocene fossils from Rancho El Refugio, Baja California, and Cerralvo Island, Mexico: California Academy of Sciences Proceedings, ser. 4, v. 30, no. 14, p. 265–284.

Instituto Nacional de Estadística Geografía e Informática, 1983a, Carta topográfica Cabo Pulmo, F12B35, Baja California Sur, México, scale 1:50,000.

Instituto Nacional de Estadística Geografía e Informática, 1983b, Carta topográfica La Ribera, Baja F12B25 California Sur, México, scale 1:50,000.

Instituto Nacional de Estadística Geografía e Informática, 1983c, Carta topográfica Las Cuevas, F12B24 Baja California Sur, México, scale 1:50,000.

Instituto Nacional de Estadística Geografía e Informática, 1983d, Carta topográfica Santiago, Baja F12B34 California Sur, México, scale 1:50,000.

Instituto Nacional de Estadística Geografía e Informática, 1983e, Carta topográfica Palo Escopeta, F12B45 Baja California Sur, México, scale 1:50,000.

Instituto Nacional de Estadística Geografía e Informática, 1983f, Carta topográfica San José del F12B44 Cabo, Baja California Sur, México, scale 1:50,000.

Instituto Nacional de Estadística Geografía e Informática, 1987, Carta geológica San José del Cabo, Baja California Sur, México, scale 1:250,000.

Lonsdale, P., 1991, Structural patterns of the Pacific floor offshore of peninsular California, *in* Dauphin, J. P., and Simoneit, B. R. T., eds., The Gulf and Peninsular Province of the Californias: American Association of Petroleum Geologists Memoir 47, p. 87–125.

López-Ramos, E., 1973, Carta geológica del territorio de Baja California: Comité de la Carta Geológica de México, scale 1:500,000.

Lyle, M., and Ness, G. E., 1991, The opening of the southern Gulf of California, *in* Dauphin, J. P., and Simoneit, B. R. T., eds., The Gulf and Peninsular Province of the Californias: American Association of Petroleum Geologists Memoir 47, p. 403–423.

Martínez-Gutiérrez, G., 1986, Excursión geológica La Paz–Cabo San Lucas, B.C.S., México, *in* Segura, L. R., ed., Mem., Foro Nacional de Escuelas de Ciencias de la Tierra, 4th, La Paz, Baja California Sur: La Paz, Universidad Autónoma de Baja California Sur, p. 85–97.

Martínez-Gutiérrez, G., 1991, Elementos morfoestructurales de la Sierra La Trinidad, B.C.S., México: Proceedings, First International Meeting on Geology of the Baja California Peninsula, La Paz, Baja California Sur, Universidad Autónoma de Baja California Sur, p. 45.

Martínez-Gutiérrez, G., 1994, Sedimentary facies of the Buena Vista–San José del Cabo Basin, B.C.S., Mexico [M.S. thesis]: Raleigh, North Carolina State University, 125 p.

McCloy, C., 1984, Stratigraphy and depositional history of the San Jose del Cabo trough, Baja California Sur, Mexico, *in* Frizzell, V. A., Jr., ed., Geology of the Baja California peninsula, Field Trip Guidebook: San Diego, Society of Economic, Paleontologists and Mineralogists, Pacific Section, v. 39, p. 237–267.

McCloy, C., 1987, Neogene biostratigraphy and depositional history of the southern Gulf of California: Geological Society of America Abstracts with Programs, v. 19, p. 764.

Miller, W. E., 1980, The late Pliocene Las Tunas local fauna from southernmost Baja California, Mexico: Journal of Paleontology, v. 54, p. 762–805.

Mina, V. F., 1957, Bosquejo geológico del territorio sur de la Baja California: Boletín de la Asociación Mexicana de Geólogos Petroléros, v. 9, p. 129–269.

Moore, D. G., and Curray, J. R., 1982, Geologic and tectonic history of the Gulf of California: Initial Reports of the Deep Sea Drilling Project, Volume 64: Washington, D.C., U.S. Government Printing Office, p. 1279–1294.

Murillo-Muñetón, G., 1991, Análisis petrológico y edades K-Ar de las rocas metamórficas e ígneas precenozoicas de la región de La Paz–Los Cabos, B.C.S., México: Proceedings, First International Meeting on Geology of the Baja California Peninsula, La Paz, Baja California Sur, Universidad Autónoma de Baja California Sur, p. 55–56.

Ortega-Gutiérrez, F., 1982, Evolución magmática y metamórfica del complejo cristalino de La Paz, B.C.S.: Resúmenes, Sociedad Geológica Mexicana Convención Geológica Nacional, 6th, p. 90.

Pantoja-Alor, J., and Carrillo-Bravo, J., 1966, Bosquejo de la región de Santiago–San Jose del Cabo, Baja California: Boletín de la Asociación Mexicana de Geólogos Petroléros, v. 58, p. 1–14.

Rodríguez-Quintana, R., 1988, Estudio paleontológico de la clase gasterópoda (mollusca) de la formación Trinidad, Baja California Sur, México [B.S. thesis]: La Paz, Universidad Autónoma de Baja California Sur, 97 p.

Sedlock, R., Ortega-Gutiérrez, F., and Speed, R. C., 1993, Tectonostratigraphic terranes and tectonic evolution of Mexico: Geological Society of America Special Paper 278, Part I, p. 2–73.

Smith, J. T., 1989, Contrasting megafunal and sedimentary records from opposite ends of the Gulf of California: Implications for interpreting its Tertiary history, *in* Abbott, P. L., ed., Geologic studies in Baja California: Society of Economic Paleontologists and Mineralogists, Pacific Section, v. 63, p. 27–36.

Smith, J. T., 1991, Cenozoic marine mollusks and paleogeography of the Gulf of California, *in* Dauphin, J. P., and Simoneit, B. R. T., eds., The Gulf and Peninsular Province of the Californias: American Association of Petroleum Geologists Memoir 47, p. 637–666.

Spencer, J. E., and Normark, W. R., 1979, Tosco-Abreojos fault zone: A Neogene transform plate boundary within the Pacific margin of southern Baja California, Mexico: Geology, v. 7, p. 554–557.

Stock, J. M., and Hodges, K. V., 1989, Pre-Pliocene extension around the Gulf of California and the transfer of Baja California to the Pacific plate: Tectonics, v. 8, p. 99–115.

Stock, J. M., and Hodges, K. V., 1990, Miocene to Recent structural development of an extensional accommodation zone, northeastern Baja California, Mexico: Journal of Structural Geology, v. 12, p. 315–328.

Umhoefer, P. J., Dorsey, R. J., and Renne, P., 1994, Tectonics of the Pliocene Loreto basin, Baja California Sur, Mexico, and evolution of the Gulf of California: Geology, v. 22, p. 649–652.

Wallace, P., Carmichael, I. S. E., Righter, K., and Becker, T. A., 1992, Volcanism and tectonism in western Mexico: A contrast of style and substance: Geology, v. 20, p. 625–628.

Manuscript Accepted by the Society December 2, 1996

Printed in U.S.A.

Index

[Italic page numbers indicate major references]

A

accommodation zones, on rift systems, vii, viii, *1*, 54
Aequipecten
  corteziana, 31
  dallasi, 112
  deserti, 25, 26, 31
  sverdrupi, 25, 26, 31
  sp., 65, 72
Africa, west, bryozoans, 111, 115
Agassizia scorbiculata, 25, 31, 37
Agua Blanca fault, vii
algae
  Bahía Concepción, 52
  Gulf of California, 131
  Loreto Basin, 87
  Punta el Bajo section, 125
  red. *See* rhodoliths
Amonia peruviana, 153
Anadara
  multicostata, 33
  sp., 51, 150
andesine, Puertecitos Formation, 13, 15
Anomia, 48
Aquepectin sp., 65, 71
arc shell, Santa Ines Archipelago, 33, 37
Arca pacifica, 36, 51
Architectonica nobilis, 19
Argopecten
  abietis, 112
  calli, 153
  circularis, 31, 33, 50, 52, 137, 138
  revellei, 112
Arroyo Arce, vii, 111, 112. *See also* Arroyo de Arce
  uplift, viii
arroyo Buenos Aires, 156
Arroyo Cadeje, 49
Arroyo Canelo, 4, 10, 12, 15, 18, 20
Arroyo Canelo section, Matomí Sandstone Member, 9. *See also* Arroyo El Canelo section; El Canelo section
Arroyo Cayuquitos, 59, 71
Arroyo de Arce, 86. *See also* Arroyo Arce
Arroyo de Gua, 112
Arroyo El Canelo, 15
Arroyo El Canelo section, Matomí Sandstone Member, 11, 12, 15
arroyo El Datilar, 153
Arroyo El Mono, 59, 62, 63, 66, 70, 71, 75, 76, 77
arroyo El Sauce, 145, 148
Arroyo El Tordillo, 45
Arroyo La Calera, 145, 146
Arroyo La Cantera, 4, 5, 8, 9, 12, 18
Arroyo La Enramada, 45
arroyo La Trinidad, 145, 150
Arroyo Los Heme, 12, 18
Arroyo Los Heme Norte. *See* Arroyo La Cantera
Arroyo Matomí area, 4, 8, 18, 22
Arroyo Matomí fan delta, 21. *See also* Matomí alluvial fan
arroyo San Bartolo, 153
arroyo San Dionisio, 159
arthropods, Refugio Formation, 153
Atlantic Ocean, bryozoans, 115, 116
augite, Puertecitos Formation, 15
Avicennia, 61, 62, 76, 78

B

Bahía Concepción, v, vii, viii, 3, 26, 36, *39*, 57, 65, 68, 73, 92, 122, 125
  as modern analog for Loreto Basin, *98*, 99
  fan development, 43
  faults, 41, 43, 69, 74
  flooding, 43
  fossils, 137
  hot springs, 69
  marine sediment, 41, 53
  rhodoliths, 128, 129, 130, 132, 134, 136, 137
Bahía Concepción Fault Zone, 74
Bahía Concepción Member, Infierno Formation, 57, 59, 65, 66, 69, *70*
Bahía Coyote, 48, 49
Bahía de La Paz, 131, 136
Bahía de Los Angeles, 136
Bahía San Gabriel, 131
Bahía Santa Ines, 26, 32
Baja California peninsula, formation of, 141
bajada, 43
balanus, Arroyo La Cantera, 7
Balanus sp., 7, 65
Baraboo region, Wisconsin, paleoislands, 26
Barbatia reeveana, 29
barnacles
  Arroyo Arce, 112
  Loreto Basin, 87
  Punta el Bajo section, 122, 123, 124, 125
  Santa Ines Archipelago, 36
Basin and Range Province, 74
  geomorphology, 22
basin, evolution of, 2
benthic foraminifera
  Puertecitos Formation, 7
  Punta el Bajo section, 125
Biflustra sp., 115
biotas
  Santa Ines Archipelago, *36*
biotite
  Puertecitos Formation, 13, 15
  Trinidad Formation, 148
bioturbation
  Infierno Formation, 78
bioturbation (continued)
  Loreto Basin, 86, 90
  Puertecitos Formation, 5, 7, 9, 19
  Punta el Bajo section, 121, 124
bivalves
  Arroyo Arce, 112
  Bahía Concepción, 41, 50, 51
  Loreto Basin, 86, 87, 90, 111
  Peninsula Concepción, 65, 70, 71
  Punta el Bajo section, 121, 125
  Santa Ines Archipelago, 36, 37
Boleo Formation, 3, 31
Bouse Formation, 4
brachiopods, Arroyo La Cantera, 5
Brazil, bryozoans, 111
Brazilian shelf, rhodoliths, 125
Bristol Channel, South Wales, paleoislands, 26
brittle stars, Gulf of California, 131
bryoliths, defined, 111. *See also* bryozoans
bryozoan macroids, defined, 111. *See also* bryozoans
bryozoans
  Loreto Basin, 87, *111*
  Punta el Bajo section, 124, 125
  Santa Ines Archipelago, 37
Bulimina sp., 8
Buliminella sp., 8
bullseye electric rays, Gulf of California, 131

C

Calabaza Member, Infierno Formation, 57, 59, 65, 66, 67, *70*
Calerita, 128, 131, 134, 136
  fossils, 136, 137
California current, 116
California Gulf, 73. *See also* Gulf of California
Californian Province, 19
Campo Cristina area, 15
Campo Cristina section, Matomí Sandstone Member, 9, 11
Canal de San Lorenzo, 128, 133, 136, 137
Cañón Las Cuevitas Member, El Moreno Formation, 3
carbonate minerals, Infierno Formation, 73, 75, 78, 81
Cardoncito drainage, 50
Caribbean Province, 19
Carmen Formation, 31, 33, 112
Carrizo Creek, 31
Caulerpa sertularoides, 131
Cayuquitos, 71
Cayuquitos Member, Infierno Formation, 57, 59, 66, 70
  chert deposits, *61*, 64, 66, 69
Cerithea sp., 150
Cerithidea albonodosa, 51

Cerithium stercusmuscarum, 51
Cerro Blanco, 41
Cerro Blanco stock, 41
cerro La Laguna, 153, 163
Cerro Prieto, 59, 64, 66, 67, 71, 75
 development of spreading center, 4
Cerro San Juan, viii
cheilostomes
 Gulf of California, 115
 Loreto Basin, 111, 112
chicharrones, 128. *See also* rhodoliths
Chincoteague Island, Virginia, bryozoans, 115
Chione
 californiensis, 36
 jamainiana, 19
 sp., 7, 19, 51, 52
chlorite, Puertecitos Formation, 21, 22
chlorozoan lithofacies, Punta el Bajo section, 125
cirripedium, Arroyo La Cantera, 7
clams, Puertecitos Formation, 5
Clementia dariena, 153
Clhamys tamiamiensis grewingki, 153
clinopyroxene, Puertecitos Formation, 13, 15
Clypeaster
 bowersi, 25, 26, 28, 32
 marquerensis, 65, 67, 70, 78
Codakia distinguenda, 36
Codium magnum, 131
Colorado River delta, 4
Comondú andesite, 31, 34, 35
Comondú Formation, 121, 142, 145
Comondú Group, vii, 25, 26, 27, 37, 41, 57, 59, 73, 80, 112
 volcanics, v, 86
Concepción Bay Fault, viii
Concepción Peninsula, 41, 43, 73, 80. *See also* Peninsula Concepción
conch, Peninsula Concepción, 65, 67
Conopeum
 commensale, 111, 113, 115, 116
 sp., 115
Conus
 brunneus, 28, 29
 multiliratus, 153
 sp., 51, 150
Coorong Lake, South Australia, 79, 80
coral, Santa Ines Archipelago, 22
coralgal, Punta el Bajo section, 125
coralline red algae, v, 119, 121, 128, 132. *See also* rhodoliths
Corallineales, 128. *See also* rhodoliths
corals, Santa Ines Archipelago, 31, 34, 37
cortez garden eels, Gulf of California, 131
Coyote Red Beds, 145
Crassatellites digueti, 29
cristobalite, Infierno Formation, 73, 75
Crucibulum sp., 48, 51, 52
crustaceans, Gulf of California, 131
Cyathodonta gatunenis, 153
Cyprideis sp., 7

D

Delicias Sandstone Member, Puertecitos Formation, 7, 8, *9*, 18
 stratotype, *10*
Delmarva Peninsula, Virginia, bryozoans, 111, 115
diatoms, Gulf of California, 131
Diplobatis ommata, 131
Dosinia
 ponderosa, 7, 19, 51, 65, 71
 sp., 7

E

East Pacific Rise, 164
echinoderms
 Puertecitos Formation, 5, 7
 Punta el Bajo section, 125
echinoids
 Arroyo Arce, 112
 Santa Ines Archipelago, 25, 26, 28, 32, 37
ectoproctaliths, defined, 111. *See also* bryozoans
Ecuador, bryozoans, 115
El Canelo section, Matomí Sandstone Member, 9, 11, 15
El Chorro Formation, 141, 142, 156
El Coloradito area, 5, 8, 9, 15, 18, 21
El Coyote region, 59
El Mono Member, Infierno Formation, 57, 59, 65, 66, *70*
 chert deposits, *61*, 64, 66, *73*
El Mono ridge, 63, 64, 71
El Refugio, 142, 152
El Rosarito, 156
Encope
 chaneyi, 7
 shepherdi, 25, 26, 33
 sverdrupi, 25, 26, 31, 37
 sp., 7
Ensenada, vii
Ensenada de Los Muertos, 136
Erendira, Baja California, 37
Euvola refugioensis, 153
Euvola, 153

F

Fasciolaria princeps, 28, 29
Fauchea
 mollis, 132
 sefferi, 132
fault-bounded basins, 2
faults, 1
 Bahía Concepción, 39, 41, 43, 74
 Gulf of California, 84
 Infierno Formation, 73
 Main Gulf Escarpment, 4
 Peninsula Concepción, 57, 66, 69, 70
 Puertecitos volcanic province, 4
 Punta el Bajo section, 125
 San José del Cabo Basin, 141
 Sierra San Fermín, 4
 Sierra Santa Isabel, 4
feldspar, Loreto Basin, 86
ferro-edenite, Puertecitos Formation, 13, 15
Ficus carbasea, 153
fish, Bahía Concepción, 41
Fish Creek area, 4, 18, 22
Flabellipecten
 bosei, 31
 sp., 33
Florimetis tritinana, 153
fossilization, potential for, vii
fossils
 Refugio Formation, 153
 Santa Ines Archipelago, 25

G

gastropods
 Arroyo La Cantera, 5
 Bahía Concepción, 41, 50, 51
 Loreto Basin, 86, 87
 Peninsula Concepción, 70, 71
 Punta el Bajo section, 121
 Refugio Formation, 153
 San José del Cabo Basin, 153
 Santa Ines Archipelago, 28
Gilbert-type fan deltas, *83*
glass, Puertecitos Formation, 13
Globigerina angustiumbilicata, 150
Globorotalia
 lenguaensis, 150
 mayeri, 150
Gloria Formation, 73
Glycymeris maculata, 29
Gracilaria textorii, 131
Grand Canyon, 25
Grandiarca sp., 150
Guadalupe plate, subduction under North American plate, 160
Gulf Extensional Province, vii, 1, 2, 4
Gulf of California, v, vii, 22, *25*, 41, 74, 99, 112, 119, 125. *See also* California Gulf
 bryozoans, 115
 faulting, 84
 opening of, 141, 164, 165
 rhodoliths, *127*
 tectonic evolution, 39
Gulf of California Fault System, viii
gypsum
 Fish Creek, 22
 Loreto Basin, 89

H

halite, Infierno Formation, 79
Halymenia
 californica, 131
 megaspora, 131
hornblende, Puertecitos Formation, 13, 15
Hornillos Formation, Comondú Group, 41
horse jaw fossil, San José del Cabo Basin, 150
hot springs
 Peninsula Concepción, 69

hot springs (continued)
  Playa Santispac, 80
Hydroclathrus clathratus, 131
hypersthene, Puertecitos Formation, 15

I

ichnofossils
  Arroyo La Cantera, 7
  Infierno Formation, 77, 78
  Peninsula Concepción, 62, 70, 73
illite, Puertecitos Formation, 21, 22
Imperial Formation, 4, 19, 22, 28, 31, 115
Imperial Valley, California, 2, 22
Infierno Creek, 73
Infierno Formation, 3, 31, 41, 57, 59, 66, 69, *70*, *73*
Isla Cayo, 131
Isla Coronados, 136
Isla Coyote, 69
Isla El Requesón, 36, 37. *See also* Isla Requesón
  rhodoliths, 128, 137
Isla Erendira, *37*
Isla Espirtu Santo, 131
Isla Las Cuevas, 67
Isla María Madre, 160
Isla Prieto, 67
Isla Requesón, v, vii, 41, 49, 122. *See also* Isla El Requesón
Isla San Jose, 130, 131, 133, 134, 137. *See also* San Jose Island
Isla San Juan, 67
Isla San Marcos, 26
Isla San Pedro Martir, 136
Isla San Sebastián, 67, 68
Isla Tiburón, 2
Isla Tortuga, 36, 69
Islas Santa Ines, 25, *31*, 34
  sea lion colony, 26

J

Jalisco block, 160
Jatropha cuneata, 36

K

kaolinite, Puertecitos Formation, 22
krummholz, 36

L

La Calera Formation, 141, 142, *145*, 148, 160
La Laguna, 152
La Paz, 128, 136, 137, 145
La Reforma–Santa Rosalia, vii
La Trinidad basement complex, 143, 145, 146, 160, 163, 164
La Trinidad fault, 160, 163
La Victoria basement complex, 141, 143, 160, 163, 164
Laevicardium
  elatum, 51
  elenense, 51, 52
Laguna Cayuquitos, 68
Laguna Salada area, 22
Laguna Salada basin, 2, 4
Las Cuevas, 150
leatherplant bush, Santa Ines Archipelago, 36
Leopecten backeri, 112
lithofagids, 121
Lithophyllum
  diguetii, 133
  lithophylloides, 133
  margaritae, 133
  pallescens, 133
  veleroae, 133
  sp., 133
Lithothamnion
  australe, 133
  crassiusculum, 133
  fruticulosum, 131
  sp., 133
Llano El Moreno Formation, 3
Longmynd region, Shropshire, England, paleoislands, 26
Loreto area, Baja California, 26, 84, 98
  rhodoliths, vii
Loreto Basin, vii, 2, 18, 37, 40, 54, 69, 74, *83*, 165
  faulting, 84
  lithofacies, 83, *86*
  paleocurrents, *95*
  paleogeographic evolution, *95*
  parasequence cyclicity, *98*
  rhodoliths, *119*
  sedimentation patterns, *98*
  stratigraphy, *84*, *92*
  structure, *84*
  subsidence, *98*
  tectonics, viii
Loreto fault, 84, 99, 119, 120, 121, 124
Los Barriles, 156
Los Barriles Formation, 141, 142, 152, *153*, 160, 163, 164

M

macroalgae, Gulf of California, 131
Main Gulf Escarpment, faults, 4, 22
mammalian bone fragment, Santa Ines Archipelago, 28
mangrove fossils, Concepción Peninsula, 61, 62, 73, 76, 78
mangrove swamps, Bahía Concepción, 39, 51
marine vertebrates, Refugio Formation, 153
Marquer Formation, 25, 26, 31, 33, 35, 37, 112
Massachusetts, bryozoans, 111
Matomí alluvial fan, 22. *See also* Arroyo Matomí fan delta
Matomí area, 12, 15
Matomí Mudstone Member, Puertecitos Formation, *8*, 14, 18
  stratotype, *9*
Matomí Sandstone Member, 15
Mauritania, west Africa, bryozoans, 115
Megapitaria squalida, 51
melachor lithofacies, Punta el Bajo section, 125
Melongena sp., 150
Membranipora
  arborescens, 111, 115
  commensale, 115
  tenuis, 115
  sp., 115
Merychippus-Pliohippus, 150
Mesa Atravesada. *See* Punta Chivato promontory
Mesa Barracas, 26, *33*
Mesa El Coloradito, 26, *32*
Mesa El Rosario, viii
Mesa El Tábano, 15, 20
Mesa Ensenada de Muerte, 26, *28*, 32, 34
mica
  Refugio Formation, 152
  Trinidad Formation, 148
micromolluscs, Gulf of California, 131
Minitas Formation, Comondú Group, 41
Mitra
  fultoni, 29
  tristis, 28, 29
model, wave assault, *35*
Modulus sp., 51
molluscs
  Arroyo Arce, 112
  Arroyo La Cantera, 5
  Bahía Concepción, *39*
  Bahía Concepción, 48, 49, *50*, 53
  Loreto Basin, 86, 119
  Peninsula Concepción, 65
  Punta el Bajo section, 121, 124
  San José del Cabo Basin, 142
  San José del Cabo Basin, 153
  Santa Ines Archipelago, 25
  Santa Ines Archipelago, 32, 34, 37
Mulegé village, vii, 40, 43, 58
Murex sp., 51, 150
Myrichthys maculosus, 131

N

Nassarius sp., 51
Nodipecten
  arthriticus, 33, 112
  modosus, 31
Nonion grateloupi, 8
North American plate, subducted by Guadalupe plate, 160
Norway, fossils, 138
Notch, Arroyo El Mono, 62, 70, 75, 77, 78

O

Old Red Sandstone, 26
oligoclase, Puertecitos Formation, 15
Oliva sp., 150
opal, Infierno Formation, 73, 75, 76, 77, 80, 81
opaque minerals, Puertecitos Formation, 13, 15
Ophiomorpha, 62, 70, 73, 77, 78
Oscar Range, Western Australia, 26
ostracodes, Puertecitos Formation, 7
Ostrea
  hermanni, 73
  palmula, 36, 51
  vespertina, 33
  sp., 65, 70, 71
oysters
  Bahía Concepción, 51
  Loreto Basin, 86, 87, 88, 89, 90
  Peninsula Concepción, 65, 70
  Puertecitos Formation, 5, 10, 20
  Punta el Bajo section, 121, 122
  San José del Cabo Basin, 150
  Santa Ines Archipelago, 25
  Santa Ines Archipelago, 31, 33, 34, 36, 37

P

Pacific plate, 53
paleo hot springs, Peninsula Concepción, 57, 66
paleocurrents, Loreto Basin, *95*
paleoecology, Loreto Basin, *111*
paleoislands, 26, 37
Palm Spring Formation, 2, 4
Panama Isthmus, closure of, *111*, 115
Panopea sp., 19
Panopeas sp., 7
Pecten
  aletes, 153
  calli, 153
  keepi, 153
pectens
  Loreto Basin, 86, 87, 88, 89, 90
  Peninsula Concepción, 70
  Punto el Bajo section, 121, 122
  Santa Ines Archipelago, 25
  Santa Ines Archipelago, 26, 28, 31, 33, 35, 37
pelecypods, Refugio Formation, 153
Pelones Formation, Comondú Group, 41, 59, 66
Peninsula Concepción, vii
  alluvial fans, 70
  basins, *57*, 80
  faults, 70
Peninsular Geological Society, v
Periglypta multicostata, 36
Perissocytheridea meyerabichi, 7
phytoplankton, Bahía Concepción, 41
Pilares Formation, Comondú Group, 41
Placunanomia cumingii, 112
plagioclase, Puertecitos Formation, 5, 13, 15
Plagioctenium, 153
Playa La Palmita, 26, 32, 33, 36, 37
Playa Santispac, 49, 69, 80
polychaetes, Bahía Concepción, 41, 52
porcellanite, Infierno Formation, 74
Porites sp., 71
potassium feldspar
  Puertecitos Formation, 5, 22
  Refugio Formation, 152
  Trinidad Formation, 148
Prague peneplain, Bohemia, paleoislands, 26
proto-Gulf stage, 2, 4
Puertecitos accommodation zone, 2
Puertecitos area, vii, 19, 22
Puertecitos Formation, *1*
  fossil assemblage, *5*
  paleoenvironment, *18*
  sedimentary facies, *5*, *18*
  sedimentation, *15*, *20*
  stratigraphy, *8*
  volcanic horizons, *11*
  volcanism, 4, *20*
Puertecitos Volcanic Province, 1, 2, 22
Punta Arena, 49
Punta Bajo, 128, 131, 133, 134, 136. *See also* Punta el Bajo section
  fossils, *136*
Punta Banda, uplift at, vii, viii
Punta Cacarizo, 28, 31, 32
Punta Chivato area, vii, *25*, 69, 125, 128
  fossils, *136*
Punta Chivato promontory, Gulf of California, 25, 26, *28*, 33, 34, 35, 37
Punta El Coloradito, 34
Punta el Bajo de Tierre Firme, 120. *See also* Punta el Bajo section
Punta el Bajo section, *119*. *See also* Punta Bajo
  description, *121*
  faults, 125
  sedimentology, *121*
Punta Los Monos, 67
Punta Prieto, 66
Purísima-Iray-Magdalena basin, 142

Q

quartz
  Infierno Formation, 74, 77
  Puertecitos Formation, 5, 22
  Refugio Formation, 152
  Trinidad Formation, 148, 150

R

Raeta undulata, 153
rancho El Chorro, 159
rancho El Machete, 150
rancho El Refugio, 153
rancho El Refugito, 153
rancho El Rosario, 150
rancho El Torete, 150
rancho La Calabaza, 150
rancho La Soledad, 150
rancho La Trinidad area, 148, 150
rancho Los Algodones, 152
Rancho Santa Rosaliita, vii, 58, 59, 75. *See also* Santa Rosaliita ranch
Rattlesnake ridge, 71
Refugio area, 152
Refugio Formation, 141, 142, 148, *150*, 156, 160, 162, 163, 164
  fossils, 153
Refugito area, 152
Requeson Formation. *See* Ricasón Formation
rhizolith, Peninsula Concepción, 61
Rhizophora, 61
rhodoliths
  anthropogenic effects on, *137*
  Bahía Concepción, v, 41, 49
  defined, 121
  distribution, *130*
  growth forms, *137*
  Gulf of California, *127*
  Isla Requesón, v
  Loreto area, vii
  Loreto Basin, *119*
  Santa Ines Archipelago, 33
  taxonomy, *133*
Rhodophyta, 128. *See also* rhodoliths
Ricasón Formation, Comondú Group, 41, 59, 63, 65, 67
rudists, Santa Ines Archipelago, 37

S

St. David's Archipelago, paleoislands, 26
Salada Formation, 142, 153. *See also* Refugio Formation
Salto Formation, Comondú Group, 41
Salton Sea, 31, 115
Salton Trough region, 2, 53
  development of spreading center, 4
San Benito fault, vii
San Felipe, 3, 4, 19, 22
San Jose Island, 133. *See also* Isla San Jose
San José del Cabo Basin, vii, *141*
  tectonism, 141
San José del Cabo fault, 141, 156, 160, 164
San José del Cabo region, 142, 163. *See also* San José del Cabo Basin
San José del Cabo trough, 142. *See also* San José del Cabo Basin
San José–Los Planes trough, 2
San Juan de Los Planes Basin, 143
San Marcos Formation, 25, 26, 28, 29, 31, 34, 35, 37, 112
San Nicolas region, 59
San Pedro Mártir area, 22
San Pedro Mártir fault, 4
San Sebastián, 71
sand dollars
  Peninsula Concepción, 65, 67, 70
  Puertecitos Formation, 5
  Santa Ines Archipelago, 25, 26, 31, 33, 34, 37

Santa Catarina area, 8
Santa Ines Archipelago, Gulf of California, *25*
  tectonics, 37
Santa Rosalia area, 3, 26
Santa Rosalia Basin, 165
Santa Rosalia Formation, 37
Santa Rosaliita, vii, 59
  uplift at, vii
Santa Rosaliita ranch, 73. *See also* Rancho Santa Rosaliita
Santiago, 148
Santiago–San José del Cabo region, 142. *See also* San José del Cabo Basin
scallops
  Bahía Concepción, 51
  Gulf of California, 137, 138
  San José del Cabo Basin, 150
Scinaia
  johnstoniae, 132
  latifrons, 132
sea anemones, Gulf of California, 131
sea-level changes
  Bahía Concepción, 43
  determination of relative, v
  history of relative, vii
  San José del Cabo Basin, 141
sea-level rise, 27
  Peninsula Concepción, 69
  Santa Ines Archipelago, 37
sea urchin, Gulf of California, 131
sedimentation patterns, Bahía Concepción, v
shrimp, Bahía Concepción, 41
Sierra La Giganta, 142
Sierra La Trinidad, 142, 150
Sierra La Trinidad basement complex, 153, 160
Sierra La Trinidad ridge, 143, 148
Sierra La Victoria, 142
Sierra La Victoria basement complex, 156, 162
Sierra La Victoria ridge, 143, 148
Sierra Las Tinajas–Sierra Pinta, 2
Sierra San Fermín, 5, 8, 9, 11, 12, 14, 15, 18, 20, 22
  faults, 4
Sierra San Pedro Mártir, uplift, 2
Sierra Santa Isabel, faults, 4
Sierra Santa Rosa, 2, 22
silica, Infierno Formation, 78, 79, 80, 81
silicification, Infierno Formation, 75, 81
Sinaloa coast, 160
smectite, Puertecitos Formation, 4, 21, 22
Solenastrea fairbanksi, 25, 31, 37
Southern Province, 19
sponges
  Bahía Concepción, 52
  Gulf of California, 131
Striotsrea sp., 153
Strombina sp., 51
Strombus
  gracilor, 51
  obliteratus, 153
  subgracilior, 65, 67, 70, 71
  sp., 150

T

Taeniconger digueti, 131
Tagelus californianus, 65, 70, 71
Tehuitzingo Formation, Oaxaca, Mexico, 79, 80
Tellina orhracea, 29
Theodoxus luteofasciatus, 50
tiger snake eels, Gulf of California, 131
Tosco-Abreojos fault, 162
Toxpneustes roseus, 131
Trigonicardia biangulata, 51
Trinidad basement complex, 143, 146
Trinidad Formation, 141, 142, 146, *147*, 152, 153, 160, 162
Trópico de Cancer, 153
Tuff of El Canelo, 4
Tuff of Los Heme, 12
Tuff of Mesa El Tábano, 8, 10, 11, 12, 15, 21
Tuff of Valle Curbina, 5, 8, 9, *12*, 14, 15, 18, 20, 21
tunicates, Gulf of California, 131
Turritela
  abrupta fredeai, 153
  imperialis, 73
turritelids, Puertecitos Formation, 10, 19
Turritella
  planigyratra, 7
  sp., 7

U

Undulostrea megadon, 112, 153

V

Valle Curbina, 5, 8, 9, 12, 15, 18
Valle de Camalu, viii
Valle de Maneadero, viii
Valle San Felipe, 22
Valle San Felipe–Valle Chico basin, 2
Vallecitos area, 4, 18, 22
Virginia, bryozoans, 111, 115
Vizcaino peninsula, uplift at, vii
volcanic glass, Infierno Formation, 73
volcanics
  Comondú Group, v
  Puertecitos Formation, 4, *11*
  Santa Rosalia, vii
volcanism, vii, 2
  explosive, 1, 20
  Puertecitos Formation, 4, *20*
  relationship to extension, 2
  relationship to sedimentation, 2

W

Wallops Island, Virginia, bryozoans, 115
waves, Santa Ines Archipelago, *36*
whales, Santa Ines Archipelago, 25, 28
winds, Santa Ines Archipelago, *36*

Typeset and printed in U.S.A. by Johnson Printing, Boulder, Colorado